Astrophysical Jets and Their Engines

NATO ASI Series

Advanced Science Institutes Series

A series presenting the results of activities sponsored by the NATO Science Committee, which aims at the dissemination of advanced scientific and technological knowledge, with a view to strengthening links between scientific communities.

The series is published by an international board of publishers in conjunction with the NATO Scientific Affairs Division

A Life Sciences B Physics	Plenum Publishing Corporation London and New York
C Mathematical and Physical Sciences	D. Reidel Publishing Company Dordrecht, Boston, Lancaster and Tokyo
D Behavioural and Social Sciences E Engineering and Materials Sciences	Martinus Nijhoff Publishers Dordrecht, Boston and Lancaster
F Computer and Systems Sciences G Ecological Sciences H Cell Biology	Springer-Verlag Berlin, Heidelberg, New York, London, Paris, and Tokyo

Series C: Mathematical and Physical Sciences Vol. 208

Astrophysical Jets and Their Engines

edited by

Wolfgang Kundt
Institut für Astrophysik der Universität Bonn, F.R.G.

D. Reidel Publishing Company

Dordrecht / Boston / Lancaster / Tokyo

Published in cooperation with NATO Scientific Affairs Division

Proceedings of the NATO Advanced Study Institute on
Astrophysical Jets and Their Engines
Erice, Sicily, Italy
17-25 September 1986

Library of Congress Cataloging in Publication Data

NATO Advanced Study Institute on Astrophysical Jets and their Engines (1986: Erice, Sicily)
Astrophysical jets and their engines.

(NATO ASI series. Series C, Mathematical and physical sciences; vol. 208)
"Proceedings of the NATO Advanced Study Institute on Astrophysical Jets and their Engines, Erice, Sicily, Italy, September 1986"—Verso t.p.
"Published in cooperation with NATO Scientific Affairs Division."
Includes index.
1. Astrophysical jets—Congresses. 2. Neutron stars—Congresses. 3. Active galaxies—Congresses. I. Kundt, Wolfgang. II. North Atlantic Treaty Organization. Scientific Affairs Division. III. Title. IV. Series: NATO ASI series. Series C, Mathematical and physical sciences; vol. 208.
QB466.J46N37 1986 523.01 87–12936

DOI: 10.1007/978-94-009-3927-1

Published by D. Reidel Publishing Company
P.O. Box 17, 3300 AA Dordrecht, Holland

D. Reidel Publishing Company is a member of the Kluwer Academic Publishers Group

---

TABLE OF CONTENTS

PREFACE

This volume is the documentation of the first Course on 'Neutron Stars, Active Galactic Nuclei and Jets', of an Erice School with a wide astrophysical scope.

The choice of the subject was made because of an apparent similarity - stressed already at earlier meetings - of four classes of astrophysical jet sources: Active Galactic Nuclei, Young Stellar Objects, Binary Neutron Stars and Binary White Dwarfs. They share important properties such as their morphology, high variability and large velocity gradients as well as - with some inference - their broad spectrum, hypersonic outflow and core/lobe power ratio.

Despite this apparent similarity of the four source classes, quite different models have been put forward for their description:
(i) The central engine of active galactic nuclei has been generally thought to be a black hole, in contrast to the central engine of young stellar objects and cometary nebulae which apparently is a pre-T-Tauri star, some six orders of magnitude less compact, and to the central engine of planetary nebulae which may or may not be a binary white dwarf.
(ii) The elongated lobes, or flow patterns, have been often interpreted as highly directional stellar wind outflows whereas in a few well-mapped cases, the elongated flow appears to be 'pumped up' through a much narrower channel, or jet, both in the extragalactic and stellar sources.
(iii) The jet substance has been mostly assumed to be of local galactic composition, i.e. to be mainly hydrogen; but since the fall of 1986 one finds a developing consensus that instead, the extragalactic jets may be composed of relativistic electron-positron pair plasma. A similar consensus has probably not yet been reached for the composition of the stellar jets, in particular of the jets of SS 433, though the St. Louis meeting in December 1986, organized by Jonathan Katz, has weakened the faith in the hydrogen jets.

A different and more uniform interpretation of the jet set has been pushed before and during this Advanced Study Institute: All jet sources contain a fast-rotating, magnetised (super-) massive core at the center of a feeding accretion disk. According to Martin Gaskell, this core can have binary nature. In the core's corotating magnetosphere, localised electric discharges accelerate charges to relativistic energies which convert into pair plasma on collision with stellar photons. The magnetosphere acts like a centrifuge generating an overpressure bubble which bulges out along the density gradient, to both sides of the heavy disk, in the form of two relativistic, supersonic jets.

This unified scheme, even though not strictly new, has not been uniformly appreciated by the lecturers of the ASI as can be gathered from the various contributions to this volume. It has created a very special atmosphere, with intense discussions before, during and after the lectures, on the beaches, around the castle and in the hospitable cellar of San Marco. Several lecture sequences were summarized with a short story, viz about a yellow fish in a cage (Hakki Ögelman), a white Indian elephant (Hagai Netzer) and a chair with an infinite number of legs (Sterl Phinney). Another sequence (Sam Falle) ended with an application of Murphy's law proving that almost all the claims made on page 10 of this book are right.

For me as the organizer and active participant, the Course was a highlight in scientific life. My thanks go to Martin Rees as whose guest in 1977 I was introduced to the subject, to Antonino Zichichi who declared me a director after a written and an oral examination in Rome and St. Vincent, to Craig Sinclair from whose hands I received the NATO ASI award and to the European Physical Society and National Science Foundation for additional support. My thanks go likewise to Alberto Gabriele and Pinola Savalli of CCSEM for their care before, during and after the Course.

Bonn, February 1987

Wolfgang Kundt
Institut für Astrophysik
der Universität

LIST OF PARTICIPANTS

Achterberg, Abraham	Sterrewacht "Zonnenbergh", Utrecht, NL
Banhatti, Dilip	Turku Univ. Observ., Finland
Banhatti, Radha	Turku Univ. Observ., Finland
Benford, Gregory	Univ. of California, Irvine, USA
Blome, Hans-Joachim	IfA, Bonn, FRG
Cawthorne, Timothy	Brandeis Univ., Boston, USA
Cohen, Martin	Univ. of Berkeley, Calif., USA
Crusius, André	MPIfR, Bonn, FRG
Dröge, Wolfgang	MPIfR, Bonn, FRG
Dunlop, James	Dept. Astronomy, Edinburgh Univ., GB
Duschl, Wolfgang	MPIfA, München, FRG
Esendemir, Akif	Middle East Techn. Univ., Ankara, Turkey
Facondi, Silvia Rosa	Dipt. di Astronomia, Bologna, Italy
Falle, Samuel	Univ. of Leeds, GB
Gaskell, Martin	SUNY, STONY BROOK, N.Y., USA
Hiltner, Peter	MPIfA, Königstuhl, Heidelberg, FRG
Innes, Davina	MPI f. Kernphysik, Heidelberg, FRG
Jessner, Axel	MPIfR, Bonn, FRG
Kapoor, Ramesh Chander	IIA, Bangalore, India
Kössl, Dieter	MPIfA, München, FRG
Kundt, Wolfgang	IfA, Bonn, FRG
Kunte, Prabhakar	TIFR, Bombay, India
Laing, Robert	Royal Greenwich Observ., Hailsham, GB
Lesch, Harald	MPIfR, Bonn, FRG
Madau, Piero	ISAS, Trieste, Italy
Malkan, Matthew	U.C. Los Angeles, Calif., USA
Mead, Andrew	Royal Observatory, Univ. of Edinburgh, GB
Means, James	Physics Dept., Irvine, Calif., USA
Melia, Fulvio	Enrico Fermi Inst., Univ. of Chicago, USA
Morganti, Raffaella	Dipt. di Astronomia, Bologna, Italy
Netzer, Hagai	Tel Aviv Univ., Israel
Ögelman, Hakki	MPIfA, München, FRG
Özel, Mehmet	NASA, GSFC, Greenbelt, MD, USA
Peacock, John	Royal Observ. Edinburgh, Scotland
Peres, Giovanni	Observ. Astronomico, Palermo, Italy
Phinney, Sterl	Caltech, Pasadena, USA
Pietrini, Paola	Observ. Astrofisico, Firenze, Italy
Pritchard, Alison	University of Leeds, GB
Saripalli, Lakshmi	TIFR, Bangalore, India
Scheuer, Peter	Cavendish Lab., Cambridge, GB
Serabyn, Eugene	Caltech, Pasadena, USA
Sorrell, Wilfred	Univ. of Wisconsin, Madison, USA
Sun, Wei-Hsin	Dept. of Astronomy, U.C. Los Angeles, CA, USA

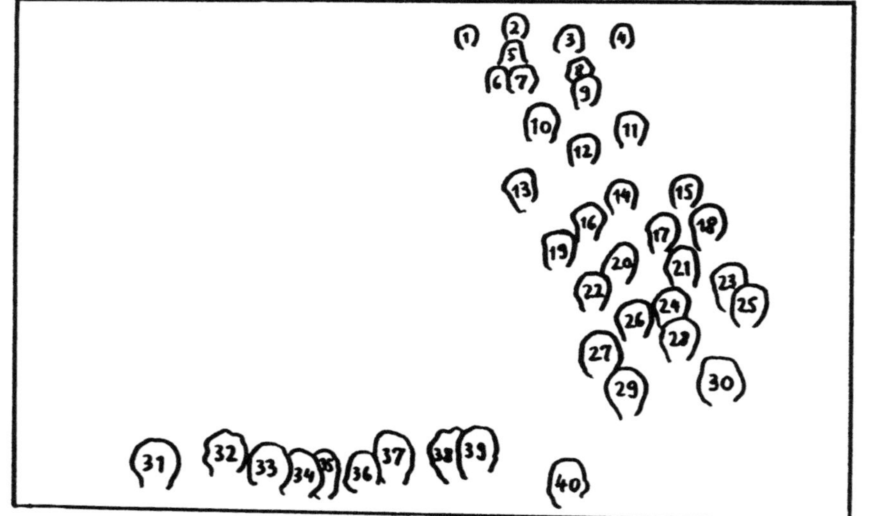

1. Peres, Giovanni
2. Melia, Fulvio
3. Morganti, Raffaella
4. Pietrini, Paola
5. Saripalli, Lakshmi
6. Banhatti, Radha
7. Sun, Wei-Hsin
8. Serabyn, Eugene
9. Madau, Piero
10. Duschl, Wolfgang
11. Malkan, Matthew
12. Benford, Gregory
13. Kundt, Wolfgang
14. Kössl, Dieter
15. Crusius, André
16. Cohen, Martin
17. Innes, Davina
18. Means, James
19. Blome, Hans-Joachim
20. Mead, Andrew
21. Dunlop, James
22. Kapoor, Ramesh Chander
23. Cawthorne, Timothy
24. Özel, Mehmet
25. Hiltner, Peter Reinhard
26. Ögelman, Hakki
27. Facondi, Silvia Rosa
28. Dröge, Wolfgang
29. Esendemir, Akif
30. Jessner, Axel
31. Sorrell, Wilfred
32. Achterberg, Abraham
33. Banhatti, Dilip
34. Laing, Robert
35. Lesch, Harald
36. Netzer, Hagai
37. Peacock, John
38. Scheuer, Peter
39. Gaskell, Martin
40. Kunte, Prabhakar

Here absent:

41. Falle, Samuel
42. Phinney, Sterl
43. Pritchard, Alison

THE SOURCES AND THEIR MODELS

W. Kundt
Institut für Astrophysik
der Universität Bonn
Auf dem Hügel 71
D 5300 Bonn 1, F.R.G.

ABSTRACT: The similarities are stressed between the different astrophysical jet sources, or Bipolar Flows, which apparently make up four classes: Active Galactic Nuclei (=AGN), Young Stellar Objects (=YSO), Binary Neutron Stars and Binary White Dwarfs. This similarity is remarkable in view of the vastly different environments and contrasts with the heterogenous models which have been advanced for their description.

1. PRESENTATION OF THE JET SET

Jets are often defined as sources showing an elongated morphology (Bridle & Perley, 1984); but the presence of an exciting compact source should be guaranteed, in particular in the stellar case , in order to avoid confusion with quasi-static sources. Elongated flow patterns are not easily established outside of planets: they need thin, supersonic pistons, supersonic w.r.t. the ambient medium, in order to guarantee that the sideways expansion (thrust) fall behind the forward expansion. An elongated morphology is therefore indicative of a highly supersonic flow.

The astrophysical jet sources fall into the following four classes:

(1) Active Galactic Nuclei: Perhaps as many as one third of all large galaxies show signs of nuclear activity (Balick & Heckman, 1982; Gaskell, this volume). This high percentage may or may not involve confusion with starburst galaxies (Terlevich & Melnick, 1985), but even 'normal' galaxies like our own appear to have an active 'monster' at their center. Nuclear activity appears to imply jet formation, at least on the scale of kiloparsecs. Remarkably, no more than one pair of jets appears to form in any one galaxy. (An exception could be NGC 1097 with its two pairs of optical jets; cf. Wolstencroft et al, 1984). A suggestive site for the spatial origin of the jets is the rotational center of the gaseous galactic disk (Gunn, 1979). Examples of extragalactic jets (of different ages) are shown in figures 1 to 4, all of them hosted by elliptical galaxies. The jets in spiral galaxies don't appear to reach the outer edge of their halo, perhaps because of entrainment of initially neutral intruders. Figure 5 shows that our own galaxy may conform to this scheme.

W. Kundt (ed.), Astrophysical Jets and Their Engines, 1–28.

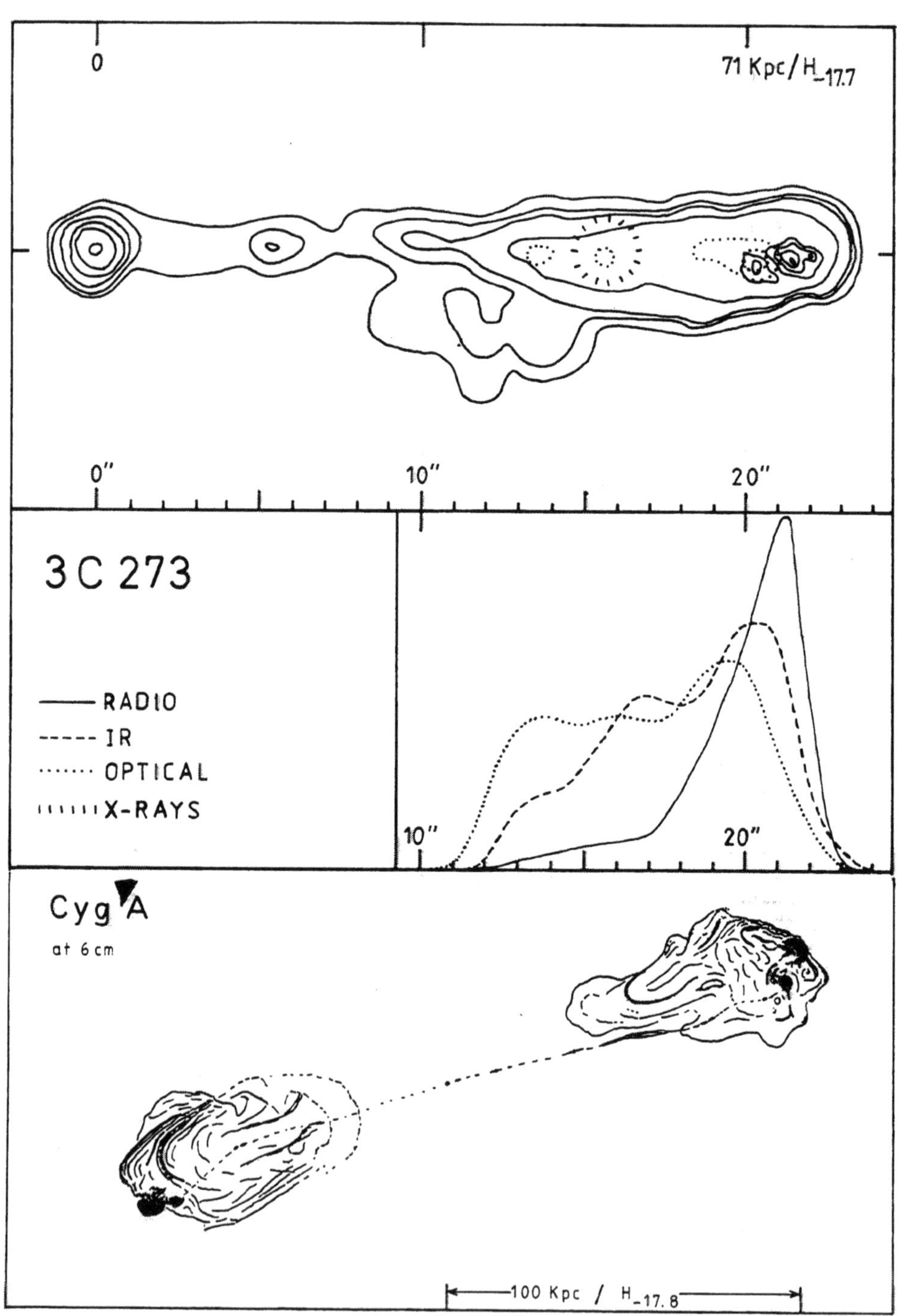
0
71 Kpc/H$_{-17.7}$
0"
10"
20"
3C 273
RADIO
IR
OPTICAL
X-RAYS
10"
20"
Cyg A
at 6 cm
100 Kpc / H$_{-17.8}$

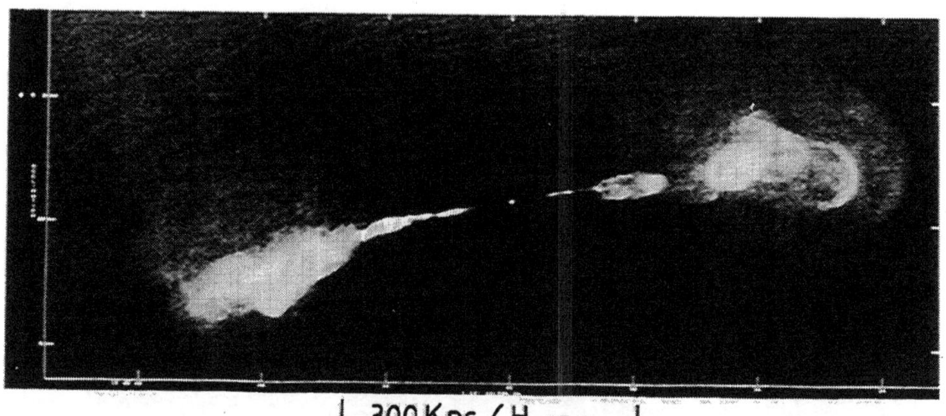

Figure 3. 6 cm radio map of Her A (= 3C 348), $z = 0.154$, taken from Dreher & Feigelson (1984). $H_{-17.8} := H_0/50$ Km s^{-1} Mpc^{-1}.

Figure 4. 1.4 GHz radio isophotes of the (nearest) radio galaxy Cen A (= NGC 5128) whose distance has been recently scaled down from 5 Mpc to $\lesssim$ 3 Mpc, based on supernova 1986g. The inner radio lobes are not resolved on this scale. The optical jet, of length 25 Kpc (d/3 Mpc), has been drawn in. For references see Gopal-Krishna & Saripalli (1984).

Figure 1. Contour map and intensity profile of the one-sided jet of 3C 273, of redshift $z = 0.158$, from radio to optical frequencies, redrawn from Kundt & Gopal-Krishna (1986) and the references given therein. $H_{-17.7} := H_0/65$ Km s^{-1} Mpc^{-1}. The jet is rotated in position angle through 48°.

Figure 2. Symbolical drawing of the 6 cm map of the radio galaxy Cyg A, $z = 0.0565$, taken from Kundt & Saripalli (1987).

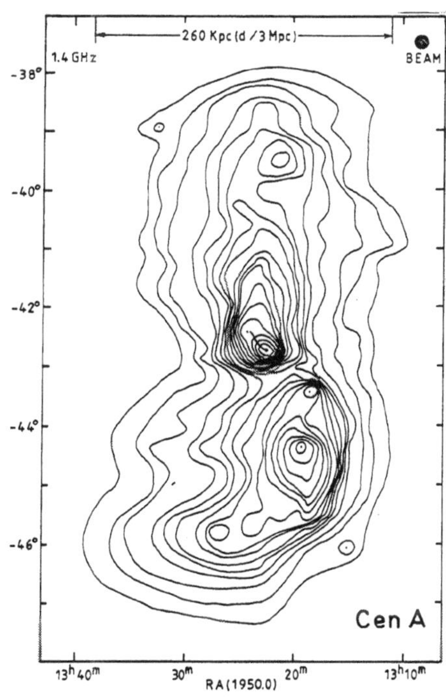

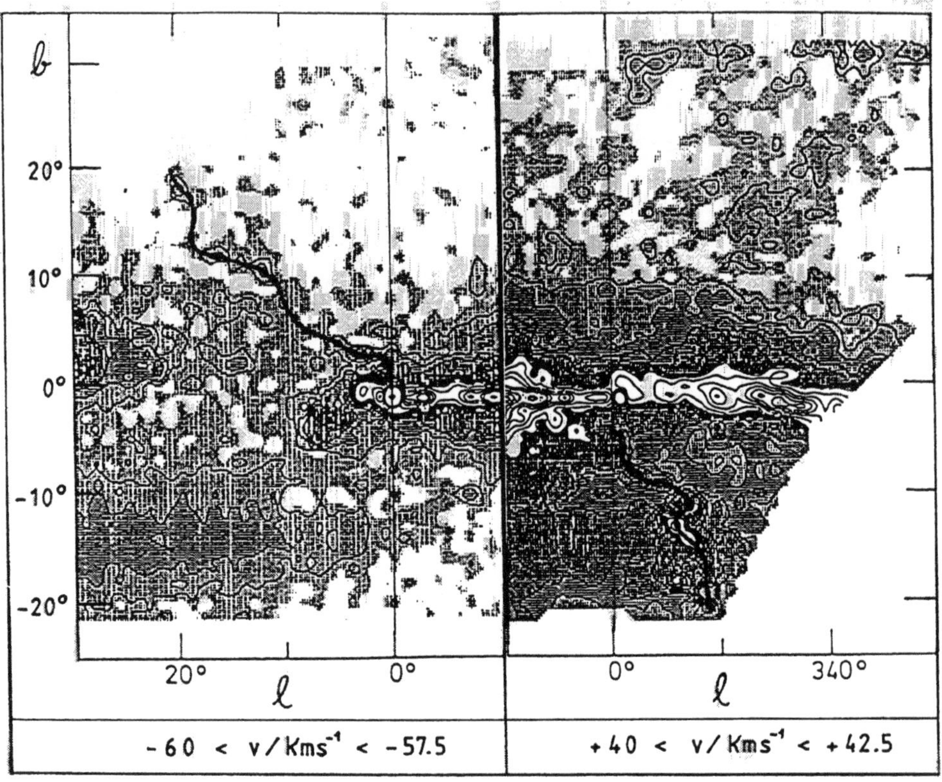

Figure 5. Our Galactic twin-jet, mapped in the 21 cm line of HI at different radial velocities v; taken from Kundt (1987a) where the original references are given. Its northern continuation, to longitudes $l \gtrsim 160^{\circ}$, can also be traced at IRAS frequencies. $10^{\circ} \simeq 1.2$ Kpc(d/7Kpc).

(2) Young Stellar Objects: Very young stars, the pre-T-Tauri or Herbig-Haro-exciting stars, are seen to drive Bipolar Flows. In a few cases, the flows are seen to be threaded by thin, knotty jets. These BPs have estimated ages of some 10^4 years. Their occurrence rate suggests that every young star passes through the BF stage (Lada, 1985; Cohen, this volume). The 'lobes' of the flows are often called 'cavities' because they look (almost) empty. Their appearance (radio or optical) depends on whether they occur inside of dense molecular clouds or inside of the warm ISM; they are sometimes 'seen' as cometary nebulae. Figures 6 and 7 show two well-mapped BFs.

(3) Binary Neutron Stars: We know only three or four binary-neutron-star candidates with jets: Sco X-1, SS 433, Cyg X-3 and possibly Circinus X-1. Sco X-1 looks like an extragalactic triple radio source (Geldzahler & Fomalont, 1986), SS 433 interacts with the (sometimes questioned) supernova remnant W 50 (figure 8), Cyg X-3 is further away

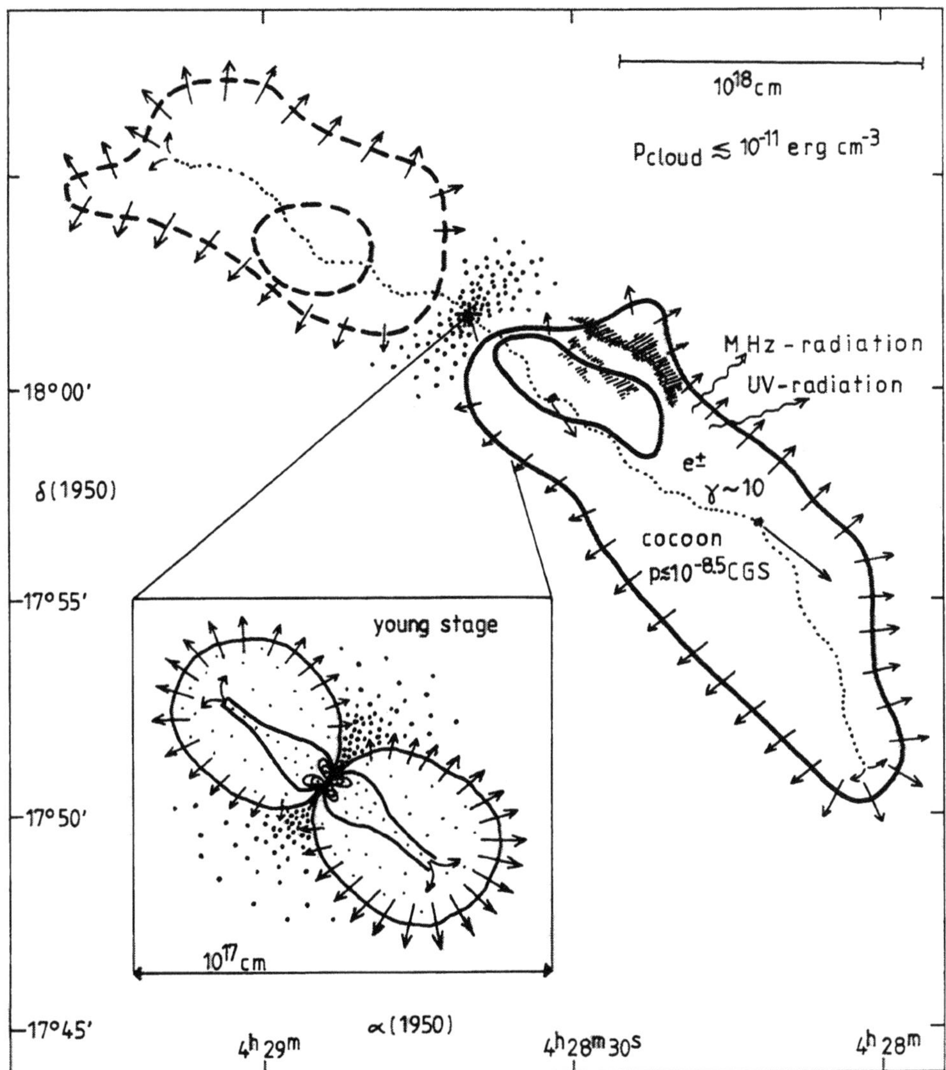

Figure 6. An interpretation of the well-mapped bipolar flow L 1551/IRS 5 given by Blome & Kundt (1987). The contours are CO-line radio isophotes on the large scale and based on I-band in the enlargement. UV radiation results from inverse-Compton collisions of the relativistic pair plasma on the infrared and optical background; together with the low-frequency synchrotron radiation, it heats the ambient medium and thus gives rise to the extended IRAS lobes (not drawn in).

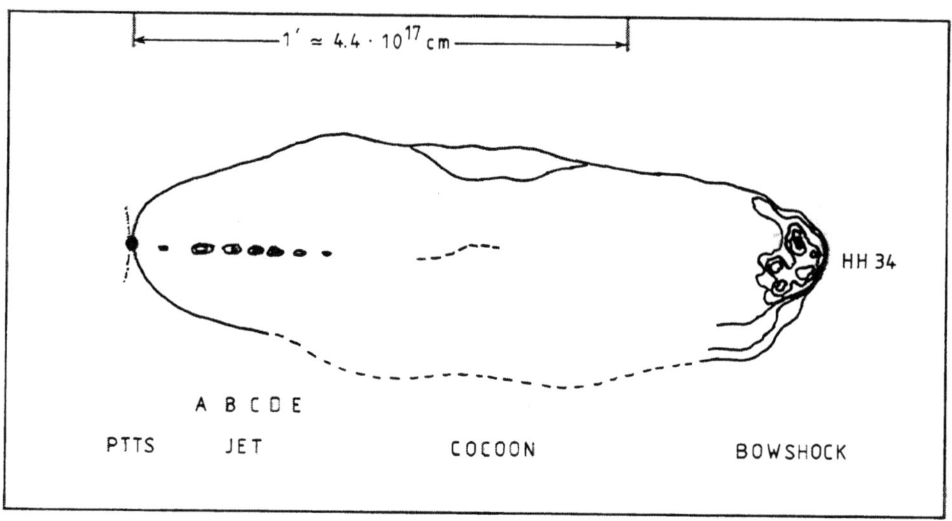

Figure 7. The young bipolar flow associated with the Herbig-Haro object HH 34, redrawn from an optical CCD image by Reipurth et al (1986). The jet is resolved into the five knots A to E with monotonic spectral properties. PTTS = exciting pre-T-Tauri star.

than the Galactic center, and Cir X-1 has sometimes been called a black-hole candidate in the past (Haynes et al, 1986). The 'missing jets' in all the other binary neutron stars may be consequences of inverse-Compton losses on photons of the companion star (Kundt, 1985b). Models at variance with each other have been proposed (Sco X-1: Achterberg et al, 1983; Kundt & Gopal-Krishna, 1984; SS 433: see Kundt, 1985a, 1987b; Cyg X-3: see Kundt, 1986). The majority of writers agree that the jets originate from a magnetised neutron star inside a stellar accretion disk, but already the strength of the magnetic dipole moment is controversial. Strongly controversial - and important to know - is whether the jets of SS 433 consist (mainly) of hydrogen or of relativistic pair plasma.

(4) Binary White Dwarfs: Bipolar nebulae are often associated with young binary white dwarfs; (others are associated with YSOs). The white dwarf is often outshone by a red-giant companion, but to my knowledge it is never missing. At least some of the bipolar nebulae host high-velocity jets along their symmetry axis, among them the Eskimo nebula (Balick, 1987). Other jet sources involving a white dwarf are the symbiotic stars CH Cyg and R Aqr. Even in 1918's well-studied nova V 603 Aql, a twin-jet is indicated by two antipodal holes blown out of the shell at the end of the first day, accompanied by post acceleration (Weaver, 1974). Inhomogeneous as the evidences yet are , they point at a fast-spinning, magnetised, accreting white dwarf as a common building block.

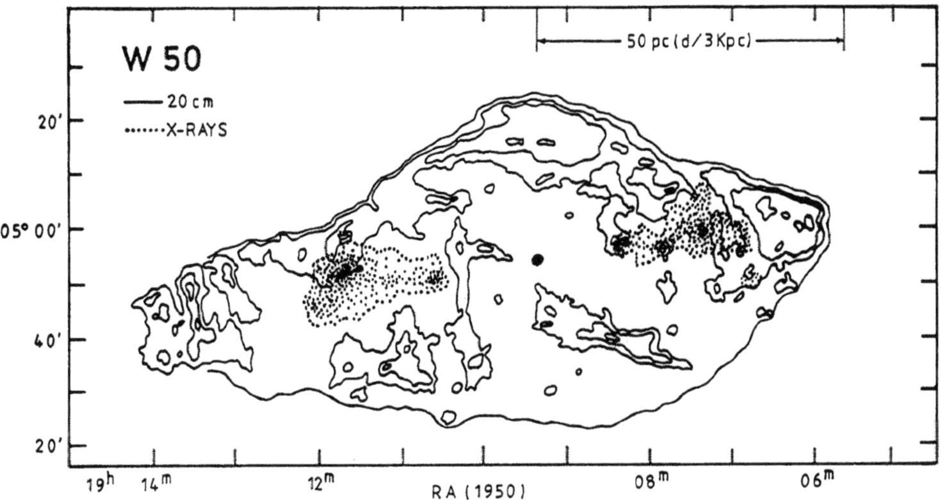

Figure 8. Radio/X-ray overlay map of the old supernova remnant W50 and its 'ansae', blown by the jets from SS 433. For references see Kundt (1987b).

2. COMMON PROPERTIES OF THE JET SET

All the sources introduced in the preceding section have the following properties (and interpretations) in common:

(1) MORPHOLOGY: They are elongated (in 3-space; instead of being spherical). In many cases, two narrow jets emerge from a compact center and end in two elongated 'lobes' or 'cavities' which are (often) interpreted as cocoons. The compact core is likely to contain a magnetised rotator at the center of a (feeding) disk. Well-resolved jets are knotty and braided and have a paraboloidal head containing one or more hotspots with considerable fine structure. For the stellar sources, the knots in the jets are often known as (compact) Herbig-Haro objects.

(2) SPECTRUM: The spectra of the compact centers (cores) - wherever known - tend to be very broad, ranging from radio frequencies through IR and UV frequencies out to X-rays, γ-rays and (occasionally) UHE γ-rays. For different sources, the spectral power per logarithmic frequency, νS_ν, has its peak either near 1 MeV, or at hard UV frequencies ($10^{16\pm1}$ Hz), or near $10^{12.5}$ Hz (cf. Ward et al, 1986). Here the UV bump may reveal the temperature ($\approx 10^{5.5}$ K) of the central engine, the MeV bump may result from dense relativistic pair plasma, and peaks below 10^{14} Hz may be due to screening by (hot or warm) dust. The upper

bound set by the cosmic X-ray background tells us that typical sources at high redshift have their spectral peaks below X-ray energies, in contrast to many 'present-day' sources; i.e. there is spectral evolution.- Even the spectra of the jets range often from the radio to the X-ray regime.- Linear polarization of cores and jets indicates a) a nonthermal origin, b) an organised structure of the central engine, and c) a paucity of low-energy electrons in the emission region.

(3) VARIABILITY: Cores and central jets of the extragalactic sources show rapid variability on all timescales down to days or even minutes, indicating near-relativistic motion of the non-thermal components and extreme compactness of the central engine. As the cores are unresolved, their variability cannot simply be due to fast-moving blobs but must be due to (coherent) beaming and/or switchon. For the stellar sources, scattering by an envelopping windzone and/or progenitor cloud is likely to damp the variability on all short timescales.

(4) SUPERSONIC OUTFLOW: High outflow velocities of interstellar matter from the cores are encountered throughout, in form of the Broad Line Region (=BLR) for AGN, and molecular masers for massive young stars. The broad emission lines for the BLR could alternatively be interpreted as due to quasi-Keplerian rotation or infall. But the following fine structure supports the outflow hypothesis: a) The motion of the BLR can be continuously pursued into the (slower) motion of the surrounding NLR (Filippenko, 1986) and even distant halo (Kundt & Krause, 1985; Turnshek, 1986); b) Stimulated hydrogen recombination lines (H 83α and H99α) have been seen blueshifted in front of Mrk 668 (=OQ 208), (Bell & Seaquist, 1980); c) The red wings of some narrow emission lines are seen attenuated in Mrk 509 (Phillips et al, 1983); d) Broad blue-shifted absorption lines occur in some 10% of all QSOs (Hartig & Baldwin, 1986); e) Time variability and emission-line shapes in three nearby Seyferts have been interpreted as outflows by Stoner & Ptak (1985); f) Gaskell (1987) interprets delayed blue emission-line wings (of C IV and Mg II) as infall, but delayed shock ionisation of the outer parts of the approaching filaments may be an alternative interpretation (see also Gaskell & Sparke, 1986).

(5) JET VELOCITIES: Emission lines from the jets of both extragalactic and galactic jets have been seen blueshifted (and/or sometimes redshifted) corresponding to radial velocities in the range from 10 to 10^3 Km s^{-1}. Such velocities reveal the motion of warm thermal inclusions in the cocoon, the 'channel wall', and should not be confused with the speed of the jet fluid (Kundt & Gopal-Krishna, 1981; Kundt, 1982, 1984a). The latter may well be extremely relativistic ($\gamma > 10$) in all cases (Noerdlinger, 1978; Kundt & Gopal-Krishna, 1981; Blome & Kundt, 1987), including SS 433 (Kundt, 1985a, 1987b).

(6) CORE-LOBE RATIOS: Rough estimates for the (total) power ratios between central engine and jet-fed outflow yield: L(core)/L(flow) = $10^{2\pm2}$ for both the extragalactic (Kundt & Gopal-Krishna, 1981) and stellar sources (Lada, 1985, perhaps to be modified by inclusion of the IR luminosities of the flows). The large scatter may be due to a longterm variability of the core. If confirmed with better analysis and statistics, this result means that jet formation occurs with an efficiency of order 1%. The corresponding ratio for the radial momen-

tum rates $\dot{P}$ has only been assessed for the stellar sources; it reads: $\dot{P}(\text{core})/\dot{P}(\text{flow}) = 10^{-2\pm2}$ and has been a major difficulty for stellar-wind interpretations (Kundt, 1984a).

3. DISCRIMINATING PROPERTIES OF THE JET SET

Whereas all the jet sources share the qualitative properties discussed in the last section, they differ vastly in their quantitative (extensive) properties.

(1) The power L of the different sources sweeps through the following ranges:

$$\begin{Bmatrix} 10^{39} \\ 10^{32} \end{Bmatrix} \lesssim L/\text{erg s}^{-1} \lesssim \begin{Bmatrix} 10^{48}, & \text{extragalactic} \\ 10^{39}, & \text{stellar} \end{Bmatrix}.$$

(2) The mass density ρ of the ambient IGM or ISM, into which the jets ram their channels, ranges as follows:

$$\begin{Bmatrix} 10^{-29} \\ 10^{-25} \end{Bmatrix} \lesssim \rho/\text{g cm}^{-3} \lesssim \begin{Bmatrix} 10^{-25}, & \text{extragalactic} \\ 10^{-19}, & \text{stellar} \end{Bmatrix}.$$

(3) The length ℓ and age t of the jets satisfy:

$$\begin{Bmatrix} 10^{21} \\ 10^{17} \end{Bmatrix} \lesssim \ell/\text{cm} \lesssim \begin{Bmatrix} 10^{25}, & \text{extragalactic} \\ 10^{19}, & \text{stellar} \end{Bmatrix}.$$

$$\begin{Bmatrix} 10^{6} \\ 10^{3} \end{Bmatrix} \lesssim t/\text{yr} \lesssim \begin{Bmatrix} 10^{8}, & \text{extragalactic} \\ 10^{4.5}, & \text{stellar} \end{Bmatrix}.$$

(4) Compactness $u := GM/Rc^2 \lesssim 1$ of the central engine: In the stellar case, u ranges between 10^{-6} (for a low-mass star) and $10^{-0.7}$ (for a neutron star) whereas in the extragalactic case, the estimates for u range between 10^{-5} and $10^{-2.5}$ in the SMC model, or u=1 in the black-hole model (see next lecture). Again there is a wide spread.

4. AMBIGUITIES IN THE MODEL ASSUMPTIONS

In the literature, the following alternative assumptions have been made for the description of the various jet sources:

(1) The BEAM is $\begin{Bmatrix} \text{HARD} \\ \text{SOFT} \end{Bmatrix}$ AS IN A $\begin{Bmatrix} \text{LAWN SPRINKLER} \\ \text{HAIR DRIER} \end{Bmatrix}$.

(2) The BEAMS are composed of $\begin{Bmatrix} \text{HYDROGEN} \\ \text{RELATIVISTIC PAIR PLASMA} \end{Bmatrix}$.

(3) The BEAM VELOCITY is $\begin{Bmatrix} \text{NON-RELATIVISTIC} \\ \text{EXTREMELY RELATIVISTIC} \end{Bmatrix}$.

(4) The BEAM-PARTICLE motion is $\left\{\begin{array}{l}\text{PRESSURE-BOUND (GASDYNAMIC)}\\ \text{FIELD-GUIDED}\end{array}\right\}$.

(5) The OUTFLOW region of a BF is a $\left\{\begin{array}{l}\text{WINDZONE}\\ \text{EXPANDING COCOON}\end{array}\right\}$.

(6) The observed HEAD of a BF is a $\left\{\begin{array}{l}\text{MACH CONE (i.e. WAVE)}\\ \text{COCOON (i.e. BALLOON)}\end{array}\right\}$.

(7) IN-SITU ACCELERATION in the knots is $\left\{\begin{array}{l}\text{ABSOLUTELY NECESSARY}\\ \text{IGNORABLE}\end{array}\right\}$.

(8) The BEAMING pattern in the jets is due to $\left\{\begin{array}{l}\text{BULK VELOCITY}\\ \text{SPREAD IN TANGENTS}\end{array}\right\}$.

(9) The BRIGHT KNOTS $\left\{\begin{array}{c}\text{are NOT}\\ \text{are}\end{array}\right\}$ PRESSURE-CONFINED.

(10) The UV-SOURCE is the $\left\{\begin{array}{l}\text{INNER DISK}\\ \text{SUPER-MASSIVE CORE}\end{array}\right\}$.

(11) The BLR is filled with $\left\{\begin{array}{l}\text{HOT HYDROGEN}\\ \text{RELATIVISTIC PAIR PLASMA}\end{array}\right\}$.

(12) The MOTION in the BLR is $\left\{\begin{array}{l}\text{INFALL / ROTATION}\\ \text{OUTFLOW}\end{array}\right\}$.

(13) The jet in 3C 273 is intrinsically $\left\{\begin{array}{l}\text{ONE-SIDED}\\ \text{TWO-SIDED}\end{array}\right\}$.

(14) The LOBES OF SCO X-1 are $\left\{\begin{array}{l}\text{MAGNETICALLY/RAM-PRESSURE-}\\ \text{STATICALLY}\end{array}\right\}$ CONFINED.

(15) The MOVING EMISSION LINES OF SS 433 are emitted by the $\left\{\begin{array}{l}\text{JETS}\\ \text{DISK}\end{array}\right\}$.

Needless to say, different sources may realize different alternatives, or the truth may be somewhere in between the two listed possibilities. But in all cases except possibly (9) (cf. Falle, this volume), I definitely prefer the second alternative. A few explaining remarks may be in order:

(1): The soft-beam model is widely preferred in the literature except for SS 433 and for the interpretation of 'precessing' sources, compare Icke (1986). Figure 9 shows what can be interpreted in two very different ways: an S-shape via straightline motion of bullets or via almost weightless blobs following a channel.

(4): If the beams consist of relativistic pair plasma, their densities are so low that neither collisions nor wave-wave interactions can cause the particles to feel a density gradient, i.e. to behave like a fluid medium. On the other hand, if the charges $\vec{E}\times\vec{B}$-drift in equi-pres-

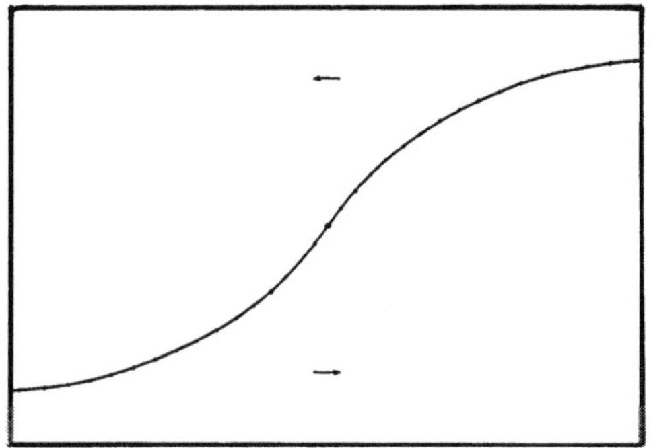

Figure 9. This shape can be interpreted in two ways: as a sequence of 'bullets' moving radially away from the center, in successively different directions (hard beam, realized by lawn sprinkler), or as the channel of a light flow dragged along by a rotating medium (soft beam, realized by hair drier).

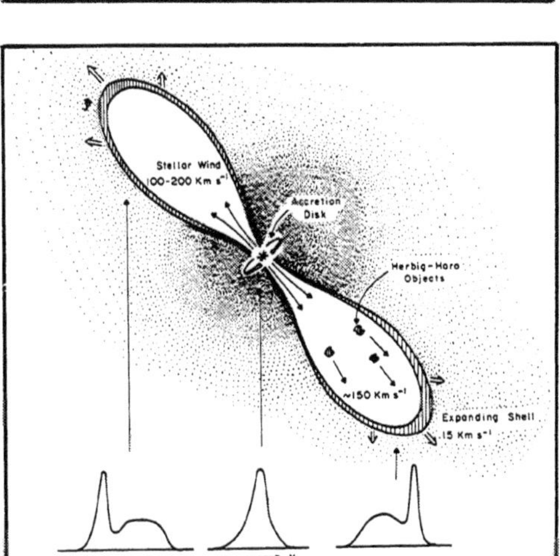

Figure 10. Interpretation of L 1551/IRS 5 by Snell et al (1980), which differs significantly from that given in figure 6.

sure (crossed) fields, the fields may force them to move at an essentially uniform speed in a (locally) uniform direction. These fields - if strong enough - can prevent the charges from moving like free particles.

(5,6): Figure 10 (from Snell et al, 1980) demonstrates how many authors have visualized the bipolar-flow phenomenon: viz. as ejection of a stellar wind into a spherical opening angle of (small but) finite size. Apart from the above-mentioned momentum problem, such models face the continual-focussing problem: why do the flows stay narrow? I therefore prefer the interpretation sketched in figure 6. In this interpretation,the 'cavity' is blown like a balloon through an imaginary straw whose far end is always kept near the far end of the balloon.

(7,8): The logical alternative to in-situ acceleration is in-situ deceleration, (essentially) realized in every TV screen: Only a small subset of the streaming particles hits obstacles (of small volume-filling factor) and is deflected through large angles whereas the majority of the charges continue their loss-free $\vec{E}\times\vec{B}$-drift. The resulting beaming pattern (↔distribution of instantaneous tangents) of the scattered charges may be similar to what is sketched in figure 11 (Kundt, 1982a, 1984b; Lind & Blandford, 1985). This alternative interpretation is in

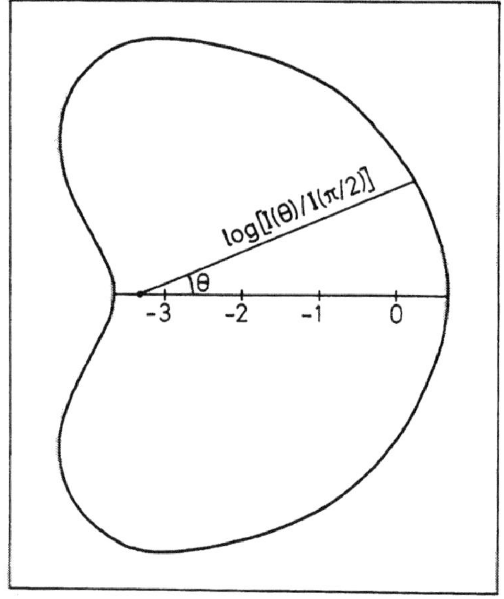

Figure 11. Empirical beamed intensity $I(\theta)$ as a function of inclination angle θ, plotted logarithmically. The plot is gleaned from the statistics of extragalactic one-sided and two-sided jets. Quite likely, its precise shape varies with the beam power and the density of the ambient medium.

line with the Scheuer & Readhead approach (1979) without running into the difficulty that the bulk Lorentz factors derived in the various analyses come out mutually different (for the same set of objects).

It is often maintained that relativistic particles in the solar wind owe their high energies to stochastic in-situ acceleration; the opposite view is defended by Sarris & Krimigis (1985).

(9): Once it is realized that there may be two populations of charges in the jets, viz the unscattered and the scattered ones, all estimates of local pressures become insecure and may have to be lowered. I shall return to this problem in my last lecture.

(10,11,12): Will be covered by my second lecture.

(13): 3C 273 is an extreme case (Kundt & Gopal-Krishna, 1986). If its receding jet were non-existent, our insight into the working scheme of the central engine would receive a serious modification: the huge overpressure in the core should discharge to both sides of a feeding disk, in accord with the almost universal presence of two lobes (rather than one) around an AGN.

(14,15), Sco X-1 and SS 433: Again it is important to understand the working scheme of these well-studied, nearby mini-quasars; is it the same as that of their extragalactic 'big brothers'? I think it is (Kundt & Gopal-Krishna, 1986; Kundt, 1985a, 1987b).

ACKNOWLEDGEMENTS

When I first expressed the conviction that even the stellar jet sources may be powered by high-pressure pair plasma, my acknowledgements got lost in the proofsheets (1984a). They were extended to John Bally, Peter Biermann, Ken Johnston, Gopal Krishna and in particular to Wilfred Sorrell. To these, I wish to add thanks to my young collaborators Hajo Blome, Axel Jessner and Reinhold Schaaf as well as to Mehmet Özel and to my senior advisor Wolfgang Priester for their continual encouragement and help. I also acknowledge help from Peter Scheuer and Bill Sherwood.

THE CENTRAL ENGINE

ABSTRACT. The literature of the past 10 years has favoured black holes over non-collapsed (super-)massive objects as the central engines of AGN behaviour. But bipolar flows and jets from compact stars are similar and driven by the cores of accretion disks. Some difficulties of the black-hole model are discussed, in particular the problem of the missing ashes. The (super-)massive magnetised core model of an accretion disk (=SMC) is revisited.

1. BRIEF HISTORY

To my knowledge, the first attempt at modelling the central engine of a quasar was by Hoyle & Fowler (1963), in the form of a supermassive rotator. Their ideas were followed up more quantitatively by Fowler (1966) who stressed the stabilizing importance of stellar rotation and by Ozernoy (1966) who also stressed the importance of magnetic fields. The discovery of pulsars caused Morrison (1969) to discuss the magnetised rotator model for AGN.

At about the same time, the search for black holes in the Universe homed in on QSOs: Lynden-Bell (1969) assessed their candidacy as the central engines of quasar activity. But even four years later, Pacini & Rees (1973) pursued the magnetised rotator (spinar) model, whose properties were further analysed by Ozernoy & Usov (1977) under the name of 'magnetoid'.

At the 1976 Texas Symposium, Martin Rees elaborated the virtues of a black-hole engine: A black hole is, among others, the stable asymptotic state towards which supermassive bodies evolve (Rees, 1977; Begelman et al, 1984). (Quantum-mechanical black-hole evaporation is ignorable on astrophysically interesting time scales and in universes of higher than $10^{-7}(M_{\odot}/M)$K temperature).

At the 1977 quasar symposium in Cambridge, a new building block was introduced by Gunn (1979), viz the disk which is needed for 'feeding the monster' throughout cosmic times. Pursuing this requirement, I got the impression (1979) that the cores of (super-) massive disks may combine the virtues of magnetised rotating stars with a) an absence of strict periodicity, b) an avoidance of gravitational collapse (with the implied unrealistically large hidden masses in the centers of present-day galaxies), and c) a nuclear enrichment of the IGM. Similar scenarios have been explored more recently by Camenzind & Courvoisier (1983) and by Krügel & Tutukov (1986); see also Terlevich & Melnick (1985) who stress the similarity of AGN activity to that caused by a very massive star.

2. PROPERTIES OF THE CENTRAL ENGINE

Assuming there is a common working scheme to all astrophysical jet sources, one can find necessary conditions on their central engine (=CE):

(1) A CE sits at the center of every massive disk.

(2) The CE blows off a fast, clumpy wind of small filling factor; (cf. alternative (12) of my first lecture).

(3) $\dot{M}_{in} \approx \dot{M}_{out}$ holds, i.e. there is no ultimate accumulation of mass in the CE; (Stoner & Ptak, 1985, in the case of AGN).

(4) Hydrogen burning can decompose a supermassive CE.

(5) The output of the CE is not periodic.

(6) The CE creates an overpressure bubble of relativistic pair plasma.

(7) The CE can be realized by an AGN, a pre-T-Tauri star, a binary neutron star, and a binary white dwarf.

(8) The overall linear and circular polarization of a CE's output need not vanish.

(9) AGN luminosity fades exponentially with cosmic age, a phenomenon suggestive of decreasing compactness, or mass, of the CE (eqs. (8) through (12) below).

Whereas the black-hole model clearly fails to satisfy properties (3), (4), and (7), the model of the (Super-) massive Magnetised Core (SMC) of a (super-) massive disk stands good chances to satisfy all nine properties. Before we discuss it, let us focus on the "generally accepted" black hole as a model for (at least) the AGN sources.

3. REASONS FAVOURING THE BLACK HOLE MODEL

Seven such reasons have been recently discussed by Blandford (1986). They are:

(1) Rapid variability suggests a very compact engine.

(2) Alignment of nuclear and extended jets suggests the presence of a stable symmetry axis.

(3) High unresolved core masses ($\sim 3 \cdot 10^9\ M_\odot$) of nearby galaxies are indicative of a black hole.

(4) The broad emission lines suggest high central masses.

(5) Superluminal expansion suggests a relativistic gravitational potential well.

(6) The high luminosities imply high efficiencies of conversion of fuel into output, higher than allowed by nuclear burning.

(7) Dense star clusters or spinars evolve towards black holes during their active cycle.

Blandford emphasizes, however, "that although this is all good circumstantial evidence, we have no proof that black holes exist in active galactic nuclei or indeed anywhere else." More recently, Trimble & Woltjer (1986) reinforce the black hole model by reformulating five of the above reasons - omitting (4) and (6) - and adding two new ones:

(8) Occasional outbursts have energies as large as the rest-energy of a star.

(9) Lack of credible alternatives (to the black hole), in spite of the enormous number of theorist-hours that have gone into the problem.

How cogent are the above nine arguments?

(1): Rapid variability is certainly a signature of compactness, but the (Keplerian) rotation rates of massive black holes are predicted similar to those of the - less massive and less compact - SMCs. Of course, Eddington's luminosity constraint sets a lower limit on the mass of the central engine which cannot be largely bypassed by having separate infall and outflow sectors. On the other hand, an SMC can anchor a transverse magnetic dipole whose ability to modulate the output is certainly superior to that of an accretion disk. The Sun shows how variable even a quasi-static star can be.

(2): The angular momentum of the (feeding) galactic disk is certainly larger - and hence more stable - than that of the putative nuclear black hole.

(3): In fact, the highest (reported) unresolved core mass M_c among nearby galaxies is that of M87 which was initially claimed to be of order $3 \cdot 10^9$ $M_\odot$ but revised downward to $M_c \lesssim 10^4$ $M_\odot$ by taking proper account of plausible deviations from spherical symmetry (of the central star cluster: Duncan & Wheeler, 1980). For our own Galaxy, Allen & Sanders (1986) and Ozernoy (1986) suggest $M_c \lesssim 10^2$ $M_\odot$. For some 65 nearby galaxies, Kormendy (1987) finds unresolved core masses of order 10^6 $M_\odot$, with only two possible mild exceptions (M 31 and NGC 3115). The very absence of supermassive unresolved cores ($M_c > 10^{10}$ $M_\odot$) in all nearby galaxies makes me question the black-hole interpretation.

(4): The mass ejection velocities of AGN are not much higher than those of hot stars.

(5): Anisotropic particle acceleration to relativistic velocities happens at the Earth' bowshock; it does not require a relativistic potential well.

(6,8): Estimates of black-hole efficiencies (in converting accreted rest-mass energy to radiated energy) have reached and exceeded 10%, but only for suitable fuel in suitable ranges of black-hole mass and accretion rate. Moreover, theoretical estimates of efficiencies tend to be overestimates (Kundt, 1984b). If black holes are to eject matter, they have to be fed supercritically, and the efficiency is likely to sink below the 1% level - which is the level of hydrogen burning.

(7): Star clusters and spinars end as black holes, but not necessarily SMCs: Their winds can be strong enough for disintegration (Krügel & Tutukov, 1986).

(9): If the authors remembered how difficult it is to get a non-black-hole paper a) through the referees, b) to the desk, c) into a review, or d) through an observer's application, they would come to the opposite conclusion; cf. Wiita (1985).

Let us return once more to reason (6), that of the high needed average efficiency ε. Peacock in his accompanying lecture has repeated Soltan's (1982) estimate of the amount of ashes left over from the primordial quasar fires, with the result that a galaxy like ours has a share (in fuelling the quasars) of $\gtrsim \varepsilon^{-1} 10^8 M_\odot$. If our galaxy had never housed a quasar, neighbouring galaxies would have an even higher share in the total fuel supply. Under the assumption, therefore, that quasar activity is powered by black holes, we find for their (average) efficiency:

$$\varepsilon \gtrsim 10^8 M_\odot/M_c \tag{1}$$

where M_c is the present-day unexplained core mass. Inserting core masses between 10^2 and 10^6 $M_\odot$ (for the estimated unresolved core masses of nearby galaxies, cf. (3) above) yields needed average efficiencies between 10^6 and 10^2, in contrast to the expected efficiencies of $\varepsilon \lesssim 10^{-2}$. Apparently we have made an unrealistic assumption. The discrepancy (by a factor of $\gtrsim 10^4$) goes away when the blackhole assumption is replaced by the condition that the AGN ashes are ejected (rather than swallowed). This discrepancy has been recently aggravated by realizing that black holes enveloped by pair plasma would radiate at $\approx 10^3$ times their Eddington luminosity, i.e. would have to be some 10^3 times more massive than estimated earlier.

4. THE SMC MODEL FOR THE CENTRAL ENGINE

In sections 2 and 3, properties have been listed and arguments given that the central engine of the jet sources may be the (super-) massive magnetised core of a (super-) massive disk. The model assumes that disks are not strictly stationary, i.e. that viscosity forces matter to spiral inward and accumulate in the central core. The reason for the (high) viscosity is probably the quasi-toroidal magnetic field which gets enhanced by the shear (of the non-rigid flow) and whose stresses would grow faster with increasing approach to the center ($\sim r^{-4}$) than turbulent stresses unless magnetic flux were continually destroyed (expelled and/or burnt). The field will therefore be always near equipressure, hence strong enough to enforce spiral-in. In the case of our Galaxy, this view is supported by the fact that on average, the molecular-cloud complexes move inward w.r.t. the Local System of Rest with a speed of 5 Km s^{-1}, corresponding to $\dot{M} \lesssim M_\odot$/yr (Blitz et al, 1983). More precisely, Blitz finds $-(4.5\pm1)$ Km s^{-1} w.r.t. the LSR for the clouds in anticenter direction and (5.5 ± 1) Km s^{-1} for the ones in center direction (private communication, 1986).

By accretion, the core of the feeding disk will always be kept near maximal (Keplerian) rotation; in its interior it is expected to burn hydrogen. Unlike a spinar, an SMC is continually refuelled. It will be fully convective and radiate thermally near its Eddington luminosity, whence (Sorrell, 1981):

$$\left\{\begin{array}{l} L_{therm} \approx L_{Edd} = 4\pi G\, M\, m_p\, c/\sigma_T \approx 10^{44} \text{ erg s}^{-1}\, M_{(6)} \\ M_{(6)} \gtrsim R_{14.5}^{\;2.1}\, Z_o^{\;-1/8} \end{array}\right\} \quad \begin{array}{r}(2)\\(3)\end{array}$$

where $M_{(6)} := M/10^6\, M_\odot$, and $Z_o := Z/Z_\odot$ is its metallicity in solar units. Unless the Eddington limit can be largely exceeded, a thermal power of 10^{46} erg s^{-1} therefore asks for a core mass not much below $10^8\, M_\odot$.

The new property of an SMC - when compared with a slowly rotating star - is its likely equi-pressure magnetic field B, inherited from the feeding disk and/or regenerated via turbulence. Clearly, it cannot exceed the equi-pressure limit of

$$B \lesssim (8\pi\varrho v_t^{\;2})^{1/2} \lesssim 10^{6.7}\text{ G } (\varrho_{-1}\, T_5)^{1/2}, \tag{4}$$

in which the turbulent velocity v_t is assumed of the order of sound speed ($v_8 \approx \sqrt{T_8}$), and the mass density ϱ is inserted in units of 10^{-1} g cm^{-3}.

If an SMC is strongly magnetised, with a significant dipole component, it has the properties of a magnetoid all of whose energies are near equipartition (Ozernoy & Usov, 1977):

$$\alpha\, |E_{grav}| = E_{rot} \gtrsim E_{turb} \gtrsim E_{mag}\ ,\ E_{rot} \approx E_{therm}\ . \tag{5}$$

Conservation of energy reads

$$\beta\, |E_{grav}| = E_{rot} + E_{turb} + E_{mag} + E_{therm} \lesssim 4\, E_{rot}\ , \tag{6}$$

whence $\beta \lesssim 4\alpha$. Simple estimates yield: $E_{grav} \approx -(5/6)\, GM^2/R$, $E_{rot} \approx M(R\Omega)^2/6 = GM^2/6R$, $E_{mag} = B^2R^3/6$, hence $\alpha \approx 1/5$. With the compactness parameter u defined as

$$u := GM/Rc^2 = 10^{-5.7}\ (M/M_\odot), \tag{7}$$

we thus get:

$$\left\{\begin{array}{l} R = 10^{14.2}\text{ cm } M_{(6)}/u_{-3}\ ,\ R\Omega/c = \sqrt{u}\ , \\ \Omega = c^3\, u^{3/2}/GM = 10^{-5.2}\ s^{-1}\ u_{-3}^{\;3/2}/M_{(6)} \\ B \lesssim c^4\, u^2/G^{3/2}\, M = 10^{7.4}\text{ G } u_{-3}^{\;2}/M_{(6)} \end{array}\right\}. \quad \begin{array}{r}(8)\\(9)\\(10)\end{array}$$

The radiated magnetic dipole power $L_D = 2\,\ddot{D}_\perp^2/3c^3$, with $D = BR^3$, $D_\perp =: \eta_D\, D$ = transverse magnetic dipole moment, ($\eta_D < 1$), amounts to

$$L_D = \eta_D\ 2\, c^5\, u^4/3G = \eta_D\ 10^{47.4}\text{ erg s}^{-1}\ u_{-3}^{\;4}. \tag{11}$$

These expressions give estimates for the size 2R, rotation period $P = 2\pi/\Omega$, surface magnetic field strength B and magnetic dipole power of the central engine whereby the 'dipole efficiency' factor η_D is probably much smaller than unity. Still, the highest observed QSO power, of order 10^{48} erg s^{-1}, may be achievable with cores whose compactness u is distinctly smaller than 10^{-2}, i.e. which can be disintegrated by hydrogen burning. Note also that for $u = 10^{-3}$, the core is 30 times smaller than its speed-of-light radius c/Ω.

When equations (7) through (11) are combined with equations (2) and (3) for the thermal structure, we find that

$$u_{-3} \approx R_{14} \, Z_o^{-1/8} , \tag{12}$$

i.e. that the compactness of an SMC grows with its radius, further that its spin period

$$P = 2\pi/\Omega \approx 10^6 \text{ s } R_{14}^{1/2} \tag{13}$$

is of the order of week for the strong sources, and that $B \sim Z_o^{-1/8}$ is very nearly independent of mass (and/or radius), probably of order Mega Gauß. These estimates look reasonable for the whole range of masses, from a solar mass (YSOs) upward to the most luminous AGN, of $M/M_\odot \approx 10^8$. The energy equipartition approximation may therefore not be overly unrealistic.

An application of these formulae to our present galactic center, with its (IR-) output of some 10^{39} erg s^{-1}, leads to $u \gtrsim 10^{-4.7}$, $R \approx 10^{12}$ cm, $c/\Omega \approx 10^{14.4}$ cm, and $M \approx 10^2 \, M_\odot$, in agreement with the suggestion by Allen & Sanders (1986) and Ozernoy (1986). In particular, if this model can be trusted, the resolved VLBI structure of Sgr A* falls inside the speed-of-light cylinder of the central engine!

A glance at the list of necessary conditions collected in section 2 for the CE shows that we have not yet indicated how the pair plasma comes into existence and how the stellar wind gets squeezed into low-filling-factor cloudlets, or filaments. A mechanism is suggested by Earth in whose corotating magnetosphere we observe occasional (small-scale: Debye length) double-layer discharges of high voltage, see figure 12. If high voltages are induced between different magnetic field lines of an SMC, e.g. in the vicinity of its speed-of-light cylinder, these voltages will give rise to field-aligned electric currents. The currents in turn can convect these voltages to a small region of high resistivity where the whole electric energy is transformed into relativistic charges. These relativistic charges create $e^{\pm}$-pairs on collision with stellar photons as soon as their energies exceed some 10^{12} eV = 1 erg (for electrons and positrons) or 10^{15} eV (for protons), respectively (cf. Mastichiadis et al., 1986). In this way, the whole magnetosphere will be filled with relativistic pair plasma once its voltage has risen above threshold. The pairs in turn will slide away from the rotation axis, along the magnetic dipole field lines, and thus be post-accelerated (Kundt, 1986). In this way, the magnetosphere resembles a centrifugal pump which creates a high pressure in its environment. Pressure equilibrium with this pair-plasma will squeeze the thermal wind of the SMC into filaments of low filling-factor.

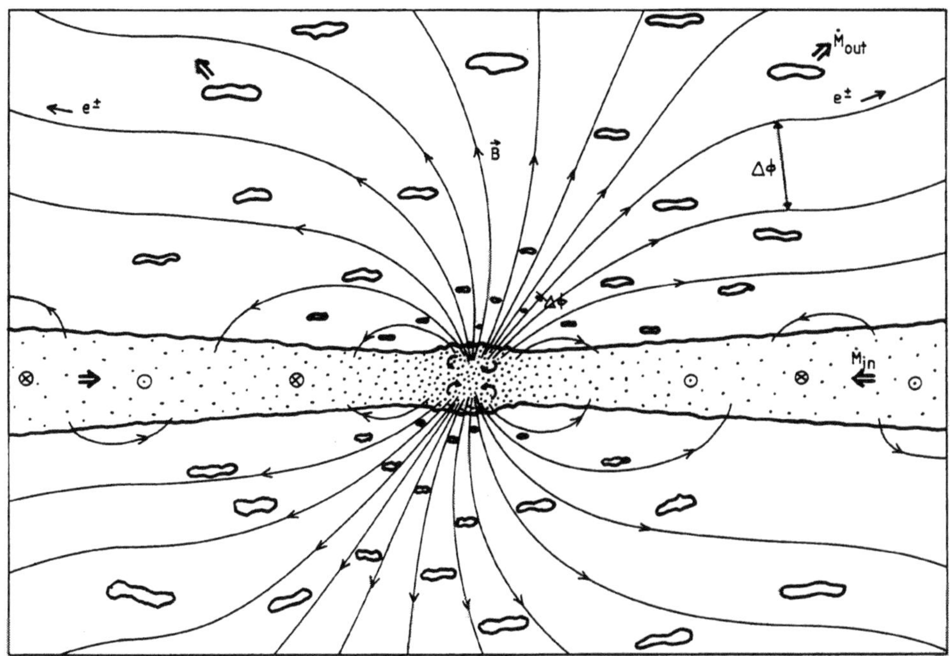

Figure 12. Sketch of an SMC: The maximally rotating core of a (feeding) disk is strongly magnetised ($B \gtrsim$ M Gauß). Near its speed-of-light cylinder, voltages $\Delta\Phi$ are induced which discharge in localized double layers (inside of the corotating magnetosphere), near the core's surface. High-energy charges collide with stellar photons and produce $e^{\pm}$-pairs which are centrifugally expelled, resulting in an overpressure bubble. The core's thermal wind takes the form of magnetically bandaged, small-filling-factor filaments, perhaps in the shape of smoke rings.

In order for this hypothetical scheme to work, it must be guaranteed that electric voltages Φ in the magnetosphere of an SMC exceed the pair-formation threshold. We get from equs. (7) to (10):

$$e\Phi \approx e\int(\vec{\beta}\times\vec{B})\cdot d\vec{x} \lesssim e\, c^2u^2/G^{1/2} = 10^{9.2} \text{ erg } u_{-3}^{2} \qquad (14)$$

where the integral has been (pessimistically) evaluated at $r = c/\Omega$, with $B(\mathbf{r}) \sim r^{-3}$. This formula predicts a magnetospheric particle energy way in excess of 1 erg for all SMC masses down to one solar mass ($u = 10^{-5.7}$). According to equ. (4), B and Φ are somewhat overestimated at the high-u end. Yet there seems to be a sufficient margin for pair production even by newly formed solar-mass stars. On the other hand, Φ decreases quickly with spin frequency, in proportion to Ω^2. One therefore expects the bipolar-flow phenomenon of young stars to terminate as soon as the latter have been spun down to the observed angular frequencies of T-Tauri stars, in agreement with the observations.

A number of recent studies have found that the hard parts of QSO spectra may well be understood as produced by relativistic pair plasma, among them those by Kazanas (1984), Fukue (1986), Kardashev et al (1986), and Blandford (1986). These findings lend further credibility to the model.

5. BINARY SMCs

As stressed by Gaskell since 1982, significant peak-redshift offsets between the BLR and NLR or redshift splittings of the peaks of the broad emission lines are known which suggest binary orbital motion at velocities $\gtrsim 10^3$ km s^{-1} of two central engines. An orbital velocity v of the two CEs would correspond to a binary separation

$$a = 10^{17} \text{ cm } M_{(7)} / v_8^{\,2} \tag{15}$$

for cores of total systemic mass $M = 10^7 M_\odot$, just larger than their BLRs (Gaskell, this volume). Quasars with such "complex" BLR profiles appear to correlate with extended radio structure, i.e. to give rise to particularly long-lived or strong jets.

A binary nature of the CE does not seem to conflict with the SMC model: On the contrary, binaries are the rule rather than the exception in the formation of massive stars from protostellar accretion disks (Blair & Candy, 1985; Kundt, 1985b). The binary cores form as a result of the bar mode instability of the disk.

A possible worry concerns the interpretation of the two high-velocity outflows seen in the form of two displaced BLRs. But if their filling factors are sufficiently small, collisions between individual filaments are of unnoticeably low power.

Why should binary AGN make longer jets than single ones? To begin with, it is not clear whether there are single AGN at all. Assuming there are, i.e. that large jets do come from binary AGN whereas short jets do not, a possible reason is a longer active cycle of the former due to a larger amount of accessible disk mass (for accretion). Another possible reason is a more completely ionised environment so that entrainment (of neutral intruders into the beam) is prohibited.

THE JETS

ABSTRACT. Jets may simply be the outlets of overpressure bubbles filled with relativistic pair plasma. They ram channels through media of various densities but can be destroyed by neutral intruders, such as photons and hydrogen atoms. The possibility of in-situ acceleration of electrons in their hotspots, the correct form of the relativistic beaming law and the confinement problem are discussed. It is argued that all jets are supersonic, that the flow in the cocoon depends on the ambient density distribution and that many extragalactic jets interact visibly with massive shells in the halos of their host galaxies.

1. UNIPOLAR FLOWS, JET FORMATION, CHANNELS, AND COCOONS

Before assessing the problem of jet formation and focussing, I should like to point at the morphological similarity between bipolar flows and certain peculiarities of supernova shells. (Remember that BLRs look similar to the unresolved Crab nebula). Two such 'bursting' supernova shells (remnants) are shown in figures 13 and 14: the Crab nebula and the galactic radio source Kes 32 = G 332.4+0.1 (Roger et al, 1985). They are not unique: roughly half of all supernova shells show similar 'chimneys', or 'spurs', or 'outlets', or 'bumps' on their circumference. More precisely, one sees zero or one bump per shell, in accord with the expectation that in about half of all cases, 'the' bump should not be visible in profile. Besides the Crab and Kes 32, impressive cases are Cas A, RCW 86 and the Cygnus Loop.

Why do 'all' supernova shells apparently have chimneys? A common explanation could be that relativistic pair plasma escapes from an overpressure storage volume, the supernova bubble (Kundt, 1983). In extreme cases like Kes 32, the overpressure has been high enough to create a 'unipolar flow' which looks almost indistinguishable from half a head-tail galaxy.

If this interpretation is right, the only difference of bipolar flows (from unipolar flows) would be that the overpressure at their center relaxes to both sides of the massive, feeding disk. All that is needed for jet formation is an overpressure bubble filled with relativistic pair plasma! The weakest spot of the bubble's edge serves as an outlet. Once a chimney has formed, its walls serve as a nozzle, particle velocities align, the flow turns supersonic, and forward (ram) pressures exceed sideways (static) pressures by large factors.

In this way the flow stays narrow, i.e. becomes a jet. At its downstream end where the growing jet rams into the ambient medium, ram pressures are reconverted into static pressures across a terminal shock: The jet blows a 'cocoon' - called 'lobe' by observers - cf. figures 2,4,6 and 7. In the case of an AGN, the pressure decreases from $\gtrsim 10^{-1}$ dyn cm^{-2} in the inner BLR to $\lesssim 10^{-8}$ dyn cm^{-2} in the outer hotspots, through a factor of $\gtrsim 10^{7}$.

The proposed scheme, that jets spring forth from overpressure bubbles, may even be recognisable in our galactic center where a

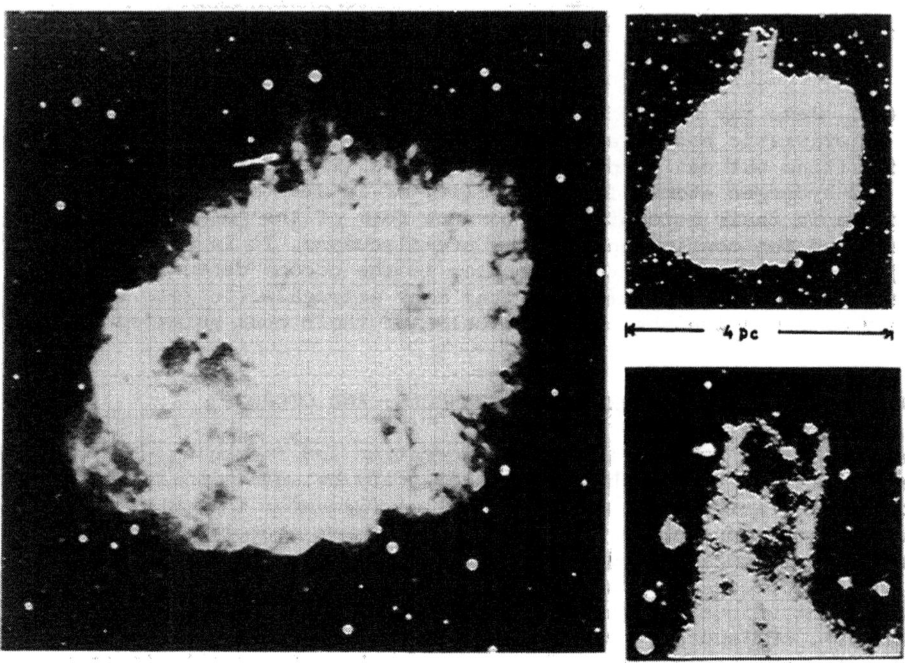

Figure 13. Deep optical emission-line map of the Crab nebula and its 'chimney', see Fesen & Gull (1986). The map is similar to the (polarized) radio morphology.

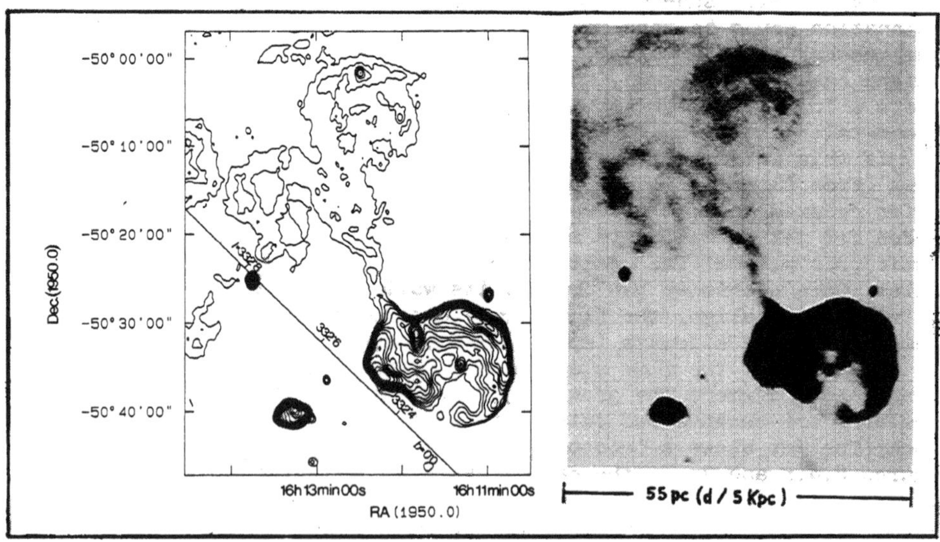

Figure 14. 0.84 GHz map of the 'outgassing' supernova shell Kes 32 (=G 332.4 + 0.1), taken from Roger et al (1985).

southern and a northern jet depart from different ends of the Sgr A East bubble (Yusef-Zadeh et al, 1986).

At the head of a flow, near its downstream termination shock, the contact discontinuity towards the ambient medium is Rayleigh-Taylor unstable during the (switchon) phase of growing pressure. The ambient medium will be squeezed into high-density, low-filling-factor filaments which later-on mark the edge of a channel, the 'channel wall'. Such channel walls can be recognised in figures 2,4,5 and 7, directly through their optical (line) radiation and/or indirectly through the synchrotron (radio) radiation emitted by the jet fluid, probably as a consequence of effective 'friction' on the walls. (Another possible explanation are internal instabilities of the supersonic flow, see Falle's accompanying lectures). In old flows, the channel walls can serve as cold traps and give rise to HI condensations, see figure 5.

As a cocoon expands, its pressure relaxes. This second phase of cocoon formation is Rayleigh-Taylor stable and leads to smooth shapes. In particular, the bowshocks of young flows (such as HH34, 3C 33 South and others) have been mapped, see figure 7. They look paraboloidal, which is expected as a consequence of 4-momentum conservation (Blome & Kundt, 1987): The curvature radius R_b of the bow shock equals the jet radius R_j times the fourth root of the density contrast:

$$R_b \approx R_j(\rho_a/\rho_j)^{1/4} \qquad (16)$$

where ρ_a is the ambient mass density and ρ_j that of the jet fluid.

2. IN-SITU ACCELERATION, CONFINEMENT, AND BEAMING

This section is devoted to three tenacious problems: Where do the radiating relativistic electrons receive their extremely high energies ($10^2 \lesssim \gamma \lesssim 10^7$), what is the beaming pattern of their synchrotron radiation, and how high is the local pressure?

In the literature it is generally argued that the electrons radiating in the hotspots must be accelerated in situ because their synchrotron loss times are short compared to their hotspot-crossing times. This conclusion is based on the tacit assumption (of homogeneity) that there is only one population of relativistic electrons. It is also influenced by the feeling that highly relativistic electrons ought not be able to traverse Mpc-distances almost lossfree, even though it is accepted that an $\vec{E} \times \vec{B}$-drift is a loss-free exact solution of the transport problem.

The proposal of in-situ acceleraton conflicts with other constraints (Kundt & Gopal-Krishna, 1981): a) The particles that transport the energy, say protons, would have larger inertia than tolerable by the inferred ram pressures; b) the energy carried by the invisible protons would escape the - already tight - energy estimates; and c) the accelerating shocks would have to be extremely efficient in transferring the energy from the marginally relativistic protons to the hyper-relativistic electrons (Kundt, 1984b). Moreover, whatever shock-acceleration process were at work in the hotspots would be more strongly at work, by many orders of magnitude, in the corotating magnetosphere of the massive central rotator.

Another case against in-situ acceleration is the jet of 3C 273 whose spectrum turns monotonically softer with distance, from optical through IR to radio frequencies, without a discontinuity at the termination shock; see figure 1 and Kundt & Gopal-Krishna (1986).

The logical alternative to in-situ acceleration is in-situ deceleration: If the hotspots contain obstacles of small filling factor, say thermal filaments, then most electrons (and positrons) will continue their loss-free $\vec{E} \times \vec{B}$-drift through the hotspots whereas a small minority of them will be scattered through large angles and thereby cross the electromagnetic fields convected by the others. The scattered electrons have high (synchrotron) radiation losses but not the others.

How high is the pressure in the scattering region? There is a classical lemma stating that whenever a synchrotron source is considerably polarized, its magnetic energy density is very near its particle energy density (Ginzburg & Syrovatskii, 1969; Singal, 1986). The lemma follows from the two constraints that field pressures cannot exceed the pressures of their anchoring sources - they would relax to a vacuum configuration - and that on the other hand, if the magnetic fields were weak, they would be screened and thus randomized by the induced magnetic fields of the gyrating charges.

This equipartition lemma should of course be applied to the whole ensemble, not just to the scattered charges. All we know about the unscattered charges is that their power must energize the downstream phenomena. They anchor the fields that make the scattered charges radiate. The 'minimum pressure' evaluated for the radiating charges via $p_{min} = n \gamma m_e c^2/3 \approx B^2/8\pi$ will therefore overestimate their pressure: It ignores the presence of the field-anchoring charges. In reality, the local fields B may be higher than calculated for equipartition and the Lorentz factors - and hence pressures - of the radiating charges correspondingly lower. This remark can weaken the confinement problem which emerged with the Potash & Wardle jet (1980) and with the bright knots in M87. Of course, the local fields cannot be stronger than allowed by the energy density of the flow. But it is difficult to estimate sideways pressures, as opposed to the forward (ram) pressures.

What is the beaming pattern of the radiating electrons (and positrons)? It has been widely assumed in the literature that there was always a Lorentz frame w.r.t. which the particles radiate isotropically. This assumption is inconsistent with a power-law energy distribution of the radiating charges (Kundt & Gopal-Krishna, 1986): In the comoving frame of a relativistic power-law distribution, only very few particles - in forward direction - are relativistic whereas all backward-moving ones are non-relativistic. The beaming pattern of a power-law ensemble cannot be obtained by boosting an isotropic radiation field (Kundt, 1982a). It may look similar to figure 11 which has been gleaned from jet statistics. It is therefore impossible to infer the jet speed from beaming considerations (Lind & Blandford, 1985).

3. BEAM DYNAMICS AND MORPHOLOGY

Let us begin with the question why only elliptical galaxies are able to become radio galaxies, i.e. produce large-scale jets, as opposed to spiral galaxies. (Seyfert galaxies are known to have Kpc-scale jets, not jets of length $\gtrsim 10^2$ Kpc). If jets exist in media whose densities range from 10^{-29} to 10^{-19} g cm^{-3}, how can the comparatively low-density halos of Seyferts prevent their growth?

If jets consist of relativistic electrons and positrons drifting jointly with transverse electric and magnetic fields at high velocities ($\gamma_{bulk} > 10^2$), they cannot easily be destroyed by an ambient plasma which drifts across their channel: the channel would simply be bent, mediated by magnetic fields. Neutral particles, however, can enter the channel. Such neutral particles are the photons of the background radiation which prevent jets from getting much longer than 3 Mpc. Starlight photons have a comparable energy density in the ISM but not in the halo. More serious, however, are hydrogen atoms which can enter the channel and get entrained as soon as they are re-ionised. $\vec{E} \times \vec{B}$-drifting charges have the common velocity $v = cE/B$, hence one proton can absorb the momentum of $m_p/m_e = 10^{3.3}$ electrons. If an AGN ejects an energy of, say, $E = 10^{59}$ erg in the form of two relativistic jets during one of its active cycles, the corresponding ejected rest mass is $M = E/c^2\gamma = 10^2 M_\odot$ for an average Lorentz factor $\gamma = 500$ of the electrons. Consequently, a jet can be seriously weakened if only one solar mass of neutral hydrogen gets entrained during the $\lesssim 10^8$ yr-long cycle. This is easily achieved in the gas-rich halos of spiral galaxies.

Confirming evidence for beam destruction by neutral particles is the fact that only young bipolar flows show bright beam heads, such as HH 34 (figure 7) or 3C 273 and Cyg A (figures 1,2), whereas old flows are lost in the maps before they end, such as L1515/IRS 5 (figure 6) or Cen A (figure 4).

The next item of interest is whether or not the speed v_j of the jet fluid is always supersonic (i.e. $v_j > c/\sqrt{3}$ for a relativistic fluid). I think it is, both because of the enormous available pressure contrast between the core and the head, and mainly because once the flow turns subsonic, its pressure will equalize on a sound-crossing time whereupon the cocoon assumes a spherical shape. The very fact that all jet sources look elongated throughout, instead of having dumb-bell shape, speaks in favour of globally supersonic jets. And so does the sidedness. But it must be stressed that what we see is usually the cocoon, not the jet. Motion in the cocoon is subsonic, hence quasi-isotropic and unbeamed. Only supersonic flows are beamed.

Another item of debate is whether there is considerable backflow of jet material in the cocoons, as occurs in several numerical simulations. Here the answer will critically depend on the boundary conditions: As the flow in the cocoon is subsonic, possibly damped by filamentary thermal inclusions (like by wad in a gas tube), it will not tolerate large pressure gradients. A cocoon expands according to $p = \rho v^2$; for constant pressure p its local expansion speed v scales therefore as $\rho^{-1/2}$. If ρ drops quickly with r, the cocoon will preferably grow at its distant end. On the other hand, if ρ is essentially constant, the cocoon's quasi-conical volume grows faster at its near end, and considerable backflow is necessary to fill the newly conquered volume; cf. figure 7.

Jets, like rivers, have the property to never split. Yet there are the inner radio lobes of Cen A, unresolved in figure 4, which are young and drain on the jets without (probably) completely exhausting them. How can they be understood? Gopal-Krishna & Saripalli (1984) have shown that the northern inner radio lobe begins where the jet pierces a thin optical-emission shell, at X-ray knot G. Apparently, the jets interact with the shells. Similar interactions of the jets with massive shells ($\approx 10^{10}$ $M_{\odot}$) may have given rise to the radio bubbles in Her A (figure 3), to the radio-dark spots in Cyg A, and to linear depolarization zones in 3C 277.3 (Kundt & Saripalli, 1987). In this way we can learn both about the properties of the apparently ubiquitious massive shells in the halos of galaxies and about how a jet overcomes high-density obstacles.

REFERENCES

Achterberg, A., Blandford, R.D. & Goldreich, P. (1983): Nature 304, 607
Allen, D.A. & Sanders, R.H. (1986): Nature 319, 191
Balick, B. (1987): Sky & Telescope, February, 125
Balick, B. & Heckman, T.M. (1982): Ann.Rev.Astron.Astrophys. 20, 431
Begelman, M.C., Blandford, R.D. & Rees, M.J. (1984): Rev. Mod. Phys. 56, 255
Bell, M.B & Seaquist, E.R. (1980): Astrophys. J. 238, 818
Blair, D.G. & Candy, B.N. (1985): Mon.Not.Roy.astr.Soc. 212, 219
Blandford, R.D. (1986): IAU 119 (Quasars, eds. G. Swarup & V.K. Kapahi, Reidel), p. 359
Blitz, L., Fich, M. & Stark, A.A. (1983): Astrophys. J. Suppl. 49, 183
Blome, H.-J. & Kundt, W. (1987): IAU 122 (Circumstellar Matter, ed. I. Appenzeller, Heidelberg, June) in press
Bridle, A.H. & Perley, R.A. (1984): Ann. Rev. Astron. Astrophys. 22, 319
Camenzind, M. & Courvoisier, T.J.-L. (1983): Astrophys. J. 266, L83
Cohen, M. (1987): IAU 122 (Circumstellar Matter, ed. I. Appenzeller, Heidelberg, June) in press
Duncan, M.J. & Wheeler, J.C. (1980): Astrophys. J. 237, L27
Fesen, R.A. & Gull, T.R. (1986): Astrophys. J. 306, 259
Filippenko, A.V. (1986): IAU 119 (Quasars, eds. G. Swarup & V.K. Kapahi, Reidel), p. 289
Fowler, W.A. (1966): Astrophys. J. 144, 180
Fukue, J. (1986): Publ. Astron. Soc. Japan 38, 167
Gaskell, C.M. (1982): Astrophys.J. 263, 79
Gaskell, C.M. (1987), in: 'Supermassive Black Holes', ed. M. Kafatos, Cambridge Univ. Press
Gaskell, C.M. & Sparke, L.S. (1986): Astrophys.J. 305, 175
Geldzahler, B.J. & Fomalont, E.B. (1986): Astrophys.J. 311, 805
Ginzburg, V.L. & Syrovatskii, S.I. (1969): Ann.Rev.Astron.Astrophys. 7, 375
Gopal-Krishna & Saripalli, L. (1984): Astron. Astrophys. 141, 61

Gunn, J.E. (1979): NATO ASI 'Active Galactic Nuclei' (eds. C. Hazard & S. Mitton, Cambridge Univ. Press), p. 213
Hartig, G.F. & Baldwin, J.A. (1986): Astrophys. J. 302, 64
Haynes, R.F., Komesaroff, M.M., Little, A.G., Jauncey, D.L. Caswell, J.L., Milne, D.K., Kesteven, M.J., Wellington, K.J. & Preston, R.A. (1986): Nature 324, 233
Hoyle, F. & Fowler, W.A. (1963): Mon. Not. Royal astron. Soc. 125, 169
Icke, V. (1986): Astron. Astrophys. submitted
Kardashev, N.S., Novikov, I.D. & Stern, B.E. (1986): IAU 119 (Quasars, eds.G. Swarup & V.K. Kapahi, Reidel), p. 383
Kazanas, D. (1984): Astrophys. J. 287, 112
Kormendy, J. (1987): IAU 127 (Structure and Dynamics of Elliptical Galaxies) ed. T. de Zeeuw, Reidel
Krügel, E. & Tutukov, A. (1986): Astron. Astrophys. 158, 367
Kundt, W. (1979): Astrophys. Sp. Sci. 62, 335
Kundt, W. (1982a): IAU 97 (Extragalactic Radio Sources, eds. D.S. Heeschen & C.M. Wade, Reidel) p. 265
Kundt, W. (1982b): Mitt. Astron. Ges. 57, 65
Kundt, W. (1983): Astron. Astrophys. 121, L 15
Kundt, W. (1984a): Astrophys. Sp. Sci. 98, 275
Kundt, W. (1984b): J. Astrophys. Astron. 5, 277
Kundt, W. (1985a): Astron. Astrophys. 150, 276
Kundt, W. (1985b): Bull. Astron. Soc. India 13, 12
Kundt, W. (1986): NATO ASI C 162 (Cosmic Radiation, ed. M.M. Shapiro, Reidel) pp. 57, 67
Kundt, W. (1987a): Astrophys. Sp. Sci.Lett. 129, 195
Kundt, W. (1987b): Astrophys.Sp.Sci.Lett.
Kundt, W. & Gopal-Krishna (1981): Astrophys. Sp. Sci. 75, 257
Kundt, W. & Gopal-Krishna (1984): Astron. Astrophys. 136, 167
Kundt, W. & Gopal-Krishna (1986): J. Astrophys. Astron. 7, 225
Kundt, W. & Krause, M. (1985): Astron. Astrophys. 142, 150
Kundt, W. & Saripalli, L. (1987): J. Astrophys. Astron.
Lada, Ch.F. (1985): Ann. Rev. Astron. Astrophys. 23, 267
Lind, K.R. & Blandford, R.D. (1985): Astrophys. J. 295, 358
Lynden-Bell, D. (1969): Nature 223, 690
Mastichiadis, A., Marsher, A.P. & Brecher, K. (1986): Astrophys. J. 300, 178
Morrison, P. (1969): Astrophys. J. 157, L73
Noerdlinger, P.D. (1978): Phys.Rev.Lett. 41, 135
Ozernoy, L.M. (1966): Sovjet Astronomy 10, 241
Ozernoy, L.M. (1986): contribution to 13th Texas Symposium on Relativistic Astrophysics, Chicago, December
Ozernoy, L.M. & Usov, V.V. (1977): Astron. Astrophys. 56, 163
Pacini, F. & Rees, M.J. (1973): Scientific American 228, 98
Phillips, M.M., Baldwin, J.A., Atwood, B. & Carswell, R.F. (1983): Astrophys. J. 274, 558
Potash, R.I. & Wardle, J.F.C. (1980): Astrophys. J. 239, 42
Rees, M.J. (1977): Ann. N.Y.Acad.Sci. 302, 613
Reipurth, B., Bally, J., Graham, J.A., Lane, A.P. & Zealey, W.J. (1986): Astron. Astrophys. 164, 51

Roger, R.S., Milne, D.K., Kesteven, M.J., Haynes, R.F. & Wellington, K.J. (1985) Nature 316, 44
Sarris, E.T. & Krimigis, S.M. (1985): Astrophys. J. 298, 676
Scheuer, P.A.G. & Readhead, A.C.S. (1979): Nature 277, 182
Singal, A.K. (1986): Astron. Astrophys. 155, 242
Snell, R.L., Loren, R.B. & Plambeck, R.L. (1980): Astrophys. J. 239, L17
Soltan, A. (1982): Mon. Not. Roy. astr. Soc. 200, 115
Sorrell, W.H. (1981): Nature 291, 394
Stoner, R. & Ptak, R. (1985): Astrophys. J. 297, 611
Terlevich, R. & Melnick, J. (1985): Mon. Not. R. astr. Soc. 213, 841
Trimble, V. & Woltjer, L. (1986): Science 234, 155
Turnshek, D.A. (1986): IAU 119 (Quasars, eds. G. Swarup & V.K. Kapahi, Reidel), p. 317
Yusef-Zadeh, F., Morris, M., Slee, O.B. & Nelson, G.J. (1986): Astrophys.J. 300, L47
Ward, M.J., Carleton, N.P., Elvis, M., Fabbiano, G., Willner, S.P. & Lawrence, A. (1986): preprint Seattle
Weaver, H. (1974): Highlights of Astronomy 3, 509
Wiita, P.J. (1985): Physics Reports 123, 117
Wolstencroft, R.D., Tully, R.B. & Perley, R.A. (1984): Mon.Not.Royal. Astron.Soc. 207, 889.

FURTHER READING

Cowling, T.G. (1985): Ann. Rev. Astron. Astrophys. 23, 1
Lyttleton, R.A. (1979): The Gold Effect, in: The Encyclopedia of Delusions, eds. Simon & Schuster, New York, p. 182
Öpik, E.J. (1977): Ann. Rev. Astron. Astrophys. 15, 1

THE QUASAR FAMILY - AN INTRODUCTION AND TAXONOMY

C. Martin Gaskell
Astronomy Program
Earth & Space Sciences
SUNY, Stony Brook
NY 11794-2100, USA

ABSTRACT. This lecture offers a definition of the quasar phenomenon and gives a description of the taxonomy of the quasar family and an explanation of some of the common terminology. The basic unity of all kinds of quasar activity is emphasised. Some differences between various family members are given and some interrelationships pointed out. Finally, it is emphasised that quasar activity at low levels is extremely common.

1. INTRODUCTION

There are a large number of names in use to describe objects showing quasar activity. The newcomer to the field is frequently quite bewildered by the nomenclature and by such names as: OVV, HPQ, Seyfert 1.95, BLRG, NLXRG, BL Lac, FR II etc. In this lecture I want to attempt to clarify this nomenclature. I will start by giving my own definition of the quasar phenomenon:

2. DEFINITION OF A QUASAR

An object is a member of the quasar family if it puts out > 10^{38} ergs per second continuously from a small region ($r \ll 0.1$pc) with a (roughly) ν^{-1} spectrum ($F_\nu \propto \nu^{-1}$) over the spectral range from 100 microns to 1 MeV.

Some remarks about this are worth making. First, on this criterion there is a "mini-quasar" in the nucleus of our galaxy (the compact non-thermal radio source). Second, 10^{38} ergs/sec has been chosen as the lower limit for purely practical reasons. It is the point at which confusion begins to set in at most wavelengths. Even in the case of our own galactic center (see lectures by E. Serabyn, this conference) there is much confusion. Below 10^{38} ergs/sec we begin to pick up stellar X-ray sources, IR emission from luminous stars and radio emission from supernova remnants and it becomes hard to tell what is

W. Kundt (ed.), Astrophysical Jets and Their Engines, 29–46.

really caused by the quasar. I personally see no reason why there could not be quasars with luminosities less than 10^{38} ergs/sec. The final remark I would like to make is that so far everything satisfying my quasar definition is in or very near the nucleus of a galaxy. Also, I believe that many galaxies have more than one quasar in them (see Gaskell 1983).

3. BASES OF CLASSIFICATION

Objects which satisfy my criterion have been found in many ways. The unusual spectral shape (which is very different from a black body) leads to "excesses" at radio, IR, UV and X-ray wavelengths (excesses compared with black bodies with a range of temperatures typical of a galactic stellar population). Quasars* are readily found from surveys of radio, ultraviolet and X-ray emission from galaxies. The hard spectrum photoionizes gas and produces emission lines. These give another way of recognising quasars. The confusion of terminology in the quasar field is largely a result of the many ways in which quasars are found. I will list here some of the main bases of classification.

3.1 Optical Appearance

I believe that at present this is one of the worst ways of classifying the quasar family because the optical appearance depends on the distance of the quasar and the quality of imaging. Historically most classification has been done by radio astronomers using the Palomar Sky Survey (which does at least give some uniformity of image quality). The main classes are:

3.1.1 "Blank Field" - nothing visible to some limiting magnitude (e.g. the sky survey limit). An object gets removed form this class as soon as an optical image is detected!

3.1.2 "[----] galaxy" - if an object looks obviously like a galaxy on the sky survey plates then the word "galaxy" gets appended to it (e.g. "Radio galaxy", "Seyfert galaxy", "X-ray galaxy" etc.).

3.1.3 "Quasi-stellar object"/"QSO" - if the object looks stellar on the sky survey plates. These quasars tend to be the higher luminosity ones (so that the quasar light drowns out the galaxy) and ones at higher redshift (so that it is harder to resolve the galaxy)

It is obvious that these classes are very distance and instrument dependent.

* *Throughout this lecture I will use the word "Quasar" to refer to all members of the quasar family and not just the highest luminosity ones.*

3.2 Radio Luminosity

If a quasar was detected in a 1960s vintage radio survey (such as the Ohio State survey) at what is now regarded as a fairly low frequency (a GigaHertz or less) and to a flux limit of a Jansky or so then it would have been called a "Radio [----]" (e.g. "radio galaxy" or "radio-loud QSO"). With better radio telescopes many formerly "radio quiet" objects have now been detected at wavelengths of a few centimeters, but the distinction between "radio-loud" and "radio-quiet" remains a useful one and one that can be refined (e.g. to refer to some absolute luminosity or to an optical to radio flux ratio).

3.3 Radio Morphology and Radio Spectra

If radio emission is detectable it permits additional sub-classification. Perhaps the most fundamental classification is into "extended" and "compact" sources. In the days before it was easy to get good high resolution radio maps it was easier to simply divide radio sources into "steep" (radio) spectra sources ($\alpha > 0.5$ in $F_\nu \propto \nu^{-\alpha}$) and "flat" spectrum sources ($\alpha < 0.5$). Since extended structure tends to have a steeper spectral index than compact structure (which is influenced by synchrotron self-absorption) the division into "steep" and "flat" spectrum sources corresponds roughly to "extended" and "compact". There is however not a perfect correspondance (there are so-called "Steep Spectrum Core" (SSC) sources). Often, as in the case of the optical image, faint extended radio emission is drowned out by a bright core and rather than just speaking about whether the source is extended or compact it is also useful to talk about "core-dominated" and "lobe-dominated" sources.

In addition to the relative size or strength of the extended radio component its detailed morphology can also be used to classify quasars. Probably the most important such classification is that due to Fanaroff and Riley (1974). Their class I sources (FR I) have the peak radio emission in their lobes closer than 50% of the way out to the edge of the lobes (they are also sometimes spoken of as "edge-darkened" sources). The FR II sources have the peak lobe luminosity (or "hot-spot") further out near the edge of each lobe.

3.4 Optical - Ultraviolet spectrum

This gives the greatest scope for classification since the optical-UV region contains a wealth of readily accessible spectroscopic information.

3.4.1 Only Stellar Absorption Lines Visible - in this case you call the object by what drew it to your attention (e.g. Radio Galaxy). The majority (55%) of radio galaxies show only stellar lines from the galaxy. However if you search hard enough in a galaxy spectrum (e.g. look in the red where extinction is lower and do a careful subtraction of the underlying galaxy continuum) you will frequently discover an emission line spectrum (see for example Keel 1983b). These spectra often show the presence of a "LINER" (see next section).

3.4.2 Emission Lines. Nuclear emission lines are subdivided into two categories of broad and narrow lines (see section 4). When emission lines are seen the so-called "narrow" lines (from a region of low density gas called the Narrow Line Region (NLR)) are almost always seen. In addition broad permitted lines might also be seen. If the broad lines are absent then the object might be described as a narrow line radio galaxy (NLRG) or a narrow-line X-ray galaxy (NLXRG) depending on how it was discovered. Optically identified galaxies with strong nuclear emission lines caused by non-thermal photoionization are called "Seyfert galaxies" (after the man who first recognized the class). Seyfert galaxies lacking broad permitted lines are called Seyfert 2 galaxies. Notice that there is nothing to stop a Seyfert galaxy being a radio galaxy (in fact Seyfert's original list included NGC 1275 which was later identified as the radio galaxy Per A). Photoionization by a non-thermal continuum is of course not the only way to produce emission lines, but fortunately photoionization by such a "hard" spectrum produces recognizable line ratios and by consideration of line ratio diagrams one can recognize normal H II regions etc. Many galactic nuclei show what is called LINER activity (Low Ionization Nuclear Emission Regions). Although it was initially thought that these had shock heated gas, it is now widely accepted that photoionization by a hard spectrum is also responsible for these emissions too. The role of shocks in galactic nuclei is still,however,a topic in need of more investigation.

When strong broad lines are present an object is called a broad line radio galaxy (BLRG) or a Seyfert 1 galaxy (if it was optically selected) or (if it is too far away for the underlying galaxy to have been seen clearly) a quasi-stellar object (QSO). It is worth noting that narrow line quasars are confined to the lower end of the luminosity range with just a few exceptions.

Osterbrock (1981) has introduced the refinement of decimal Seyfert subclasses (e.g. Seyfert 1.5, 1.8, 1.95, 1.98 etc). These have well defined meanings in terms of the relative prominence of the broad components of Hα and Hβ. Because of continuum variability (and hence broad line variability) a Seyfert can change its sub-classification significantly from month to month.

3.4.3 Absorption lines. When absorption lines are seen in bright quasars they are usually associated with intervening galaxies along the line of sight. The only cases where we can be unequivocally sure that the absorbing gas is associated with the quasar are when we see broad absorption lines blueshifted from the emission-line redshift (implying that it was ejected) or when when the absorption line column densities vary with the quasar's continuum (implying that the gas is close enough to the quasar that the quasar's radiation field dominates the ionization balance). Quasars showing broad blueshifted absorption troughs are called "broad absorption line" quasars (BAL quasars)

3.4.4 Featureless continuum - "BL Lac object" (named after the first member of the class identified). This is a tricky class to define both in principle and in practice, because emission lines usually are present at some level (see Miller, French & Hawley 1978) and their prominence changes as the continuum varies. Of course, a common reason for not seeing lines can simply be that the spectrum is of a low signal to noise ratio. A subsequent observer with better data might then classify the object as something else.

3.4.5 Optical Variability. Probably all quasars are variable at some level in the optical and ultraviolet but if a quasar is particularly variable ($\Delta m > 1$ mag.) and on a short timescale (months, weeks or even days) then it is labelled an "Optically Violently Variable" quasar (or OVV).

3.4.6 Polarization. If the broad band polarization of a quasar is more than a few percent the quasar is called a "Highly Polarized Quasar" (or HPQ).

The sets of HPQs and OVVs have a large overlap. BL Lac objects also show high polarization and OVV behavior (in fact polarization and variability should probably be used as part of the definition of a BL Lac object - see Miller 1978). BL Lacs, HPQs and OVVs are therefore frequently lumped together and called "Blazars" (although the high polarization of some HPQs is probably quite unrelated to that of blazars).

It can be seen from the above that the classification of quasars is multifaceted. By showing the historical and observer-wavelength biases I hope I have left a clear impression that the different names do not necessarily imply fundamental differences in the type of quasar. After an introduction to the emission line regions (which are covered more fully in the lectures by H. Netzer) I will turn to the subject of the unity of the quasar family.

4. THE EMISSION LINE REGIONS

4.1 Narrow-Line Region (NLR)

This is a region of 'low density' photoionized gas ($N_H < 10^8 cm^{-3}$) occupying a relatively large volume of space ranging from radii of less than a parsec to (in some objects) many kiloparsecs. The NLR can be spatially resolved in nearby galaxies (see for example Keel 1983a). Because the density is low (especially in the outer regions) strong forbidden lines can be seen (e.g. λ5007 [O III]). This region was formerly called the forbidden line region, but this was something of a misnomer as strong permitted lines are seen as well (e.g. Balmer lines). The full widths at half maxima (FWHM) of the Doppler broadened lines lie in the range $150 < FWHM < 1400$ km/sec. The lower half of this range (up to 400 km/sec) is comparable with the stellar velocity dispersion in a galactic nucleus and indeed where it can be measured the stellar velocity dispersion is correlated with the NLR line width (Wilson & Heckman 1985).

4.2 Broad Line Regions (BLR)

The permitted emission lines can show very great Doppler widths. The absence or extreme weakness of high velocity wings of forbidden lines such as the strong [O III] lines implies that the density of the high velocity gas (i.e. the BLR) is much greater than that of the NLR ($N_H > 10^8 cm^{-3}$). Because the density is higher a much smaller volume of gas suffices to produce the lines (and a much lower mass). Studies of emission line variability show that most of the BLR is within a light month of the central energy source in low luminosity quasars (see Gaskell & Sparke 1986).

The FWHM of the BLR is > 400 km/sec and almost always much greater than any observed stellar velocity dispersion. In some objects the half width at zero intensity (HWZI) approaches 10% of the speed of light. Although it might be noted that the FWHM of a NLR in one quasar could exceed the FWHM of the dense gas (BLR) in another quasar, it is almost invariably the case that in any given quasar FWHM(high density) > FWHM(low density).

I believe that there are two "families" of BLR clouds. I will call these:

BLR I - fully ionized clouds, optically thin
BLR II - predominantly neutral, optically very thick.

Arguments for the necessity of these two families of BLR are given by Collin-Souffrin _et al_ (1981, 1986) and Gaskell (1986b). These two regions are probably somewhat separated spatially (Gaskell & Sparke (1986) find BLR II to be further out than BLR I). BLR I is responsible for producing the strong high ionization ultraviolet lines (C IV, N V, O VI, He II etc.) while BLR II is responsible for producing the strong low ionization lines (notably Fe II - see lecture by H. Netzer). The two BLRs show systematically different redshifts (Gaskell 1982b) with the high ionization lines being slightly blueshifted. Most quasars show both a BLR I and a BLR II but sometimes the lines from BLR II are very weak relative to BLR I or totally absent (e.g. in Mrk 359 see MacAlpine _et al_ 1985). The correlations between the BLRs and other properties of quasars will be mentioned in section 6.

5. THE UNITY OF THE QUASAR FAMILY

5.1 The Ubiquity of Broad-line Regions

I want to assert in this section that all quasars probably have broad line regions and indeed that the presence of a BLR is an "almost defining characteristic" of the quasar family. The problem is, I believe, simply one of detection. The following effects can cause a BLR to be "hidden" from the observer. It can be

(i) "drowned-out" by the Doppler-boosted continuum,
(ii) so weak that it is lost amongst the stellar absorption lines,
(iii) blocked by some obstacle,
(iv) temporarily "turned-off" because the continuum has faded.

Over the last decade new observational techniques, improvements in detectors, more sophisticated data analysis and more observations* have shown that these four effects are indeed operating. I will mention just a few examples:

(i) High signal to noise observations of BL Lac objects (esp. when the continuum is faint) can show broad lines (e.g. Miller, French & Hawley 1978).

(ii) Careful subtraction of stellar contributions to spectra can reveal broad wings to lines (e.g. Keel 1983b).

(iii) Simple reddening can be (partially) got round by observing further to the red; e.g. by surveying for the strong Hα λ6563 line (Stauffer 1982a,b; Keel 1983b). Infra-red observations of the nucleus of our galaxy have revealed a broad He I 2.058μm line (Hall, Kleinmann and Scoville 1982).

(iv) The new technique of spectropolarimetry has revealed the spectrum of a broad line region in the polarized (scattered) light in the nucleus of the Seyfert 2 galaxy NGC 1068 (Antonucci & Miller 1985). Further spectropolarimetric observations of other Seyfert 2 galaxies are revealing similar hidden BLRs (Miller 1986), although the technique can only be applied to those Seyfert 2s with high polarization.

(v) The nucleus of NGC 7603 has changed its spectral type from Seyfert 1.0 to Seyfert 1.9 (Tohline & Osterbrock 1976) in only a few months. In modern terminology the variations in NGC 3516 reported by Andrillat & Souffrin (1968) also correspond to a change between a type 1.0 and a 1.9 .

Because of all these demonstrable effects that can "hide" BLRs the hypothesis that <u>every</u> member of the quasar family has a BLR cannot be disproved and I believe that the existence of a BLR is an essential part of quasar activity.

Not only do all quasars seem to have BLRs but also the BLRs are very similar. The ratio of line luminosity to continuum luminosity is remarkably constant. Fig. 1 (next page; adapted from Shuder 1981) shows how BLR Hα luminosity is directly proportional to the luminosity of the non-thermal continuum at λ4800 for a range in luminosity of almost a million (this result was also independently reported by Yee, 1980).

The constancy of the line/continuum ratio (or "equivalent width" in traditional astronomical parlance) implies at least two things: (i) the covering factor is remarkably constant from object to object and does not vary significantly even over a luminosity range of a million and (ii) the intrinsic shape of the optical to UV continuum is also remarkably constant over the same range in optical luminosity.

*or, to summarise in the words of a famous astronomer "making the graduate students work harder!"

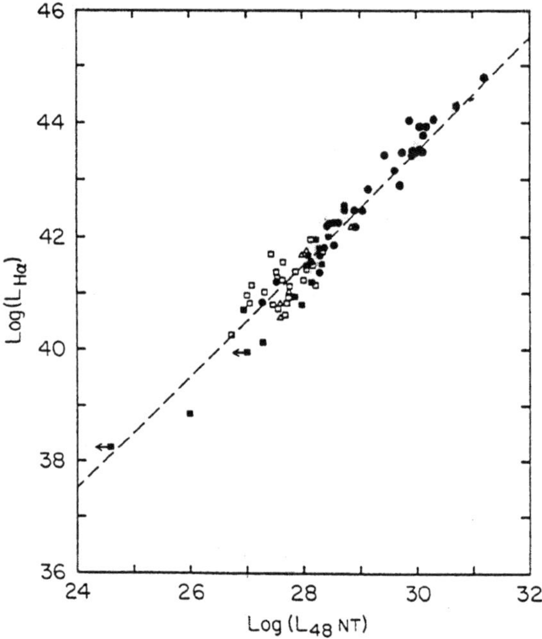

Figure 1. Continuum and Hα luminosities for broad line region dominated quasars (filled symbols) and narrow line region dominated quasars (open symbols). $L_{48}NT := dL/d\nu(\lambda=4800Å)$ / erg $s^{-1}Hz^{-1}$.

5.2 The Similarity of Continuum Shapes

With the success over the past decade of satellite missions such as the IUE, EINSTEIN, EXOSAT and IRAS and with the extended wavelength coverage of ground-based observations we have a pretty good idea of the shape of quasar spectra from radio wavelengths to γ-ray energies (although coverage approaches completeness only for one object - 3C 273). The most significant gap in our knowledge is in the extreme ultraviolet (from 14eV to 100eV) but fortunately this spectral region contains the ionization edges of H I (13.6eV), He I (24.6eV) and He II (54.4eV) and these can be used to probe the spectral shape here by counting the number of recombination photons in the H I, He I and He II lines (see Shuder 1981). As mentioned in 5.1 the constancy of the line/continuum ratio implies a constancy of shape in the extreme ultraviolet. More direct measurements in other wavebands support the inferences from the optical emission lines.

In the optical region a variety of spectral shapes are seen but the differences are probably largely due to reddening (I am assuming here that other complications such as the light of the underlying galaxy and blended emission lines have been taken care of; these are non-trivial). The topic of reddening is reviewed by MacAlpine (1985), De Zotti & Gaskell (1985) and Gaskell (1986). Known samples of quasars are biased towards low reddening. Just how low the reddening is is a matter of dispute, but I argue elsewhere that it is not zero and probably has an E(B-V) of 0.2 mag (Gaskell 1986 a). The precise value is important because of its effect on our estimate of the EUV spectrum. Whatever the reddening, the shape of the optical - UV continuum is fairly constant. Gaskell (1980, 1981) showed that the majority of core dominated radio sources have an apparent UV spectral index $\alpha \sim 0.6 - 0.7$ ($F_\nu \sim \nu^{-\alpha}$) but there is a possible luminosity or evolutionary effect (see 6.1.5 and 6.2.3 below). From the ratios of UV to optical fluxes plotted by Malkan (1984) one can see that most quasars observed with the IUE have apparent optical (λ4220) to UV (λ1460) spectral index of 0.5 to 0.7. This constancy of spectral shape implies that the intrinsic optical to UV continuum shape is (again) very constant and that there is little variation in the reddening for the IUE observed quasars. It must be mentioned however that there are known examples of quasars with their spectra rising from the optical to the UV (presumably almost unreddened) and that in samples of quasars chosen without optical bias a significant fraction can show very heavy reddening (Gaskell 1986 a).

The high energy (2 - 100 keV) X-ray spectra of quasars also show very similar spectra. With few exceptions, the spectrum is a power law with an index of 0.62 ± 0.04 (Rothschild et al 1983) and this is sometimes seen extending down to lower energies (100eV to 1 keV) but a good many objects show much "softer" (i.e. steeper) spectra at low energies (often called a "soft X-ray excess"). Mrk 841 (Arnaud et al 1985) is a good example. The soft X-ray excess probably represents the Wien cutoff of a 700,000 K black body.

The power-law spectrum from γ-rays to the sub-mm region which I took as the defining characteristic of quasars turns down between 0.1 and 0.3 mm (see for example Edelson & Malkan 1986), presumably because of synchrotron self-absorption right near the central engine. 0.1 to 0.3 mm therefore probably represents the longest wavelength at which we are seeing close to the central engine. Emission at longer wavelengths comes from more extended emission that is not always present in all objects. These extended regions can extend out to a megaparsec away from the center. The extended regions contribute some X-ray emission as well as radio emission. In addition to this extended emission some of the radiation is reprocessed near the central source. The BLR clouds contribute free-free and bound-free continua (e.g. the Balmer continuum) and a quasi-continuum around λ4500 - λ2300 (the so-called "little blue-bump") due to blended Fe II lines. Energy absorbed by dust is re-radiated in the IR.

BL Lac objects and related objects are now believed to be quasars in which the continuum is enhanced by relativistic motion towards the observer (Blandford & Rees 1978). These show the simplest and most uniform overall spectral energy distributions. Landau et al (1986) show that over 80% of such sources are well characterized by a spectrum which is a simple parabola in log F_ν vs log ν from wavelengths of 20cm to λ1400 (over six orders of magnitude in frequency). The peak is always near 10^{10} Hz (i.e. 3cm) and the shape is always similar. The relationship between these uniform continua and the spectra of more normal quasars is uncertain.

6. DIFFERENCES AND INTERRELATIONSHIPS

Although I have been stressing the similarities in continuum properties of quasars there are also significant differences. The most significant one is "radio-loudness". The formation of extended radio structure is a major topic of this conference (Kundt 1987) but I would like to issue a reminder that most quasars (90%!) are radio-quiet (by my earlier definition) and of the 10% that are radio-loud only a small percentage of those show extended double lobes. Powerful radio jets are clearly not a necessary part of the quasar phenonemon, and a theory to explain jets must also explain why they usually do not occur! In this section I will briefly summarize some of the known relationships of quasar properties with luminosity, redshift and extended radio structure.

6.1 Luminosity Effects

6.1.1 The Fanaroff-Riley Break. Fanaroff & Riley (1974) showed that there is a remarkably sharp luminosity break between their classes FR I and FR II (defined in 3.3 above). Extended sources are FR Is at powers below 2×10^{25} W Hz^{-1} s^{-1} at 178MHz (H_o = 50 km s^{-1}Mpc^{-1}) and FR IIs above that.

6.1.2 The Baldwin Effect. The equivalent width of the broad C IV emission line is anti-correlated with the continuum luminosity (Baldwin *et al* 1978; Wampler *et al* 1984). Although the Baldwin effect was originally established for core-dominated radio QSOs (in order to overcome possible biases in optically selected objects) there is no reason to believe that it does not apply to lobe-dominated quasars too (this has simply not been tested) and it has been shown that the Baldwin effect works in radio-quiet quasars (Smith 1978; Kiang, Cheng & Zhou 1983) and that it extends to low luminosity Seyfert 1 galaxies (Wampler *et al* 1984).

Wampler *et al* (1984) list a variety of possible explanations of the Baldwin effect, but most simple effects can be ruled out by considering C III] λ1909. Gaskell (1981) pointed out that the C IV/C III] ratio is anti-correlated with luminosity in bright quasars and that the equivalent width of C III] is independent of luminosity. Mushotsky & Ferland (1984) have gone further and shown that a luminosity dependence of the ionization parameter (the ratio of ionizing photon density to hydrogen density) is probably behind the Baldwin effect. The ionization parameter is higher in lower luminosity objects.

6.1.3 The Covering Factor. X-ray absorption is luminosity dependent, being greatest for low luminosity quasars (Mushotzky 1982; Lawrence & Elvis 1982; Reichert *et al* 1985). Quasars with a hard X-ray (2-10keV) luminosity less than 10^{44}ergs s^{-1} usually show soft X-ray absorption (covering fractions > 70%); those with luminosities greater than 5×10^{44}ergs s^{-1} never do (implying cover-ing fractions of less than 70%).

6.1.4 Dust. De Zotti & Gaskell (1985) find some evidence that the reddening of quasars is luminosity dependent. The lower luminosity objects are most likely to show heavy reddening. Lawrence & Elvis (1982) point out that the IR spectral index steepens for low X-ray luminosity objects; implying a larger dust re-emission contribution to the infrared in these sources.

6.1.5 Continuum Shape. Reichert et al (1982), Zamorani (1982), Avni & Tananbaum (1982) and others have found that the optical to X-ray spectral index steepens as the optical luminosity increases. The UV spectral index might also be a function of optical luminosity (see Cheng & Gaskell 1987).

6.1.6 BLR Line Width. There is a weak tendency for high luminosity quasars to have broader lines. Baldwin (1977) pointed out that the FWHM of C IV shows a probable increase with luminosity. Steiner (1981) found a correlation of the monochromatic flux at 4000 Å with the FWZI of Hα (especially if only objects with strong NLRs were considered). Shuder (1984) and Joly et al (1985) found similar correlations and noted that such correlations "saturate" at high luminosities (L_{opt} > 10^{45}ergs s^{-1}; this saturation is probably the reason Richstone et al 1980 failed to find a correlation).

6.1.7 Variability. The degree of optical variability decreases with increasing optical luminosity (Angione 1973, Pica & Smith 1983). Barr and Mushotzky (1986) show that there is a significant correlation between X-ray luminosity and the timescale of fastest X-ray variability (which ranges from less than an hour for $L_x = 10^{40}$erg s^{-1} up to several months for $L_x = 10^{46}$ergs s^{-1}). They interpret this as evidence that the emitting plasma is near the limit of being dominated by electron-positron pairs (see also Kundt 1987).

6.1.8 Narrow lines. Narrow line dominated objects tend to have lower optical and X-ray luminosities than BLR dominated objects although there are many Seyfert 2 galaxies brighter than faint Seyfert 1 ones. The brightest quasars are all BLR dominated. Although some workers have reported correlations of [O III]/Hβ with L_o or L_x (e.g. Steiner 1981), Reichert et al (1982) have shown that there are no correlations when NLR dominated objects are considered separately.

6.2 Evolutionary Effects

The luminosity evolution of quasars and evolution of the luminosity function is well established and evolution (esp. in the radio) is well covered by other lecturers but I will mention a few changes in the spectral properties that are perhaps telling us something about evolutionary changes in or near the central engine.

6.2.1 Size of radio lobes. The extended radio structure of quasars is much larger now than it was in the earlier days of the universe (Miley 1971).

6.2.2 Radio Turnover Frequency. Menon (1984) has shown that the (rest frame) turnover frequency in radio loud quasars moves to higher frequencies at large redshifts. This implies a dramatic change in size of the radio emitting region from several kiloparsecs (at low redshift) to only a few parsecs at high redshift. The environment of a quasar is thus demonstrably different at earlier times.

6.2.3 The UV and X-ray Continuum. Reichert et al (1982) find a slight increase in L_x/L_o with redshift (after allowing for the luminosity dependence). Avni and Tananbaum (1982) state that their results imply that the evolution of the X-ray luminosity function is weaker than the evolution of the optical luminosity function. Cheng and Gaskell (1987) find that the UV spectral index (λ2200 - λ1200) becomes flatter with increasing redshift. The causes of these effects is unknown.

6.3 Radio Structure Related Effects

6.3.1 Optical Morphology. From the very first identification of extra-galactic radio sources in the 1950s it was realized that strong radio sources almost always occur in elliptical galaxies, and that many of these (30%) are peculiar or disturbed. On the other hand, radio-quiet objects are almost all in disc galaxies (see for example Adams 1977, Keel 1980). Adams (1977) noted that there was a surplus of Seyfert nuclei in disturbed and interacting systems. Amongst radio galaxies, Heckman et al (1986) find FR I sources to have more normal optical morphologies. The FR II sources frequently have peculiar optical morphologies.

6.3.2 Line Widths. Osterbrock (1977) showed that broad-line radio galaxies had much broader broad lines on average than Seyfert 1 galaxies. Miley and Miller (1979) showed that lobe-dominated high luminosity quasars similarly had broad lines much broader than core dominated ones. If the difference between lobe and core dominated sources is only one of viewing angle then the cloud motions must be primarily in a plane perpendicular to the radio axis. Blandford and Königl (1979) suggested therefore that the BLR clouds might be rotating in a disc. There are, however, many problems with a rotating broad-line region (see Mathews 1982) and a more reasonable alternative is that the broad lines are indeed broader when large-scale radio structure is able to form. It is not difficult to think of reasons why this would be so and, as Setti and Woltjer (1977) suggested,the cause probably lies in the host galaxy environment. Strong support for this idea comes from Su & Simkin (1980) who find line width to be correlated with galaxy type but not with inclination (when it can be measured). Su & Simkin find a continuous relationship between spiral Seyfert 1 galaxies and "N"-type (elliptical) radio galaxies.

6.3.3 BLR II Strengths. The most obvious characteristic of BLR II (see section 4.2 for definition) is optical Fe II emission. Osterbrock (1977) noted that while strong Fe II λ4570 is (almost) always observed in Seyfert 1 galaxies it is absent in broad-line radio galaxies. Setti and Woltjer (1977) showed that this correlation with radio structure extends to higher luminosity objects. Wills & Browne (1986) suggest that the Fe II - radio structure correlation is spurious and the result of a systematic measuring error due to the FWHM - radio structure correlation (see preceeding section) because line blending of the broad lines at large FWHM artificially raises the continuum level so lowering the Fe II/Hβ ratio. The analysis of Gaskell (1985a) shows that the FWHM versus Fe II/Hβ correlation is caused by Hβ getting stronger as the lines get broader (the opposite of the "Wills effect"). Simulations by Boroson, Persson & Oke (1985) showed that the Wills effect was unimportant and an inspection of a similar simulation by Wills (1987; her figure 6) shows that after allowance for the effect there is still a real difference in optical Fe II strength. Boroson et al (1985) also show that there is a big difference in Fe II emission at the same FWHM. I believe that there is a real difference in the ratio of BLR I/BLR II that depends on radio structure. Setti & Woltjer (1977) suggested that the linking factor is the galactic environment. Gaskell (1982a), Heckman (1983) and Boroson et al (1985) used the optical spectral differences between different radio classes to argue against unified beaming models. Possible ways round some of these objections are given in Wills & Browne (1986).

6.3.4 Displaced BLR peaks. Many quasars have the peaks of their broad lines at very different redshifts from the NLR redshift (Gaskell 1982, 1983). This is probably due to the orbital motion of two supermassive objects. Gaskell (1983) points out that the best cases (with the BLR peaks displaced by > 1000 km s^{-1} relative to the NLR) seem to be in extended radio structure quasars, and suggests that the presence of a supermassive binary might be an essential ingredient in forming large-scale collimated radio jets. If true this could be an important link between jets seen in our own galaxy (e.g. SS 433) and quasar jets. Possible unified schemes for astrophysical jet sources are discussed by Kundt (1984, 1987). I believe that all so called "complex" BLR profiles are the results of binary BLRs. Miley & Miller (1979) and others have noted that complex line profiles are mostly found in quasars with extended radio structure.

6.3.5 Narrow Line Emission. Lobe-dominated quasars have stronger narrow lines (see fig. 2). The narrow lines are also stronger in quasars that do not have a strong BLR II contribution (see fig. 3). The strength of the [O III] emission is not simply caused by the continuum being weak since the [O III]/Hβ ratio shows a similar relationship to Fe II strength (see fig 1. of Steiner 1981) and there is a strong separation of very extended [O III] emission between radio-compact (or quiet) and radio-extended quasars, with the latter having much stronger extended [O III] emission (Boroson, Persson & Oke 1985).

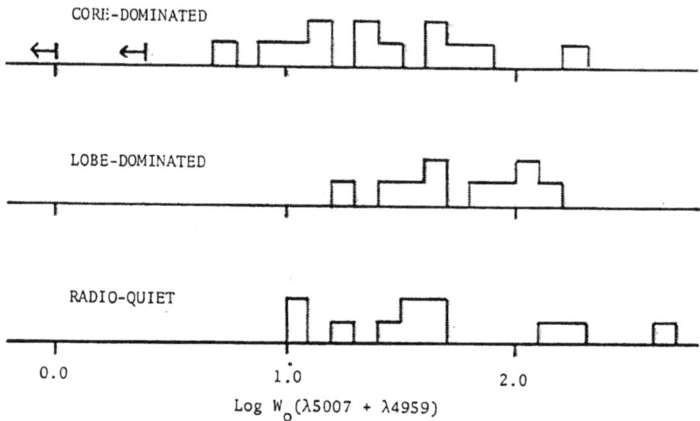

Figure 2. Distribution of equivalent withs of [O III] as a function of radio type. Data from Baldwin (1975), Gaskell (1981) and Blumenthal, Keel & Miller (1982).

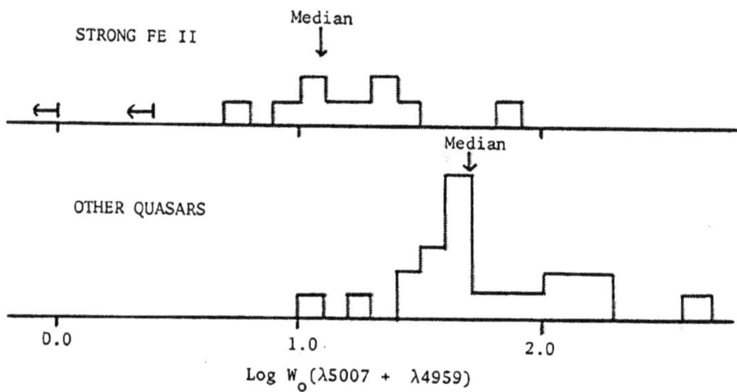

Figure 3. Distribution of equivalent widths of [O III] as a function of Fe II strength. "Strong Fe II" means Fe II λ4570/Hβ > 0.5. Data from Baldwin (1975), Gaskell (1981) and Blumenthal, Keel & Miller (1982).

I believe that these narrow-line differences are a strong reason for believing that core-dominated and lobe-dominated quasars are fundamentally different but Wills & Browne (1986) suggest that there could be a problem caused by the correlation of [O III] strength with radio power (de Bruyn & Wilson 1978; Whittle 1984). There is still a need for further work here with carefully chosen samples of quasars.

6.3.6 Optical Polarization. Stockman, Angel & Miley (1979) discovered that the optical polarization of high luminosity quasars is parallel to the quasar's extended radio structure. For Seyfert 1 galaxies the polarization is also parallel to the radio structure but for Seyfert 2's the polarizations are generally perpendicular and higher than in the Seyfert 1's (Antonucci 1983). Radio galaxies also show a similar separation into "parallel" and "perpendicular" groups (Antonucci 1984). The perpendicular ones have the weakest radio cores and the parallel ones have the strongest cores. The "perpendicular" polarizations are caused by scattering (since Hα is polarized perpendicular to the radio structure). Some sort of physically and optically-thick torus is needed to block the central radiation source. I believe that optical polarization measures provide the best evidence for disc structure in quasars, but it should be noted that the discs needed are vastly bigger than the accretion discs postulated to explain the EUV thermal excess.

The "parallel" group need a different explanation - perhaps a thinner scattering disc.

6.4 Summary of Interrelationships

The above sketch of differences between members of the quasar family (sections 6.1 - 6.3) and the relationships with luminosity, cosmic epoch and radio-structure is by no means complete but I believe it contains a lot that is not explained by currently popular theories and which has ultimately to be explained. I would particularly draw attention to the correlations which do not readily fit simple unified models that attempt to explain everything as an orientation effect with perhaps one other parameter (mass of the central engine, say). I think that the picture has to be more complicated than that and that "environmental" factors play a big role. On the other hand, this is not to say that orientation effects are unimportant. There clearly have to be some orientation effects.

7. WHAT FRACTION OF GALAXIES HAVE QUASARS IN THEIR NUCLEI?

The results of surveys of nearby galaxies to look for quasar activity are thoroughly reviewed by Keel (1985). These surveys (e.g. Heckman 1980, Stauffer 1982 and Keel 1983a) show that 80% of late type galaxies show LINER activity (see 3.4.2 for definition) down to the 10^{38}ergs s^{-1} level. There is a sharp drop-off at types Sb/Sbc to $\sim$ 20%, although the fraction of Sc galaxies showing activity rises with increasing luminosity. About 50% of ellipticals show LINER activity. There is, incidentally, little doubt now that LINERS are *bona fide* members of the quasar family since they are radio sources (Heckman 1980), X-ray sources (Halpern & Steiner 1983) and half of them show a BLR (Filippenko and Sargent 1985).

As is well known, the evolution of quasars is very pronounced and the brightest quasars were a thousand times more common at z $\sim$ 3. This means that there must be a lot of dead or dormant quasar engines in the present-day, nearby galaxies and one would expect that almost every "decent-sized" galaxy has one. The surveys of low-level activity in nearby galaxies seem to be showing that this is indeed the case.

The central engines do not need to stay dormant. Hummel (1981), Stauffer (1982) and Kennicutt & Keel (1984) found that pairs and close groups of galaxies are far more liable to show quasar activity than isolated galaxies. Kennicutt and Keel find that 38% of interacting galaxies show Seyfert activity while for isolated field galaxies the percentage is only 3 or 4%. Over a third of the brightest quasars can be seen to be interacting with other galaxies (Hutchings, Crampton and Campbell 1984), and compact companion galaxies that have probably suffered the effects of very close encounters are seen in the vicinity of ∿ 20% of bright quasars (Stockton 1982). This triggering of activity by neighbours might be caused by more gas somehow being fed into a dormant central engine by some sort of perturbation (Toomre & Toomre 1972) or perhaps by a "direct hit" on the nucleus by an unseen dwarf companion (Gaskell 1985b). Whatever the mechanism, models for the central engines of quasars need to allow for external triggering.

8. ACKNOWLEDGEMENTS

I would like to thank the organisers of the Erice International school, and especially Prof. Wolfgang Kundt for the invitation to lecture at the Advanced Study Institute and for their hospitality. Financial support from NATO is gratefully acknowledged.

REFERENCES

Adams, T. F. (1977) Astrophys J. Suppl., **33**, 19.
Andrillat, Y. & Souffrin, S. (1968), Astrophys. Letts., **1**, 111
Angione, R. J. (1973) Astron. J., **78**, 353.
Antonucci, R. R. J. (1983), Nature, **303**, 158.
Antonucci, R. R. J. (1984), Astrophys. J., **278**, 499.
Antonucci, R. R. J. & Miller, J. S. (1985), Astrophys. J., **297**, 621.
Arnaud, K.A. et al (9 authors) (1985), Mon.Not.R.astr.Soc. **217**, 105.
Avni, Y. & Tannanbaum, H. (1982) Astrophys. J. Letts., **262**, L17.
Baldwin, J. A. (1975), Astrophys. J., **201**, 26.
Baldwin, J. A. (1977), Astrophys. J., **214**, 679.
Baldwin, J. A., Burke, W. L., Gaskell, C. M. & Wampler, E. J. (1978), Nature, **273**, 431.
Barr, P. & Mushotzky, R. F. (1986), Nature, **320**, 421.
Blandford, R. D. & Könige, A, (1979), Astrophys. J., **232**, 34.
Blandford, R. D. & Rees, M. J. (1978) in Pittsburg Conference on BL Lacertae Objects, ed. A. M. Wolfe (Univ. of Pittsburg: Pittsburg) p 328.
Blumenthal, G. R., Keel, W. C. & Miller, J. S. (1982) Astrophys. J., **257**, 499.
Boroson, T. A., Persson, S. E. & Oke, J. B. (1985), Astrophys. J., **293**, 120.
Boroson, T. A., & Oke, J. B. (1985), Astrophys. J., **281**, 535.
Cheng, F. & Gaskell, C. M. (1987) in preparation.
Collin-Souffrin, S., Dumont, S, Heidmann, N. & Joly, M., Astron. & Astrophys., (1980), **83**, 190.

Collin-Souffrin, S., Dumont, S., & Tully, J. (1982), Astron. Astrophys., **106**, 362.
Collin-Souffrin, S., Dumont, S., Joly, M & Pequignot, D. (1986), Astron. Astrophys., **166**, 27.
de Bruyn, A. G. & Wilson, A. S. (1978), Astron. Astrophys., **64**, 433.
De Zotti, G. & Gaskell, C. M. (1985), Astron. Astrophys., **147**, 1.
Edelson, R. A. & Malkan, M. A. (1986), Astrophys. J., **308**, 59.
Fanaroff, B. L. & Riley, J. M. (1974), Mon. Not. R. astr. Soc., **167**, 31P.
Filippenko, A. V. & Sargent, W. L. W. (1985), Astrophy. J. Suppl. **57**, 503.
Gaskell, C. M. (1979), Bull. Am. astr. Soc., **11**, 622.
Gaskell, C. M. (1981), Ph.D. thesis, Univ. of California, Santa Cruz.
Gaskell, C. M. (1982a), Astrophys. J., **252**, 447.
Gaskell, C. M. (1982b), Astrophys. J., **263**, 79.
Gaskell, C. M. (1983), Liege Astrophysical Colloquium, **24**, 473.
Gaskell, C. M. (1985a), Astrophys. J., **291**, 112.
Gaskell, C. M. (1985b), Nature, **315**, 386.
Gaskell, C. M. (1986a) in Continuum Emission in Active Galactic Nuclei ed. M. L. Sitko (Kitt Peak National Observatory: Tucson) p 39.
Gaskell, C. M. (1986b) in IAU Symposium No. 119 "Quasars" ed. G. Swarup & V. K. Kapahi (Reidel: Dordrecht)
Gaskell, C. M. & Sparke, L. S. (1986), Astrophys. J., **305**, 175.
Hall, D. N. B., Kleinmann, S. G. & Scoville, N. Z., (1982), Astrophys. J. Letters, **260**, L53.
Halpern, J. R. & Steiner, J. E. (1983), Astrophys. J. Letts., **269**, L37.
Heckman, T. M. (1980), Astron. Astrophys., **87**, 152.
Heckman, T. M. (1983), Astrophys. J. Letts., **271**, L5.
Heckman, T. M. et al (1986), Astrophys. J., **311**, 526.
Hummel, E. (1981), Astron. Astrophys. **96**, 111.
Hutchings, J. R., Crampton, D. & Campbell, B. (1984), Astrophys. J. **280**, 4
Joly, M., Collin-Souffrin, C., Masnou, J. L. & Nottale, L. (1985), Astron. Astrophys., **152**, 282.
Keel, W. C. (1980), Astron. J., **85**, 198.
Keel, W. C. (1983a), Astrophys. J. Suppl., **52**, 229.
Keel, W. C. (1983b), Astrophys. J., **268**, 632.
Keel, W. C. (1983c), Astrophys. J., **269**, 466.
Keel, W. C. (1984), Astrophys. J., **282**, 75.
Keel, W. C. (1985) in Astrophysics of Active Galaxies and Quasi-Stellar Objects ed. J. S. Miller (University Science Books: Mill Valley) 3.
Kennicutt, R. C. & Keel, W. C. (1984), Astrophys. J. Letts., **279**, L5.
Kiang, T., Cheng, F. H. & Zhou, Y. Y. (1983), Mon. Not. R. astr. Soc., **203**, 25P.
Kundt, W. (1984), Astrophys. Sp. Sci., **98**, 275.
Kundt, W. (1987), this conference.
Landau, R. et al (1986), Astrophys. J., **308**, 78.
Lawrence, A. & Elvis, M. (1982), Astrophys. J., **256**, 410.
MacAlpine, G. M. (1985) in Astrophysics of Active Galaxies and Quasi-Stellar Objects ed J. S. Miller (university Science Books: Mill Valley) p 259.
MacAlpine, G. M., Davidson, K., Gull, T. R. & Wu, C.-C. (1985), Astrophys. J., **294**, 147.

Malkan, M. A. (1984) in X-ray and UV Emission from Active Galactic Nuclei ed W. Brinkmann & J. Trümper (European Southern Observatory: Garching) p 121.
Mathews, W. G. (1982), Astrophys. J. **258**, 425.
Menon, T. K. (1984) Astron. J., **88**, 598.
Miley, G. K. (1971), Mon. Not. R. astr. Soc., **152**, 477.
Miley, G. K. & Miller, J. S. (1979), Astrophys. J. Letts., **228**, L55.
Miller, J. S. (1978), Comments in Astrophysics, **7**, 175.
Miller, J. S. (1986), Private Communication.
Miller, J. S., French, H. B. & Hawley, S. A. (1978) in Pittsburgh Conference on BL Lac Objects ed. A. M. Wolfe (University of Pittsburgh: Pittsburgh) p 176.
Mushotzky, R. F. (1982), Astrophys. J., **256**, 92.
Mushotzky, R. F. (1984), Astrophys. J., **278**, 558.
Mushotzky, R. F. & Ferland, G. J. (1984), Astrophys. J., **278**, 558.
Osterbrock, D. E. (1977), Astrophys. J., **215**, 733.
Osterbrock, D. E. (1981), Astrophys. J., **249**, 462
Pica, A. J. & Smith, A. G. (1983), Astrophys. J., **272**, 11.
Reichert, G. A., Mason, K. O., Thorstensen, J. R. & Bowyer, S. (1982), Astrophys. J., **260**, 437.
Reichert, G. A., Mushotzky, R. F., Petre, R. & Holt, S. S. (1985), Astrophys. J. **296**, 69.
Richstone, D. O., Ratnatunga, K. & Schaeffer, J. (1980), Astrophys. J., **240**, 1.
Rothschild, R. E., Mushotzky, R. F., Baity, W. A., Gruber, D. E., Matteson, J. L. & Peterson, L. E. (1983), Astrophys. J. **269**, 423.
Setti, G. & Woltjer, L. (1977), Astrophys. J. Letts., **218**, L33.
Shuder, J. M. (1981), Astrophys. J., **244**, 12.
Shuder, J. M. (1984), Astrophys. J., **280**, 491.
Smith, M. G. (1978), Vistas in Astronomy, **22**, 321.
Stauffer, J. R. (1982a), Astrophys. J. Suppl., **50**, 517.
Stauffer, J. R. (1982b), Astrophys. J., **262**, 66.
Steiner, J. E. (1981), Astrophys. J. **250**, 469.
Stockman, H. S., Angel, J. R. P. and Miley, G. K. (1979), Astrophys. J. Letts., **277**, L55.
Stockton, A. (1982), Astrophys. J., **257**, 33.
Su, H. J. & Simkin, S. M. (1980), Astrophys. J. Letts., **298**. L1.
Tohline, J. E. & Osterbrock, D. E. (1976), Astrophys. J. Letts., **210**, L17.
Toomre, A. & Toomre, J. (1972), Astrophys. J., **178**, 623.
Wampler, E. J., Gaskell, C. M., Burke, W. L. & Baldwin, J. A. (1984), Astrophys. J., **276**, 403.
Whittle, D. M. (1984), Mon. Not. R. astr. Soc. **213**, 33.
Wills, B. J. (1987) in IAU Colloquium No. 94 "Physics of Formation of Fe II lines outside LTE" ed. R. Viotti & M. Friedjung (Reidel: Dordrecht).
Wills, B. J. & Browne, I. W. A. (1986), Astrophys. J., **302**, 56.
Wilson, A. S. & Heckman, T. M. (1985) in Astrophysics of Active Galaxies and Quasi-Stellar Objects ed. J. S. Miller (University Science Books: Mill Valley) p 39.
Yee, H. K. C. (1980), Astrophys. J., **241**, 984.
Zamorani, G. (1982), Astrophys. J. Letts., **260**, L31.

OUR GALACTIC CENTER

E. Serabyn
Max-Planck-Institut für Radioastronomie
Auf dem Hügel 69
D-5300 Bonn 1, F.R.G.

INTRODUCTION. The last few years have produced large and rapid advances in our understanding of the galactic center. New information has come via a variety of methods, including high spatial resolution radio and infrared continuum mapping, and line observations in the micron through centimeter wavelength range. These observations have addressed the distribution and kinematics of both the stellar population and the interstellar medium in the vicinity of the galactic center. This review summarizes the understanding of the central 100 pc of our galaxy which has emerged from this new body of data. Since the region divides itself naturally into two areas, those being the central 10 pc, and the region out to 100 pc, the discussion will be presented in terms of these two topics. Although the distance to the galactic center is actually ~7 to 8 kpc, a distance of 10 kpc is assumed throughout, to allow direct intercomparison of all data sets. At this distance 1' corresponds to a physical size of 3 pc.

1. THE CENTRAL 10 PARSECS

1.1 The location of the galactic center

The ionized Sgr A region (Fig. 1) was first thought to lie at the center of our galaxy because of its location close to both the kinematic center of the rotating nuclear HI disk, and to the optically estimated location of the center of the galaxy (Oort and Rougoor 1960). The latter location is not well determined, because approximately 30 magnitudes of visual extinction (Becklin et al. 1978) prevent observation of the stars in the central region at $\lambda \leq 1$ μm. However, since the 2 μm flux from the central stellar cluster was also found to peak within the Sgr A region (Becklin and Neugebauer 1968), and since emission lines from Sgr A have uniquely broad linewidths for a galactic HII region ($\gtrsim 200$ km s^{-1}; Pauls et al. 1974), the identification of Sgr A with the galactic center is well founded.

W. Kundt (ed.), Astrophysical Jets and Their Engines, 47–65.

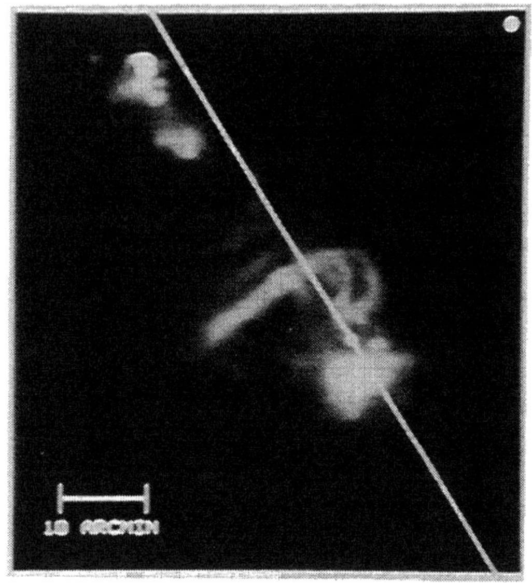

Fig. 1. 10.7 GHz continuum emission from the galactic center (Seiradakis et al. 1985). The sloping line is the galactic plane. The Sgr A region lies just below the point on this line.

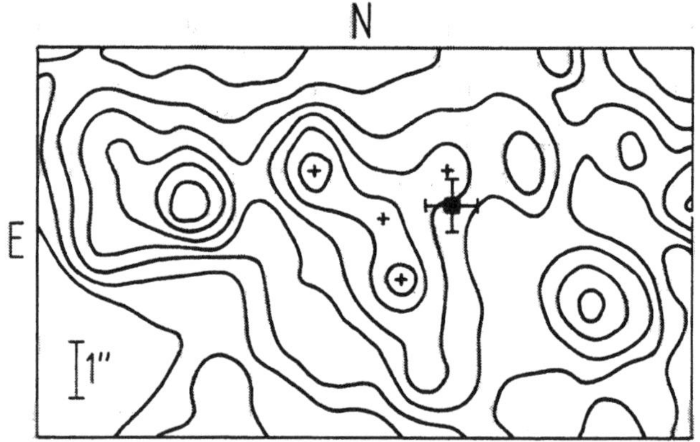

Fig. 2. 2.2 μm emission from the central few arc sec of the galaxy (Forrest et al. 1986). The four components of IRS 16 are marked with crosses. The black square marks the position of the radio point source.

While the extended Sgr A region can define the location of the galactic center at the arc minute level, the infrared observations of the central stellar cluster are able to locate the "center", in the sense of the center of symmetry of the central stellar cluster, to an accuracy of a few arc seconds. The 2 μm maps of the central few parsecs show a number of unresolved point sources superposed on a broad background of emission, near the center of which is located a slightly extended (2") source IRS 16 (Becklin and Neugebauer 1975; Allen et al. 1983). Since the 2μm background is presumably due to stars comprizing the stellar cluster at the galactic center, the extended source IRS 16 near its center has commonly been assumed to be the very center of this late-type distribution of stars.

This assumption has recently been called into question, on two grounds. First, higher resolution IR observations have succeeded in resolving IRS 16 into 4 separate components (Fig. 2; Storey and Allen 1983; Henry et al. 1984; Allen and Sanders 1986; Forrest et al. 1986), and second, all 4 of these components have IR colors much bluer than those of the surrounding sources. While the 2 μm background and most of the point sources can be characterized by a color temperature of ~4000 K, appropriate to late-type stars, the blue colors of IRS 16 indicate color temperatures of order 30,000 K (Henry, Depoy and Becklin 1984, Stein and Forrest 1986). Thus, IRS 16 apparently is not composed of late-type stars (Wollman et al. 1982).

The components of IRS 16 may be clusters of young stars near the galactic center, but since a small cluster of stars need to lie exactly at the center of symmetry of the much more massive cluster of late-type stars present, the location of such a cluster would not necessarily define the center. In addition, it is not clear which component of IRS 16 would be "central". On the other hand, the blue colors of the IRS 16 components allow the possibility that the IR radiation from one of these components is produced in an accretion disk around a massive central object. If so, the appropriate component of IRS 16 would define the center of our galaxy, in the sense of a single central object. Again, based only on the IR measurements, it is not obvious which of the 4 components could be the "central" one.

Alternatively, the extended distribution of 2 μm radiation from the central stellar cluster can be expected to have a center of symmetry. However, while showing a steep increase towards the central few arc seconds (Becklin and Neugebauer 1968; Sanders and Lowinger 1972; Bailey 1980; Allen et al. 1983), the cluster center of symmetry is not defined to better than ~5", because of the presence of numerous bright point sources, and because of variable extinction on larger (arc min) scales. Thus the extended 2μm light distribution indicates only that the center of our galaxy is located within several arc sec of IRS 16.

The other choice for the "center" of the galaxy is a bright radio point source of size $\leq$20 AU (Balick and Brown 1974; Lo et al. 1985) which is located about an arc sec from IRS 16 (see Fig. 2). This source, Sgr A^*, has spectral characteristics similar to those seen in compact sources present in the cores of several external galaxies, in particular, a positive spectral index α of ~0.25 ($F_\nu \propto \nu^\alpha$; Lo et al. 1981; Backer 1982). However, the total radio luminosity of this point source, ~10 $L_\odot$, is orders of magnitudes weaker. Its other properties include an elongation approximately perpendicular to the galactic plane (Lo et al. 1985), and a transverse velocity consistent with a location at rest at the galactic center (Backer and Sramek 1982). Since the properties of this compact radio source make it unique in our galaxy, it is conceivable that the radiation from this object is due to an exotic object at the galactic center, such as a black hole. If it is massive ($M \gtrsim 10^6$ $M_\odot$), it would define the galactic center.

The relative positions of Sgr A^* and the components of IRS 16 have been carefully investigated, in order to determine if any of the infrared sources correspond to Sgr A^* (Allen and Sanders 1986; Forrest et al. 1986). The conclusion, seen in Fig. 2, is that Sgr A^* is ~1.3" west of the central component of IRS 16. This conclusion has been reached by a number of methods, and is at a significant level. The northwest component of IRS 16, IRS 16 NW, is the closest to Sgr A*, but it is not yet known if the two are coincident or not (the observed separation of ~0".5 is not significant). Thus it may be that none of the bright IR sources are coincident with Sgr A*.

In summary, radio observations point to an unusual object, Sgr A*, lying at our galactic center, while the infrared continuum observations indicate that the center of the stellar cluster is within a number of arc sec of this source also. The very blue 2 μm source IRS 16 lies close to the radio point source, but apparently only the component IRS 16 NW lies close enough to Sgr A* to possibly be coincident. For the purposes of the following discussion, the "center" will refer to both Sgr A* and IRS 16.

1.2 The ionized gas distribution

The ~4' (12 pc) sized Sgr A region (Fig. 1) consists of a superposition of two sources: a thermal HII region Sgr A West, and a nonthermal shell of emission Sgr A East (Downes and Martin 1971; Ekers et al. 1975, 1983). Sgr A East is presumably a supernova remnant along the line-of-sight, and absorption line studies place it within a few hundred pc of the center (Güsten and Downes 1980). It is not clear exactly how close it is to the center, or if it is interacting with Sgr A West, although the coincidence of the western part of the Sgr A East shell with part of the thermal structure of Sgr A West suggests that an interaction may be taking place (Ekers et al. 1983).

The Sgr A West region, upon which the remainder of this section will focus, occupies roughly the central arcmin of our galaxy. Its structure, revealed through high resolution radio continuum maps (Brown and Johnston 1983; Lo and Claussen 1983; Ekers et al. 1983; Lo 1986; Ishiguro et al. 1986), is rather unusual, in that it appears to have a "spiral" shape (Fig. 3), surrounding the central radio point source. Although the earliest VLA maps resolved only two arms of the "spiral", later higher resolution observations have revealed a third "arm" projecting eastward. This immediately ruled out any models with bipolar symmetry, such as the twin precessing jet model of Brown (1982).

In fact, more recent observations have shown that the mini-spiral is not a single structure at all, but rather a superposition of a number of distinct features along the line-of-sight. This is shown most clearly by spectral line observations of the H76α line (van Gorkom et al. 1983) and the [Ne II] line (Serabyn and Lacy 1985). Both lines show rotation dominating the motion of the "western arc" feature (Fig. 3), with velocities varying quite regularly from ~−100 km s^{-1} at the southern end of the feature to +100 km s^{-1} at its northern end.

Near the midpoint of the western arc, where the "spiral" appears to bend eastward toward the center, the velocity of the western arc feature crosses 0 km s^{-1}, and a second velocity component, near −120 km s^{-1}, is seen. This second, higher velocity feature originates in the structure which runs eastward from this location toward the radio point source, implying a large velocity discontinuity between the western arc and the east-west feature. Thus, the velocity information allows a separation of

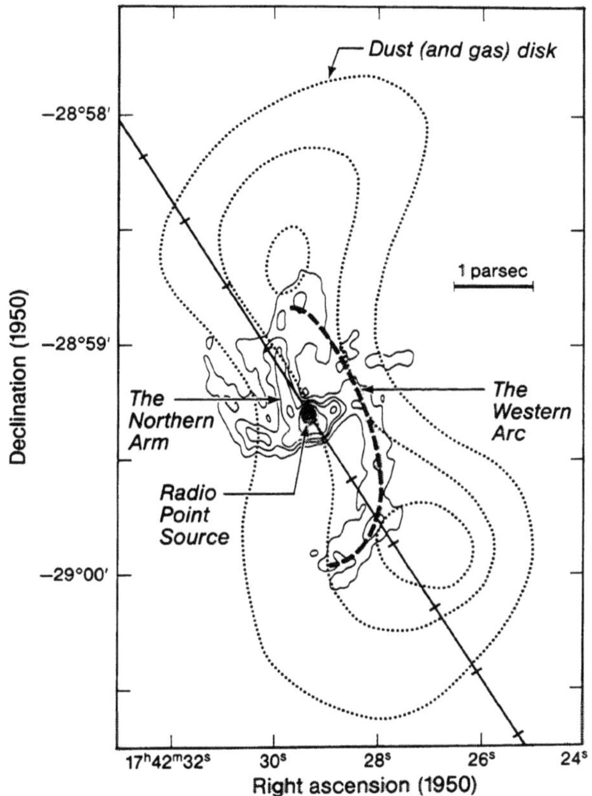

Fig. 3. *Superposition of the 15 GHz continuum emission from Sgr A West (solid contours; Ekers et al. 1983) and the far infrared dust continuum emission (Becklin et al. 1982).*

the "spiral" continuum emission pattern into two distinct gas features: the western arc, which runs along the western edge of the map, and an East-West emission ridge which crosses the midpoint of the western arc.

The continuum emission is thus composed of a number of separate features, most of which appear elongated or filamentary. The most obvious of these are the western arc and the northern and eastern arms (Fig. 3). In the central 10", just south of the radio point source, there is additional emission which may or may not relate to these more extended features. All three filamentary structures have velocities which vary regularly with position along the filaments, reaching maximum velocities of 140 and 165 km s^{-1} on the northern and eastern arms, respectively. Furthermore, all three features seem to show a zero-crossing in velocity near the position of the center (Serabyn et al. 1987), and all three pass by the center, rather than originating at a common point. The last two points argue against outflow models, of either the expanding shell or jet type, and for orbiting or infalling gas.

A number of gas orbit models have now been explored, those being tidal disruption of an infalling gas cloud (Schwarz 1984), orbits spiraling inward due to the drag of a hot intercloud medium (Quinn and Sussman

1985), and drag-free orbits of arbitrary shape about either a point mass or a stellar cluster (Serabyn and Lacy 1985, Serabyn et al. 1987). The latter approach, although based on the simplest assumptions, has yielded rather good fits to both the velocities and spatial positions of all three gas filaments. The gravity dominated orbit model also allows the mass within the orbits to be estimated, as is discussed below.

In contrast to the larger scale filaments, for which streams of gas in orbit about the center are a good description, the kinematics of the central 10" region are not yet clear. This central region contains a number of overlapping gas features, with velocities as high as ±300 km s^{-1} (Lacy et al. 1980), and velocity gradients up to 25 km s^{-1} arc sec^{-1} (Serabyn et al. 1987). The typical line-of-sight velocity spread is ~35 to 50 km s^{-1}, much smaller than previous observations (with larger beams) indicated. Several of the velocity components in the central region may be continuations of the northern and eastern arms.

The broadest velocity emission is observed in the few arc sec region surrounding IRS 16 (Hall et al. 1982; Geballe et al. 1984). Emission out to ~±700 km s^{-1} is seen in both infrared H and He lines, in the form of low level wings on the previously discussed narrower line emission. The weakness of this emission, coupled with the fact that atmospheric absorption lines affect the profiles, makes observations of this "broad line" emission rather difficult. Thus, the kinematics of this gas are not yet clear. Interpreted in terms of mass outflow from a central object, a mass outflow rate of $\lesssim 10^{-2}$ $M_\odot$ yr^{-1} is indicated (Krisciunas et al. 1985). Given the small size (~3-4") of the emission region, and its location close to IRS 16 and Sgr A^*, it is clear that this gas provides important information on the kinematics closest to the center.

1.3 Neutral and Molecular Gas

Surrounding the central ionized "spiral" is a region of neutral and molecular material, extending from ~2 pc to ~10 pc from the center (Fig. 3). This region was first detected via its far infrared thermal dust emission (Becklin et al. 1982), which shows a double lobed structure surrounding the central arc min. Because the FIR radiation is optically thin, and because each dust grain radiates more at every wavelength as its temperature is raised, there is a lack of dust in the central arc min region compared to the immediate surroundings. However, the color temperature of the dust radiation is centrally peaked, suggesting a central heating source. The picture which has emerged is that of a relatively transparent central cavity of ~2 pc radius, surrounded by dense (~10^5 cm^{-3}; Genzel et al. 1984; 1985) molecular gas and dust.

Molecular line observations (Genzel et al. 1982, 1984, 1985; Kaifu et al. 1983; Harris et al. 1985; Sandqvist et al. 1985; Lugten et al. 1986; Serabyn et al. 1986; Güsten et al. 1987) have shown that this double lobed structure forms a flattened torus or disk of gas which rotates about the center with a velocity of ~100 km s^{-1}. The existence of the western arc is then easily understood as the ionized inner edge of this rotating torus (Fig. 4). Here again, a central illuminating source and a transparent central region are indicated.

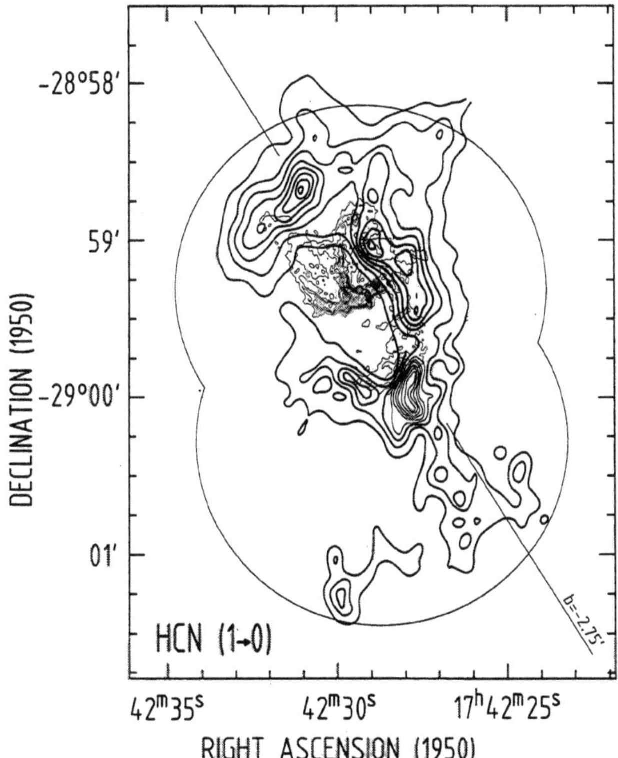

Fig. 4. The molecular HCN J=1-0 emission from the inner edge of the molecular disk (Güsten et al. 1987) superposed on the 5 GHz continuum emission from Sgr A West (Lo and Claussen 1983).

The material in the neutral ring is clumpy (clump size ~0.3-1.0 pc) and shows decreasing excitation with radius. This can be derived from a comparison of the relative intensities of the transitions of the CO molecule (Harris et al. 1985; Lugten 1986, priv. comm.) or of the CS molecule (Güsten, priv. comm.). The CO molecule indicates a decrease in the density-temperature product by a factor of ~3 between radii of 2 and 5 pc, while CS observations indicate a falloff of density. The simplest model of a uniform disk may be an oversimplification however, as an azimuthal excitation gradient may also be present.

Along the inner edge of the disk is also a narrow ring of shocked H_2 which may be excited by a central mass outflow hitting the inner edge of the molecular disk (Gatley et al. 1984; 1986). For example, if the broad H and He emission seen near IRS 16 is due to outflow from a central object, and if the ouflow reaches the molecular ring, the momentum would be more than sufficient to account for the excitation of the shocked H_2. Thus, it is conceivable that the central 2 pc cavity is a mass loss bubble. Alternatively, Genzel et al. (1985) have proposed that the sharp inner radius may be due to a Lindblad resonance, and Güsten et al. (1987) suggest a central explosion for the "clearing" agent. Some sort of agent is necessary to keep the central cavity clear, as friction between clumps in the molecular gas should cause the gas to fall inward. Indeed, this is one possible explanation for the streams of ionized gas present inside the ring (the northern and eastern arms; Güsten et al. 1987).

Observations of C^+ (Lugten et al. 1986) and CO (Serabyn et al. 1986) reveal that the emission from the molecular disk is tilted out of the galactic plane by ~25° on the plane of the sky. The direction of this tilt is opposite to that of the +22° tilt of the larger scale neutral HI disk, which extends from some tens or hundreds of pc out to ~1.5 kpc radius (Burton and Liszt 1978). Thus, between a scale size of a few pc and a few kpc, the rotation axis of the interstellar gas apparently changes direction twice. The reason for this is not clear, but may be related to the possible existence of a bar or to some triaxility in the stellar bulge at the center of our galaxy (Liszt and Burton 1980; Vietri 1986).

Measurement of a position velocity diagram along the tipped major axis of the 10 pc sized disk reveals a constant rotational velocity with radius in the CO J=1−0 line (Serabyn et al. 1986). Previous observations indicating a fall off of rotational velocity with radius (Harris et al. 1985; Lugten et al. 1986) seem to be the result of measurements off the kinematic major axis, combined with an azimuthal excitation gradient in the ring. The finer details of the kinematics have now been probed with a few arc sec resolution in the HCN J=1−0 line (Güsten et al. 1987), and further suggest that the disk may not be a completely planar structure, being somewhat warped.

1.4 Energetics

Although the extinction to the galactic center prevents the ionization source of Sgr A from being identified directly, the ionizing spectrum does have several known properties: its total luminosity is ~1−3x10^7 $L_\odot$ (Becklin et al. 1982), the number of Lyman continuum photons is ~4x10^{50} s^{-1}, and, based on the low ionization states present, T_{eff} ≤35,000 K (Lacy et al. 1980). These parameters are all consistent with those of B stars, implying that recent star formation at the galactic center could explain the ionization state of Sgr A. However, for a normal initial mass function, the O stars present would produce too many hard ionizing photons to be consistent with the low ionization state of Sgr A. To get around this problem, it is necessary to assume that a burst of star formation occurred ~4x10^6 yrs ago, and that the ensuing O stars have already evolved off the main sequence, leaving only B stars to produce the ionizing photons. This also allows an explanation of the high abundance of M type supergiants at the galactic center, by equating them with the evolved O stars (Lebofsky et al. 1982). Thus it is possible to set up a consistent picture in which a distribution of newly formed stars ionizes Sgr A West and heats the dust ring at r >2 pc, with the only disturbing aspect being that no star formation is currently going on in the central few pc.

On the other hand, it is possible that a single central source provides the ionizing radiation for Sgr A West. In this case, there is an additional constraint on the source, namely that the calculated 2 μm flux must not exceed the 2 μm flux from the source. It is not clear which of the 2 μm sources to use for the limit (section 1.1), but the 2 μm flux from any of the components of IRS 16 is sufficient to be consistent with a T=35,000 source which provides a large fraction of the flux necessary to ionize Sgr A (Henry, Depoy and Becklin 1984; Serabyn and Lacy 1985).

The central source could be a central cluster of ionizing stars as discussed previously, or possibly an accretion disk around a central object. However, a face on accretion disk of large radial extent is ruled out by the 2 μm flux from the components of IRS 16. On the other hand, a nearly edge-on disk is not ruled out, even if the assumed 2 μm flux limit is lowered a factor of ~3 below the flux from any of the IRS16 components, because of the sin i factor, where i is the inclination of the disk. Since the 10 pc sized ring around the galactic center, the kpc sized HI disk, and the galactic plane are all seen close to edge-on, it is not unreasonable for the possible central accretion disk to be close to edge-on also. Therefore the 2 μm flux limit is not a stringent enough constraint on source models to rule out a central accretion disk, especially a non-steady-state one. There thus remain a number of possibilities for the source of the ionization in Sgr A.

1.5 The mass distribution

There are two approaches to mass distribution at the center of our galaxy: through the stellar light distribution and kinematics, and through the gas kinematics. Since the stellar kinematics have only recently begun to be explored (Mc Ginn et al. 1986), most of the information comes from the stellar 2 μm flux distribution. This reveals that the stellar mass distribution is of the form $\rho^{-\alpha}$, with α ~1.7 to 1.8 (Becklin and Neugebauer 1968; Sanders and Lowinger 1972; Bailey 1980; Allen et al. 1983). However, the unknown mass-to-light ratio of the central stellar cluster leaves the normalization of this distribution uncertain. Furthermore, it has recently been suggested that at small radii (≤20") this power law distribution is produced not by the population II cluster at the galactic center, but by the presence of a few bright 2 μm sources which are not part of this population (Rieke and Lebofsky 1986). These 2 μm sources may instead be due to recent star formation at the galactic center (Lebofsky et al. 1982). Removal of these sources (IRS16 among them) implies that the central stellar cluster may actually have a flat core radius of ~1 pc.

The other approach to the mass distribution is through the gas kinematics. For this approach to be valid, the gas must be circulating about the center in orbits dominated by gravity, with nongravitational forces such as pressure and magnetic fields being negligible. In the case of the galactic center, there is good evidence that much of the gas is indeed orbiting about the center. The best example is the western arc (Fig. 3), along which the velocities closely follow the pattern expected for a circular orbit about the center. The material in the neutral disk also appears to be following relatively circular orbits about the center, although a few disturbances do seem to be present. For these cases, the mass estimates follow simply from the measured orbital velocities and radii, and yield an enclosed mass beyond r ~2 pc which increases approximately linearly with radius (Serabyn et al. 1986), in rough agreement with the stellar cluster results. The normalizations also agree if an M/L ratio of ~1 is assumed.

Inside of the molecular disk, the ionized filaments also appear to be following orbits about the center, although noncircular ones. By fitting orbits of various shapes to these features, it was found that in order to obtain good fits to the velocities, a substantial fraction of the mass enclosed within the western arc's orbit at r=1.7 pc must also be present inside of much smaller radii (Serabyn and Lacy 1985, Serabyn et al. 1987). This implies a mass distribution at small radii which is much more centrally peaked than that of the central stellar cluster, and so points toward a massive central object. This result can be seen in Fig. 5, which shows the mass estimates resulting from the kinematics of these various gas features, assuming gravity dominated orbits. All of the mass estimates in this figure rely on the presence of spatially extended orbits along which the velocity variation can be measured. Thus, none of the mass estimates rely on the presence of a single high velocity, or on a virial argument.

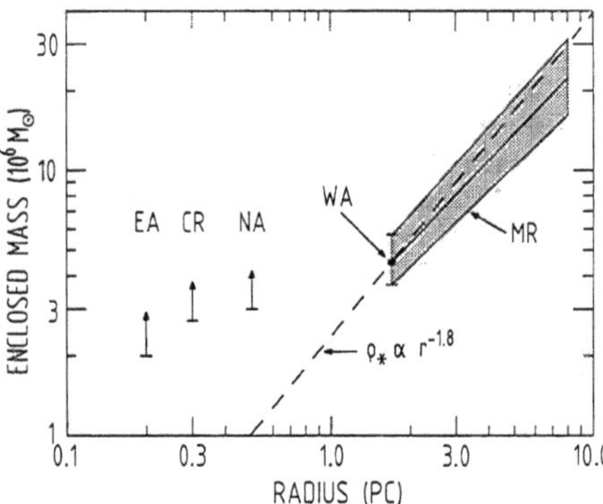

Fig. 5. Mass estimates in the central 10 pc. The hatched area is derived from the kinematics of the molecular disk at r >2 pc (Serabyn et al. 1986; Güsten et al. 1987), and the four other points and limits refer to estimates based on various ionized filaments in Sgr A West (Serabyn and Lacy 1985; Serabyn et al. 1987). The dashed line gives the stellar mass estimate (Becklin and Neugebauer 1968) for M/L=0.9.

A number of assumptions do go into modeling the orbits of the ionized filaments, however. Among these is the assumption that single particle orbits adequately represent the actual situation. Since the fitted orbits require large changes in orbital radii, this is adequate if either the origin of these gas streamers is very far away, or if the source of the gas is stationary. The latter would apply if, for example, the northern and eastern arms are infalling streamers of gas which arise as a result of a collision of a passing cloud with the molecular ring. In this case, the point of impact could be stationary even though the molecular ring rotates.

The other important assumption is that magnetic forces can be neglected. Aitken et al. (1986) estimate that B >10 mG along the northern arm. Although at this lower limit the magnetic energy density is significantly smaller than the gas kinetic energy density, larger field strengths could make them comparable.

2. ENERGETIC ACTIVITY AND THE CENTRAL 100 PARSECS

2.1 Energetic activity in the central 10 pc

The preceding section has presented a rather quiescent picture of the central 10 pc region, with orbiting material dominating much of the kinematics, and primarily molecular cloud and HII region temperatures involved. This picture, however, is incomplete. There are a number of indications of more energetic activity present, beginning with the aforementioned radio point source, which has a brightness temperature of nearly 10^9 K. This section addresses the higher temperature and more exotic phenomena seen, and extends the scope of the discussion to a radius of ~100 pc.

A direct indication of the presence of a hotter component of gas in the central parsec is the existence of KeV X-ray emission from this region (Watson et al. 1981). The angular resolution (1') of the observations does not allow a determination of whether the emission comes from the extended HII region or from the radio point source, but if it does arise in extended emission, gas temperatures of 10^7-10^8 K are required. Higher energy X-rays, with total luminosity ~3×10^{38} ergs s^{-1}, and positron annihilation radiation have also been observed in the general direction of the galactic center, but as these were measured with beams of greater than a degree (Matteson, J.L. 1982, Leventhal et al. 1986), their association with the galactic center is not assured. Nevertheless, the position annihilation radiation, which has in the last few years been too weak to be detectable, was unique in the galaxy, and so may be at the galactic center. The time variability of the emission limits the emission region to a size of ~10^{18} cm (Jacobson 1982), and the narrow linewidth indicates that the annihilation takes place in a medium with T <50,000 K, consistent with the temperature of the gas in the HII region.

Since many active galactic nuclei have jets emerging from their nuclei, it is natural to look for such a structure in our galactic center. Because of interference from material along the line-of-sight, only recently has evidence for a jet originating in the nucleus of our galaxy been obtained (Yusef-Zadeh et al. 1986). These observations, at 160 MHz, reveal a linear ridge of emission, extending at least 30 pc perpendicular to the galactic plane (the extent of the region mapped). The jet is one-sided, being visible only below the galactic plane. Its spectrum is nonthermal, with $\alpha < -0.85$ ($F_\nu \propto \nu^\alpha$). A disturbing aspect of the jet hypothesis, though, is that the emission appears to emerge from the galactic plane at an offset of ~2' to positive longitudes of the radio point source.

2.2 The galactic center arc

Moving farther afield, at ℓ ~0.°1 to 0.°2 is the curious radio continuum structure known as the galactic center arc (Figs. 1 and 6; Pauls et al. 1976; Yusef-Zadeh et al. 1984). This filamentary structure consists of two types of filaments: straight filaments perpendicular to the galactic plane, and "arched" filaments oriented somewhat along, and slightly above the

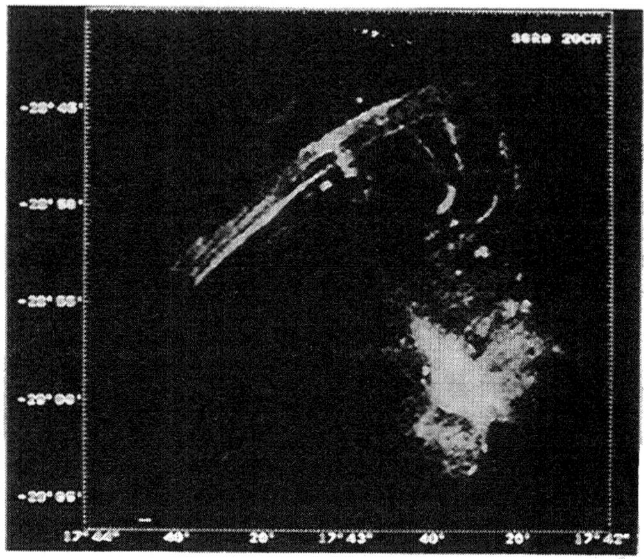

Fig. 6. 1.5 GHz emission from Sgr A (lower right), and the galactic center arc (Yusef-Zadeh et al. 1983).

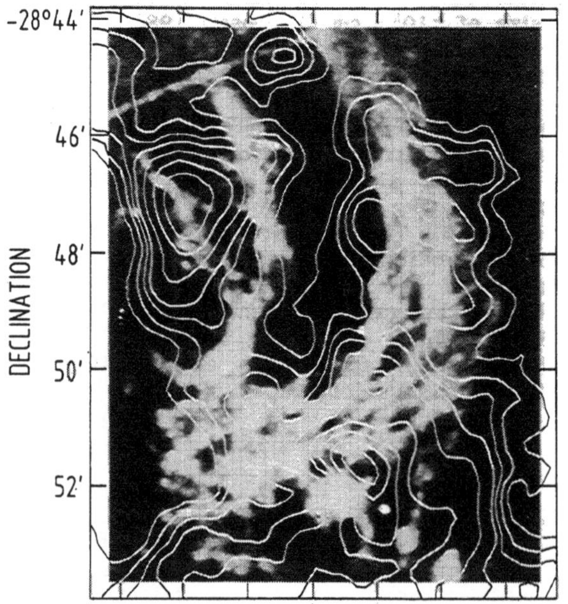

Fig. 7. Contours of constant CS J=2-1 emission intensity (Serabyn and Güsten 1987) superposed on a radiograph (Morris and Yusef-Zadeh 1985) of the arched filaments in the arc.

galactic plane. A weaker set of arched filaments is also present at negative galactic latitude (Fig. 1). The arched filaments give an impression of connecting the ends of the straight filaments to the galactic center. Observations of recombination lines and thermal dust emission (Pauls and Mezger 1980; Yusef-Zadeh 1986; Dent et al. 1982) have shown that the arched filaments are thermal in nature, while the straight filaments are nonthermal. The recombination lines also show that the velocity of the arched filament gas tends to increase from ~-50 km s^{-1} at the ends nearer the center, to ~0 km s^{-1} at the more distant ends.

The molecular gas in the vicinity of the arched filaments has now been mapped with sufficient resolution to show a clear association of molecular material with the ionized filaments (Bally et al. 1986; Serabyn and Güsten 1987). At negative velocities, a molecular gas component is present with a velocity structure very similar to that of the arched filaments. A superposition of the two gas components (Fig. 7) reveals that in many cases, the arched filaments lie along the edges of the negative velocity molecular clouds. This negative velocity molecular material also appears to abut Sgr A, giving the first direct indication that the arc is as close to the galactic center as its projected distance suggests. The ionization mechanism is not well defined, but photoionization by a source at the galactic center, or by a uniformly distributed stellar source seem ruled out by the geometrical arrangement of the neutral and ionized material. However, ionization by a few pockets of OB stars, or collisional ionization by the diffuse intercloud medium (with positive velocities of ~+100 km s^{-1}) are possibilities.

The straight nonthermal filaments remain more of a mystery. Except for a slight curvature, these features extend almost linearly for ~30 pc perpendicular to the galactic plane, with a width of $<$ 1 pc. The extreme linearity of these features leads naturally to the suggestion that their morphology is determined by a magnetic field, with synchrotron emission of electrons spiralling about the field lines being the emission mechanism (Yusef-Zadeh et al. 1984). Consistent with this emission mechanism is the existence of linear polarization along sections of the filaments (Tsuboi et al. 1985, 1986; Seiradakis et al. 1985). However, polarization is not present along the entire length of the filaments (Fig. 8), leading to the suggestion of a nonuniform distribution of foreground thermal matter which depolarizes sections of the emission. The rotation measure of the polarized emission yields a magnetic field strength of ~10 uG.

The spectral index of the emission from the nonthermal filaments is rather flat (Yusef-Zadeh et al. 1984), suggesting a young monoenergetic distribution of electrons. A small age is consistent with the expection that the age of the filaments must be small compared to the rotation period about the center (10^6 yr), in order for differential rotation not to distort the filaments. The source of the electrons present in the filaments is not at all clear, with most suggestions being rather speculative. Seiradakis et al. (1985) suggest an origin in a center of activity near the midpoint of the straight filaments, while Sofue and Fujimoto (1986) suggest an origin in material flowing out from the center.

In addition to the "organized" system of filaments constituting the galactic center arc, there are yet other filamentary structures present

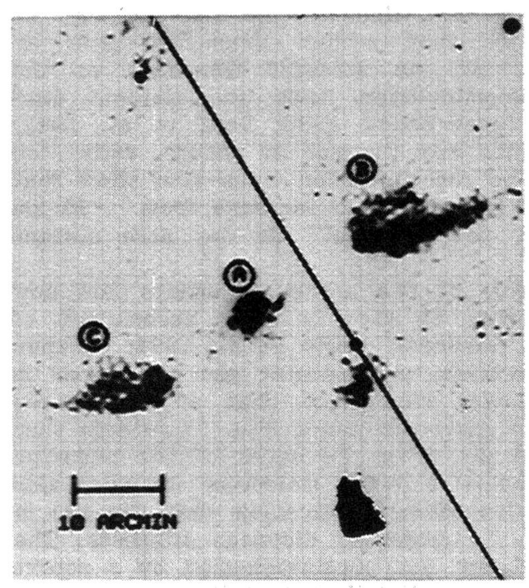

Fig. 8. Polarized intensity radiograph of the same region as in Fig. 1 (Seiradakis et al. 1985). The straight filaments in the arc lie along the line BAC.

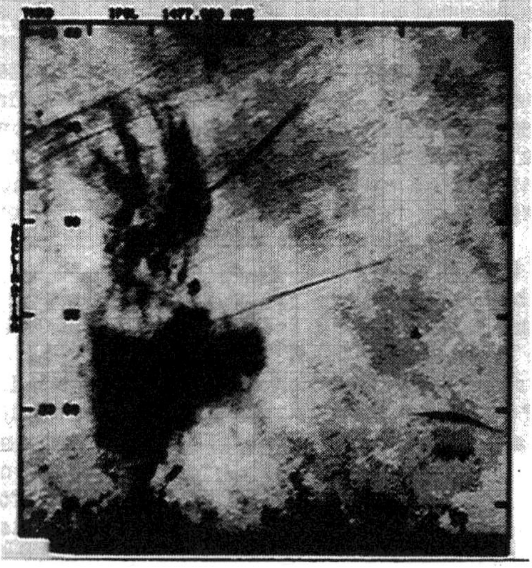

Fig. 9. 1.5 GHz emission from Sgr A, showing three isolated "threads" on the right side of the frame (Morris and Yusef-Zadeh 1985).

near the galactic center (Fig. 9). Three well isolated linear "threads" of emission are seen within ~50 pc of the galactic center. These threads are narrower than the filaments in the arc, being ~0.3 pc thick, but up to 30 pc long also. Very little is known of the nature of these threads with even the spectral index being undetermined. One thing is clear, however, that being that they do not extend in a direction which could be extrapolated back to the galactic center.

2.3 The galactic center lobe

The nonthermal filaments which are part of the arc are apparently not unique, since a similar structure perpendicular to the galactic plane is present on the opposite side of the center (Liszt 1985). Its location is not symmetric with respect to the galactic center, being at $\ell \sim -0.6^{\circ}$, as opposed to $\ell \sim 0.2^{\circ}$ for the arc.

Although both the arc and Sgr C filaments extent only ~30 pc perpendicular to the galactic plane, both may be related to much larger scale phenomena, since at both locations, spurs of radio continuum emission extent up to ~200 pc above the galactic plane (Fig. 10; Altenhoff et al. 1978; Sofue and Handa 1984; Reich et al. 1984). These two radio spurs may actually loop over to connect to each other ~200 pc above the galactic center, forming an Ω-shaped lobe (Sofue and Honda 1984). Such a structure could be due to a large-scale explosion at the galactic center or to magnetic activity similar to that in solar loop prominences. Although the weakness of the emission from the top of the Ω makes it difficult to determine whether the two spurs are actually connected, different physical conditions in the two spurs seem to argue

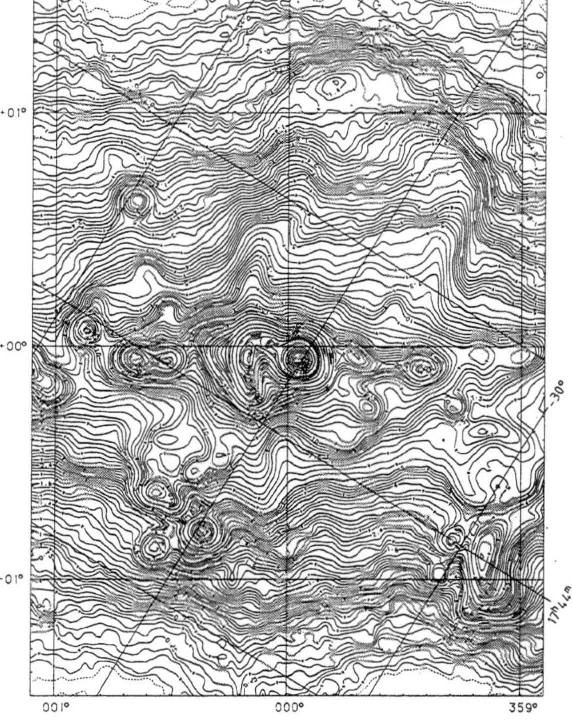

Fig. 10. 2.7 GHz emission from the central 2° of the galaxy, in (ℓ,b) coordinates (Reich et al. 1984). Sgr A is near the center of the map, the arc is at $\ell \sim 0.2^{\circ}$, and Sgr C is at $\ell \sim -0.6^{\circ}$. Above the latter two features, spurs of emission extend above the galactic plane to $b \sim 1^{\circ}$.

for different origins. These differences include polarization on the eastern, but not the western spur (Tsuboi et al. 1986), CO emission associated with the western, but not the eastern spur (Sofue 1985), and the extension below the galactic plane of the eastern, but probably not the western spur (Fig. 10).

SUMMARY. The components of the central 100 pc of our galaxy take a variety of forms. At the very center of our galaxy are located a bright radio point source, Sgr A*, and an infrared source with unusually blue colors, IRS 16. The precise nature of these sources have yet to be determined. Surrounding these are the central stellar cluster, with a mass roughly proportional to radius, and a spiral shaped HII region, Sgr A West. The filamentary features in this HII region seem to be streamers of gas in orbit about the center. This central 2 pc ionized region is in turn surrounded by a disk of molecular gas and dust which also is circulating about the center, and which extends out to a radius of ~10 pc.

On larger scales, rather exotic activity is present. A variety of filamentary structures are seen: isolated threads with a length to width ratio of ~100, groupings of linear, nonthermal (but flat spectral index) filaments which lie perpendicular to the galactic plane, and curved filaments which appear to outline the edges of molecular clouds. In addition, spurs of radio continuum emission extend ~200 pc above the galactic plane. Finally, a one-sided jet from the center seems to extend at least 30 pc below the galactic plane. The nature and excitation mechanisms of these structures is still rather uncertain, and further observations and modeling are clearly needed to clarify the processes at work.

ACKNOWLEDGEMENTS

I would like to thank R. Güsten, H. Lesch, W. Kundt, and R. Wielebinski for helpful comments and discussions. I am also grateful to R. Genzel and R. Güsten for permission to use data prior to publication.

REFERENCES

Aitken, D.K., Roche, P.F., Bailey, J.A., Briggs, G.P., Hough, J.H., and Thomas, J.A.: 1986, MNRAS **218**, 363

Allen, D.A., Hyland, A.R., and Jones, T.J.: 1983, MNRAS **204**, 1145

Allen, D.A. and Sanders, R.H.: 1986, Nature **319**, 191

Altenhoff, W.J., Downes, D., Pauls, T., and Schraml, J.: 1978, A.A. Suppl. **35**, 23

Backer, D.C.: 1982, in IAU Symposium 106, Extragalactic Radio Sources, eds. IAU Symposium 106, D.S. Heeschen and C.M. Wade, p. 389

Backer, D.C. and Sramek, R.A.: 1982, Ap.J. 260, 512

Bailey, M.E.: 1980, MNRAS **190**, 217

Balick, B. and Brown, R.L.: 1974, Ap.J. **194**, 265

Bally, J., Stark, A.A., Wilson, R.W., and Henkel, C.: 1986, preprint

Becklin, E.E., Gatley, I., and Werner, M.W.: 1982, Ap.J. **258**, 135
Becklin, E.E., Matthews, K., Neugebauer, G., and Willner, S.P.: 1978, Ap.J. **220**, 831
Becklin, E.E. and Neugebauer, G.: 1968, Ap.J. **151**, 145
Becklin, E.E. and Neugebauer, G.: 1975, Ap.J. (Letters) **200**, L71
Brown, R.L.: 1982, Ap.J. **262**, 110
Brown, R.L. and Johnston, K.J.: 1983, Ap.J. (Letters) **268**, L85
Burton, W.B. and Liszt, H.S.: 1978, Ap.J. **225**, 815
Dent, W.A., Werner, M.W., Gatley, I., Becklin, E.E., Hildebrand, R.H., Keene, J., and Whitcomb, S.E.: 1982, in The Galactic Center ed. G.R. Riegler and R.D. Blandford, p. 33
Downes, D. and Martin, A.H.M.: 1971, Nature **233**, 112
Ekers, R.D., Goss, W.M., Schwartz, U.J., Downes, D., and Rogstad, D.H.: 1975, A.A. **43**, 159
Ekers, R.D., van Gorkom, J.H., Schwartz, U.J., and Goss, W.M.: 1983, A.A. **122**, 143
Forrest, W.J., Pipher, J.L., and Stein, W.A.: 1986 Ap.J. (Letters) **301**, L49
Gatley, I., Jones, T.J., Hyland, A.R., Wade, R., Geballe, T.R., and Krisciunas, K.: 1986, MNRAS **222**, 299
Gatley, I., Jones, T.J., Hyland, A.R., Beattie, D.H., and Lee, T.J.: 1984, MNRAS **210**, 565
Geballe, T.R., Krisciunas, K., Lee, T.J., Gatley, I., Wade, R., Duncan, W.D., Garden, R., and Becklin, E.E.: 1984, Ap.J. **284**, 118
Genzel, R., Watson, D.M., Crawford, M.K., and Townes, C.H.: 1985, Ap.J. **297**, 766
Genzel, R., Watson, D.M., Townes, C.H., Dinerstein, H.L., Hollenbach, D., Lester, D.F., Werner, M.W., and Storey, J.W.V.: 1984, Ap.J. **276**, 551
Genzel, R., Watson, D.M., Townes, C.H., Lester, D., Dinerstein, H., Werner, M.W., Storey, J.W.V.: 1982, in The Galactic Center, eds, G.R. Riegler and R.D. Blandford, p. 72
Güsten, R. and Downes, D.: 1980, A.A. **87**, 6
Güsten, R., Genzel, R., Wright, M.C.H., Jaffe, D.T., Stutzki, J. and Harris, A.I.: 1987 (preprint)
Hall, D.N.B., Kleinmann, S.G. and Scoville, N.Z.: 1982, Ap.J. (Letters) **260**, L53
Harris, A.I., Jaffe, D.T., Silber, M., and Genzel, R.: 1985, Ap.J. **294**, L93
Henry, J.P., DePoy, D.L., and Becklin, E.E.: 1984, Ap.J. (Letters) **285**, L27
Ishiguro, M., Fomalont, E., Morita, K.I., Kasuga, T., Kanzawa, T., Iwashita, H., Kawabe, R., Kobayashi, H., and Okumura, S.: 1986, preprint
Jacobson, A.S.: 1982, in The Galactic Center, ed. G.R. Riegler and R.D. Blandford, p. 123
Kaifu, N., Inatani, J., Hasegawa, T., Morimoto, M.: 1983, in IAU Symposium 106, The Milky Way Galaxy, eds. H. van Woerden, W.B. Burton and R.J. Allen, p. 367
Krisciunas, K., Geballe, T.R., and Wade, R.: 1985, BAAS **16**, 980
Lacy, J.H., Townes, C.H., Geballe, T.R., and Hollenbach, D.J.: 1980, Ap.J. **241**, 132
Lebofsky, M.J., Rieke, G.H., and Tokunaga, A.T.: 1982, Ap.J. **263**, 736
Leventhal, M., MacCallum, C.J., Huters, A.F., and Stang, P.D.: Ap.J. **302**, 459

Liszt, H.S.: 1985, Ap.J. (Letters) **293**, L65
Liszt, H.S., and Burton, W.B.: 1980, Ap.J. **236**, 779
Lo, K.Y..: 1986, Science **233**, 1394
Lo, K.Y., Backer, D.C., Ekers, R.D., Kellermann, K.I., Reid, M., and Moran, J.M.: 1985, Nature **315**, 124
Lo, K.Y., Cohen, M.H., Readhead, A.S.C., and Backer, D.C.: 1981, Ap.J. 249, 504
Lo, K.Y. and Claussen, M.J.: 1983, Nature **306**, 647
Lugten, J.B., Genzel, R., Crawford, M.K., Townes, C.H.: 1986, Ap.J. **306**, 691
Matteson, J.L.: 1982, in The Galactic Center, ed. G.R. Riegler and R.D. Blandford, p. 109
McGinn, M.T., Sellgren, K., Becklin, E.E., and Hall, D.N.B.: 1986, in The Proceedings of the Symposium on the Galactic Center in honor of C.H. Townes, ed. D.C. Backer, AIP
Morris, M. and Yusef-Zadeh, F.: 1985, Astron. J. **90**, 2511
Oort, J.H. and Rougoor, G.W.: 1960, MNRAS **121**, 171
Pauls, T., Downes, D., Mezger, P.G., and Churchwell, E.: 1976, A.A. **46**, 407
Pauls, T. and Mezger, P.G.: 1980, A.A. **85**, 26
Pauls, T., Mezger, P.G., and Churchwell, E.: 1974, A.A. **34**, 327
Quinn, P.J. and Sussman, G.J.: 1985, Ap.J. **288**, 377
Reich, W., Fürst, E., Steffen, P., Reif, K., and Haslam, C.G.T.: 1984, A.A. Suppl. **58**, 197
Rieke, G.H. and Lebofsky, M.J.: 1986 in The Proceedings of the Symposium on the Galactic Center in honor of C.H. Townes, ed. D.C. Backer, AIP
Sanders, R.H. and Lowinger, T.: 1972, Astron. J. **77**, 292
Sandqvist, A., Wooten, A., and Loren, R.B.: 1985, A.A. **152**, L25
Schwarz, U.J.: 1984, Kapteyn Astronomical Institute Annual Report, p. 48
Seiradakis, J.H., Lasenby, A.N., Yusef-Zadeh, F., Wielebinski, R., and Klein, U.: 1985, Nature **317**, 697
Serabyn, E. and Güsten, R.: 1987, preprint
Serabyn, E., Güsten, R., Walmsley, C.M., Wink, J.E., and Zylka, R.: 1986, A.A. , **169**, 85
Serabyn, E. and Lacy, J.H.: 1985, Ap.J. **293**, 445
Serabyn, E., Lacy, J.H., and Townes, C.H.: 1987, in preparation
Sofue, Y.: 1985, Publ. Astron. Soc. Japan **37**, 697
Sofue, Y. and Fujimoto, M.: 1986, preprint
Sofue, Y. and Handa, T.: 1984, Nature **310**, 568
Stein, W.A. and Forrest, W.J.: 1986, preprint
Storey, J.W.V. and Allen, D.A.: 1983, MNRAS **204**, 1153
Tsuboi, M., Inoue, M., Handa, T., Tabara, H., and Kato, T.: 1985, Publ. Astron. Soc. Japan **37**, 359
Tsuboi, M., Inoue, M., Handa, T., Tabara, H., Kato, T., Sofue, Y., and Kaifu, N.: 1986, Astron. J. **92**, 818
van Gorkom, J.H., Schwarz, U.J., and Bregman, J.D.: 1983, in IAU Symposium 106, The Milky Way Galaxy, ed. H. van Woerden, W.B. Burton and R.J. Allen, p. 371
Vietri, M.: 1986, Ap.J. **306**, 48
Watson, M.G., Willingale, R., Grindlay, J.E., and Hertz, P.: 1981, Ap.J. **250**, 142
Wollman, E.R., Smith, H.A., and Larson, H.P.: 1982, Ap.J. **258**, 506
Yusef-Zadeh, F.: 1986, Ph.D. thesis, Columbia University

Yusef-Zadeh, F., Morris, M., Slee, O.B., and Nelson, G.J.: 1986, Ap.J. (Letters) **300**, L47
Yusef-Zadeh, F., Morris, M., and Chance, D.: 1984, Nature **310**, 557

underground physics

JETS IN X-RAY BINARIES

Hakkı Ögelman
Max Planck Institut für extraterrestrische Physik
D-8046, Garching bei München
Federal Republic of Germany

ABSTRACT. The three galactic X-ray binaries that show evidence for jets, Sco X-1, Cyg X-3, and SS433 are examined. SS433 is the only source that shows rich jet features in all wavelengths from radio to gamma rays; the jets of Sco X-1 and Cyg X-3 are only evident in the radio region. Observational features for the three systems, as X-ray binaries, are summarised and compared. A possible common model for the central engine, involving a fast spinning neutron star with weak surface field ($B_s \lesssim 10^9$ gauss), is outlined.

1. INTRODUCTION

At present, there exist three X-ray emitting binaries that show definite evidence for having jets; namely, Sco X-1, Cyg X-3, and SS433. A fourth possibility, 1E2259+596 has also been discussed in the literature (Gregory and Fahlman, 1980). In comparison with the extragalactic counterparts, there are two main advantages in studying these galactic jets. One advantage is that they are nearer to us thus allowing a more sensitive investigation; the second advantage is that we have a better idea about the nature of the central engine. Although the common preferance for the central object in SS433 appears to be a black hole, in principle all three sources could have a neutron star as their compact object.

My outline for this review is as follows: First, I summarize briefly the observational evidence pertaining to the galactic binary X-ray sources in general. Next, I review the properties of the three X-ray binaries with jets. Finally I discuss the prospects of a fast spinning neutron star as the central engine of the jets in X-ray binaries.

By the term X-ray binaries, I will strictly imply the class of more luminous X-ray sources ($L_x \gtrsim 10^{35}$ erg s^{-1}) that have a neutron star or a black hole as the central object, thus excluding the less luminous X-ray binaries that contain a white dwarf as the compact object, i.e. cataclysmic variables.

W. Kundt (ed.), Astrophysical Jets and Their Engines, 67–90.

2. A REVIEW OF X-RAY BINARIES

Presently there exist $\sim 10^2$ known X-ray sources that are estimated to be binary systems with neutron stars. About half of these have measured binary periods; the X-ray luminosities are in the 10^{35} to 10^{38} erg s^{-1} range. The primary energy source is the energy released by the accreted mass, as it streams from the companion onto the surface of the compact object, giving the luminosity L_x:

$$L_x \sim \frac{GM_x\dot{M}_x}{R_x}$$

where M_x, R_x, and $\dot{M}_x$ are the mass, radius and the accretion rate associated with the X-ray emitting compact object. With typical values of $M_x = 1.4\ M_\odot$, and $R_x = 10^6$ cm, a neutron star can radiate close to its Eddington luminosity $L_{Edd} \sim 10^{38}$ erg s^{-1} if $\dot{M}_x \sim 10^{-8}\ M_\odot\ yr^{-1}$. In case the central object is a black hole instead of a neutron star, a comparable amount of energy per unit mass is thought to be released before the disappearance of the material through the event horizon at $R_s = 2GM_x/c^2$.

A closer look at the galactic X-ray binaries reveals that they are divided into two main categories: massive X-ray binaries with companion masses $M_* \gtrsim 10\ M_\odot$ and the low mass systems with $M_* \lesssim 2\ M_\odot$. Due to the expected age of the stellar component, these two categories are also called type I and II respectively. Roughly 50% of the sources belong to each category. The expected X-ray lifetimes for the high-mass and the low-mass systems are $\sim 10^4 - 10^5$ and $\sim 10^8 - 10^9$ years respectively (Savonije 1983). Considering that we see about equal numbers in both categories, the formation rate of high mass X ray binaries must be about 10^4 times more. If we further take into account the smaller number of massive stars, the formation probability of an X-ray binary must be a factor $\sim 10^6$ more in high mass systems (van den Heuvel 1983).

The rotation period of the neutron star has been detected in a sizeable fraction of the massive binary systems (~ 25), and in few of the low mass systems (~ 3). These periods cover a wide range, 10^{-1} to 10^3 s. The existence of a rotation period is regarded as a firm evidence for a neutron star versus a black hole. The Doppler shift of the pulse arrival times allows accurate determination of the orbital parameters such as size, eccentricity and mass function f(M). If in addition, independent information about the normal star such as eclipse duration, luminosity class, or the semi-amplitude of the optical Doppler velocity curve, exist, we can determine all parameters of the system. About 7 such systems exist all of which yield neutron star masses compatible with the 1.2 to 1.6 $M_\odot$ expected from theory.

Orbital periods range from ~ 10 min to ~ 200 days. Generally, the orbital period is a first order indicator of the system mass, since the size of the companion must be less than or equal to the Roche lobe. Consequently, for this condition to be fulfilled, low mass systems should be closer and thus have shorter periods. Roughly, systems with orbital periods less than one day are low mass systems, and those longer are high mass systems.

Among the galactic X-ray binaries there are three for which there is an indication that they maybe black holes, namely Cyg X-1, LMC X-3, and A0620-00 (McClintock, 1986). The likely masses of the compact objects in these systems are $\gtrsim 3M_{\odot}$.

Within this general framework of X-ray binaries, I proceed to examine the observational record of the three with jets.

3. Sco X-1

Sco X-1 is the brightest X-ray source in the sky. Radio emission from the source was detected following its discovery (Andrew and Purton 1968; Ables 1969). Hjellming and Wade (1971) discovered that the radio source consisted of a variable central source roughly coincident with the position of the X-ray source, plus two separate jet-like lobes about 1.2' from the center ($\sim 10^{18}$ cm); it was this radio picture that has enabled Sco X-1 to join the "jet set". Figure 1 shows the VLA picture of Sco X-1.

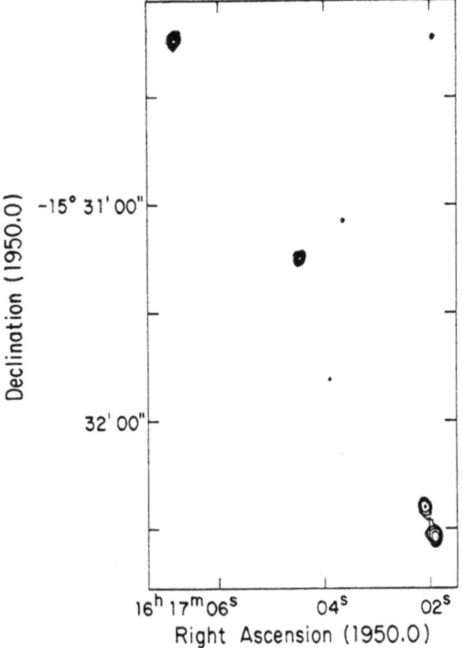

Figure 1: A VLA map of Sco X-1 at 1.465-GHz. The central source is nearly coincident with the X-ray source (from Fomalont et al. 1983).

Further studies have shown that the lobes are moving away from the central source with projected velocities of $\sim$ 35 km s^{-1} (Geldzahler and Fomalont 1985; Hjellming and Johnston 1985). Blandford and Rees (1974) have suggested a model similar to the extragalactic sources, where the energy generated at the central object, is collimated and transported along the jet axis. The relativistic particles and the fields are generated when this beam collides with the interstellar matter; i.e. the working surface. The radio luminosity of the lobes, 10^{29} erg s^{-1} , and their size, $\sim 1.5 \times 10^{15}$ cm, and the assumption that the radio emitting process is synchrotron emission of relativistic electrons in equipartition with the magnetic field, gives $\sim 10^{-6}$ erg cm^{-3} as the magnitude of the energy density. This also implies that the total energy in the lobes is of the order of $\sim 10^{40}$ ergs. In such a model, to keep the expansion velocity around the observed 35 km s^{-1}, ρ_{ext}, the density of the external medium has to be greater than:

$$\rho_{ext} \gtrsim 2.5 \times 10^{-21} \left(\frac{P_{lobe}}{10^{-6}}\right)\left(\frac{v_{ex}}{20\ km\ s^{-1}}\right)^{-2} g\ cm^{-3}$$

which is a very high value ($n_H \sim 10^3$ cm^{-3}) and may possibly exist in a supernova remnant around the source (Kundt and Gopal-Krishna 1984). Achterberg, Blandford, and Goldreich (1983) have suggested that magnetic pinching by a toroidal field may confine the jets, rather than the external pressure. They further estimate that a current of 3×10^{13} Amperes, flowing along the jet will be able to accomplish this focusing. In this model, the total power carried by the jet is $\sim 10^{32}$ erg s^{-1} , which may easily be provided by the central neutron star.

Recently, Geldzahler and Fomalont (1985) have observed that the extended components vary 10 to 30 % on time scales $\lesssim$1 year, with an indication that the variations in opposite lobes may be correlated. If we consider that these correlations are causal, it implies that the speed of the material flowing inbetween is of the order of 0.5 c.

Let us examine the properties of the central X-ray source responsible for these jets. Sco X-1, the brightest and first non-solar X-ray source discovered (Giacconi et al. 1962) is also the earliest identified with an optical counterpart (Sandage et al. 1966). It is a typical low mass X-ray binary with $L_x \sim 10^{37}$ to 10^{38} erg s^{-1} in the 2 to 10 keV band. It is very variable with timescales of minutes to hours or more (Bradt et al. 1975; Holt et al. 1976). Although it is not an eclipsing system, photometric (Gottlieb, Wright, and Liller 1975), and spectroscopic (Cowley and Crampton 1975) observations have revealed that it is a binary with 0.79 day period. Further spectroscopic observations in the optical band indicate that the inclination $i \sim 30^o$, $M_x \sim 1.3\ M_\odot$, and $M_* \sim 1\ M_\odot$ (Crampton et al. 1976). Material flows from the Roche-lobe of the red dwarf companion via an accretion disk to the neutron star and provides the X-ray luminosity. Although no rotation period for the neutron star has been observed, recently quasi periodic oscillations (QPO) in the 6 to 20 Hz range have been observed in the X-ray flux (Middleditch, and Priedhorsky 1985). It has been argued that these frequencies correspond to the Keplerian frequencies near the Alfven radius, i.e. the radius around which the disk material starts getting attached to the magnetic field of the neutron star. It is possible to relate the Alfven radius r_A to the dipole moment μ of the neutron star and the accretion rate $\dot{M}_x$ as (Henrichs 1983):

$$r_A \sim \mu^{4/7}(2GM_x)^{-1/7}\dot{M}_x^{-2/7}$$

The range of observed QPO frequencies imply that Sco X-1 has to have a surface field of $\sim 10^9$ gauss. Using the observed luminosity dependent behaviour of the QPO frequency, and the changes of the X-ray spectrum, Priedhorsky (1986) argues that when the accretion rate falls below a certain minimum, the rotational energy of the neutron star can transfer energy to the disk; normally it is the disk that transfers rotational energy to the neutron star. It has further been observed that large radio flares start during these low luminosity phases (Miyomoto and Matsuoka 1977), implying that the radio emitting plasmons, i.e. the beginning of the jetstream, are injected on these occasions.

4. Cyg X-3

Cyg X-3 is a bright variable X-ray source, with $L_x \sim 10^{38}$ erg s^{-1} ; the X-ray flux is periodic at 4.8^h. The stability of this period implies that it is the orbital period of the system. Usual assumptions about the system, a Roche-lobe filling companion with a neutron star, yields 0.5 to 4 $M_\odot$ as the mass of the stellar companion(van den Heuvel and De Loore 1973). Several models, including stellar wind (Willingale, King, and Pounds 1985), cocoon (Milgrom and Pines 1978; Bonnet-Bidaud, and van der Klis 1981), and accretion disk corona (White, and Holt 1982), have been proposed to explain the shape of the light curve, which does not show an eclipse.

It is again only the radio emission from Cyg X-3 that shows evidence for jets, i.e. asymetrically expanding, elongated lobes. Braes and Miley (1972) discovered that Cyg X-3 is a bright variable radio source whose flux density at 20 cm may vary from 20 mJy to 200 mJy on timescale of months corresponding to a radio luminosity of 5×10^{32} to 5×10^{33} erg s^{-1} at an assumed distance of 10 kpc. Gregory (1972) discovered a giant radio flare during which the flux density reached 20 Jy. The time evolution of the radio spectrum of these giant flares was consistent with a relativistically expanding synchrotron source. High resolution mapping of the flares indicated that the source becomes elongated in the weeks following the flare (Geldzahler et al. 1983). Spencer and Johnston (1986) have found that, while the central component is ~ 20 mas before expansion, it develops asymetrically into two oppositely directed lobes at a rate of 5 to 18 mas per day. At an assumed distance of 10 kpc, this corresponds to a projected velocity of 0.35 c; Figure 2 shows the radio map of these lobes. Molnar (1986) suggests that the giant flares may be just the high amplitude end of a continuum of flare sizes. Furthermore, there is a claim that these radio flares are periodic at 4.95 hour, which is $\sim 3\%$ longer than the X-ray period. It is suggested that this period difference may be caused by the precession of the lines of apsides in an eccentric orbit. It is also possible to attribute this difference to the effect of a third body in an eccentric orbit of ~ 4 day period. Further confirmation of the different radio periodicity is necessary.

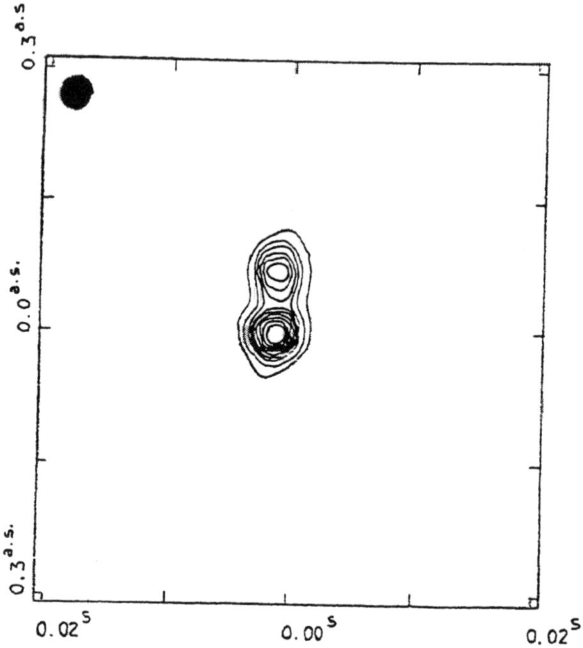

Figure 2: Map of Cyg X-3 at 6 cm, 13 days after a major flare (from Spencer and Johnston 1986).

Cyg X-3 has been observed to emit 10^{36} erg s^{-1} in the infrared ($\sim 2\mu m$); this flux is also modulated with the 4.8 hour period (Becklin et al. 1974). In the optical band, due to $\gtrsim 28^m$ of visual absorption, there is no hope of seeing anything (Weekes et al. 1981).

A series of measurements in the 10^8 to 10^{16} eV gamma-ray range indicate that Cyg X-3 may be a remarkably strong source. In the 10^8 eV band, positive detection with SAS-2 was reported by Lamb et al. (1977); however, the more sensitive exposure by COS-B failed to detect any positive flux (Swanenburg et al. 1981). In the 10^{12} eV, atmospheric Cerenkov range, positive detection has been claimed by Stepanian (1984). At even higher energies, 10^{15} to 10^{16} eV, a luminosity of $\sim 10^{37}$ erg s^{-1} , modulated at 4.8 hour period has been claimed by Samorski, and Stamm(1983), and Lloyd-Evans et al. (1983). There is even a report of 12.6 ms periodicity at $\sim 10^{12}$ eV by Chadwick et al. (1985). Observations with better statistics are needed to confirm these results.

5. SS433

SS433 is a unique galactic object that shows the jet phenomena in all wavelengths, from radio to infrared, to optical, to X-rays, to gamma-ray lines (see Margon, 1984 for a recent review). The most striking aspect of this object is revealed in the optical part of the spectrum. In addition to strong, broad Balmer and HeI emission lines at their expected wavelengths, the spectrum contained features at unfamiliar wavelengths. These lines were later interpreted as Doppler-shifted Balmer and HeI lines from oppositely directed, precessing beams of relativistic plasma moving with velocity of .26 c (Liebert et al. 1979; Margon et al. 1979; Fabian and Rees 1979; Milgrom 1979). A simple kinematic model, initially proposed by Abell and Margon (1979), appears to describe the wavelength shifts of the jets, shown in Figure 3, remarkably well.

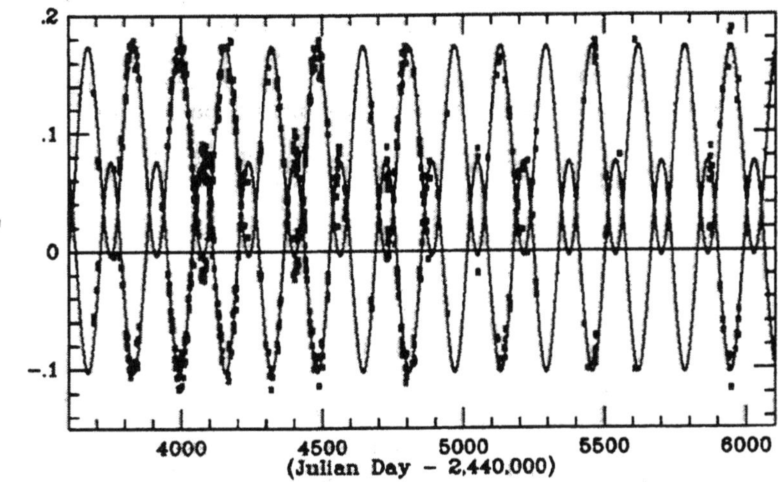

Figure 3: Doppler shifts of SS433 and the prediction of the kinematic model (from Margon 1985).

In this model, the wavelength shift $z = \frac{\Delta\lambda}{\lambda}$ can be expressed as:

$$z = \gamma(\pm v \sin\theta \sin i \cos\psi \pm v \cos\theta \cos i + 1) - 1$$

with the values $v = .2601$, $\gamma = 1.0356$, $\theta = 19.8^o$, $i = 78.8^o$. ψ, which is the precession phase, is given by:

$$\psi(t) = \psi_o + 2\pi(t - t_o)/P_{pr}$$

with $\psi_o = \cos^{-1}(-\cot i \cot\theta)$, $t_o = JD\ 2,443,562.3$, and $P_{pr} = 162.5$ days.

There are some deviations from this kinematic model. One deviation is the periodic modulation at 6.28 day and 5.83 day periods (Katz et al. 1982). The dominant 6.28 day period, at an amplitude of 5 to 10% of the 162 day variations, is interpreted as the nodding motion of the accretion disk due to the binary torque variations; the unexpected power at 5.83 day period is not yet accounted for. An additional noise-like deviation from the 162 day period also exists, implying that the precession period is not very stable. Anderson, Margon, and Grandi (1983) find a residual noise which is of the order of $\frac{<\Delta P_{162}^2>^{\frac{1}{2}}}{P_{162}} \sim 1\%$. This noise appears to propagate more like a phase noise rather than period noise which was claimed for the case of the 35 day cycle of Her X-1 (Boynton, Crosa, and Deeter 1980). However, Ögelman (1986) finds that the residual noise of Her X-1 can also be expressed as phase noise modulated by an exponential time filter.

The intensity of the Doppler-shifted lines are highly variable on timescales $\lesssim$1 day. The intensity correlations of the red and blue shifted lines on these timescales imply that the two radiating regions are seperated by a distance $\lesssim 10^{15}$ cm .

In addition to the jet lines, the optical spectrum of SS433 contains some relatively stationary lines. The 13.09 day orbital periodicity was first observed in the radial velocity variation of these lines (Crampton, Cowley, and Hutchings 1980; Crampton, and Hutchings 1981). The amplitude of these velocity variations, K_*, is quite important in determining the masses of the binary components:

$$f(M) = \frac{4\pi^2(a_x sin\ i)^3}{GP_{orb}^2} = \frac{M_* sin^3 i}{(1+q)^2}$$

where $q = \frac{M_x}{M_*}$, and:

$$q = \frac{K_* P_{orb}}{2\pi a_x sini} \quad .$$

It seems that the H and HeII velocity amplitudes give quite different results. Initially, radial velocity from Balmer lines gave an $f(M) \sim 0.5\ M_\odot$; subsequently, HeII λ4686 emission line velocities indicated an $f(M) \sim 10\ M_\odot$.

SS433 is closely associated with the diffuse nebula W50, classified as a supernova remnant. Its dimension of 60′ to 70′ corresponds to about 100 pc at the estimated distance of 5 kpc. There is an alignment of the major axis of the remnant with the direction of the jets, indicating that mass flow from the jets may be affecting the remnant (Ryle et al. 1978; Clark, and Murdin 1978).

In the radio region, SS433 has been extensively observed; it consists of a central source and extended components (Hjellming, and Johnston 1981). The spectrum of the central source is non-thermal, with power law index of - 0.6 . The intensity of this central component is very variable on timescales of hours or more. However, no periodicities of the radio flux at 162 day precession period or 13 day orbital period has been found. Recently Bonsignori-Facondi, and Braccesi (1986) have discovered a 0.15 Jy component ($\sim$ 6% of the total) of the 408 MHz flux density from the central source that shows a correlation time 16 s and is modulated by the 162 day period. They suggest that this source is located very near the compact object, has a size of $\sim 5 \times 10^{11}$ cm, and radiates coherently with a pulsar-like emission mechanism.

An important fraction of the radio flux from SS433 is in the extended components. Hjellming, and Johnston (1981) showed that the extended portions of the radio emisssion have oppositely directed proper motions of 3″ per year. Identifying this angular velocity with .26 c it was posssible to determine a distance to SS433 as 5.5 kpc. Further high resolution observations with the VLA (Hjellming, and Johnston 1985), indicate that the jets have a corkscrew pattern which fits remarkably well to the kinematic ejection picture, as shown in Figure 4. The asymmetry of angular expansion velocities allows an independent estimation of the expansion velocity as .26 c.

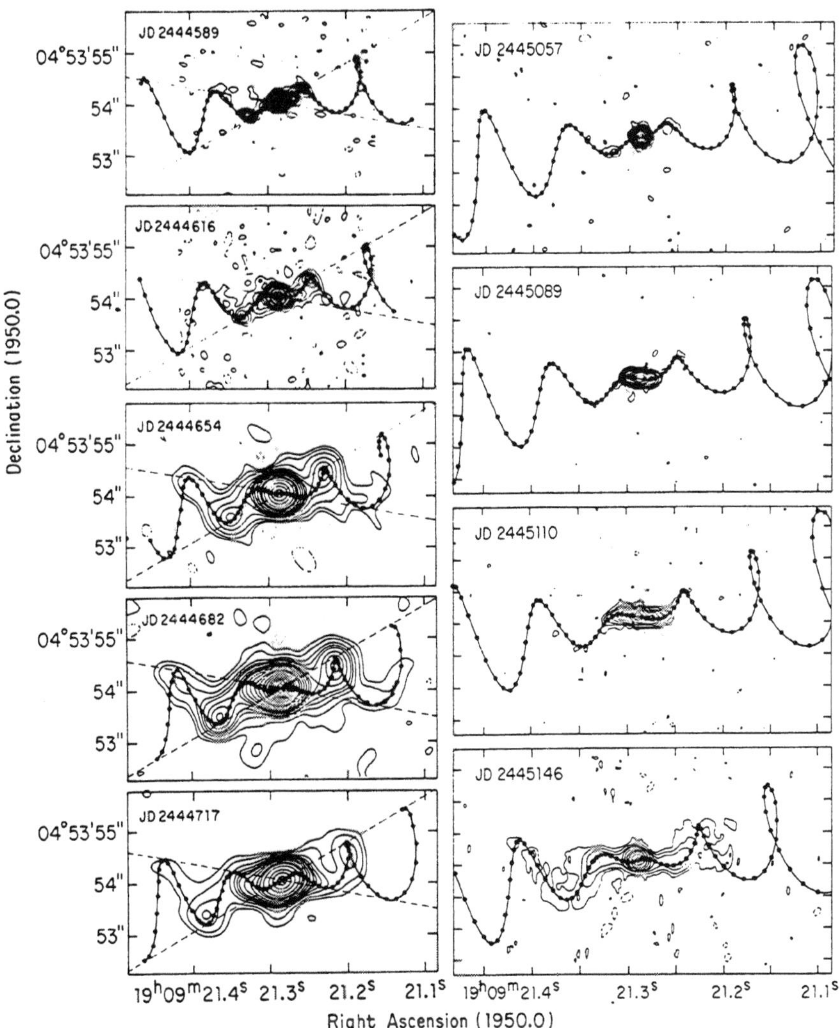

Figure 4: VLA maps of the jets of SS433 together with predicted proper motion paths (from Hjellming and Johnston 1985)

The spectral index of the expanding features is also -0.6, similar to the central source. Hjellming, and Johnston (1985) suggest that the observed jitter in the kinematic model period may be due to lateral velocities of the order of the sound speed in a 3×10^8 K plasma (i.e. $\pm$ 2000 km/s) entrained in the walls of the outgoing jet.

In contrast to the generally accepted kinematic picture, Kundt (1985) has suggested that a variety of observational difficulties can be avoided if one considers a model where the observed moving lines originate near the inner edge of an accretion disk at a distance of $\sim 10^7$ cm from a precessing neutron star. In this model the radio jets are explained as corkscrew shaped channels along a cone of average opening angle $\theta \leq 53^o$ which are being traversed by extremely relativistic matter of $v \sim c$.

At X-ray wavelengths, the central part of SS433 is not a particularly bright source with 2 to 10 keV luminosity $L_x \sim 3 \times 10^{35}$ erg s^{-1} . *EINSTEIN* observations have also revealed softer X-ray emission from the extended lobes of SS433, going out to a distance of 40′ ($\sim$ 60 pc) from the center, and having a luminosity $L_x \sim 6 \times 10^{34}$ erg s^{-1} per lobe (Watson et al. 1983). The authors favor a $T \sim 2 \times 10^7$K thermal bremsstrahlung interpretation of the lobe emission which they interpret as a supernova shell distorted by the presence of jets. Grindlay et al. (1984) have examined the *EINSTEIN* data covering the central source which accounts for more than 90% of the X-ray flux from SS433. They find that the emission is variable on timescales from a fraction of an hour to months. They also find that either a power-law type of spectrum or a thermal bremsstrahlung spectrum with $T \sim 10^8$K fits the data. No variability on timescales less than 5 minutes is observed, implying that the source is extended (10^{12} to 10^{13} cm). Recent observations of SS433 with the *EXOSAT* satellite have revealed a 6.7 keV iron line emission that is Doppler-shifted in accordance with the kinematic model of the blue-shifted jet (Watson et al. 1986). These results, together with a *TENMA* point, are shown in Figure 5. The fact that the red shifted jet is not seen can be combined with the system geometry to restrict its size to $l_x \lesssim 10^{12}$ cm (see Figure 6). Watson et al. (1986) proceed to argue as follows(see also the rewiev by Matsuoka (1986)): From the size, it is possible to calculate a cooling time:

$$\tau_j \sim \frac{l_x}{\beta_j c} \sim 10^2 \; s \quad .$$

Asuming that the emission is thermal bremsstrahlung, the cooling time can also be written as:

$$\tau_{cool} \sim 1.5 \times 10^{11} T^{\frac{1}{2}} n^{-1} \; s \quad .$$

Substituting $T \sim 10^8$K gives the number density as $n \sim 10^{13}$ cm^{-3}. One further equation, relating L_x to n and volume V is:

$$L_x \sim 1.4 \times 10^{-27} n^2 T^{\frac{1}{2}} V \sim 3 \times 10^{35} \text{ erg s}^{-1}$$

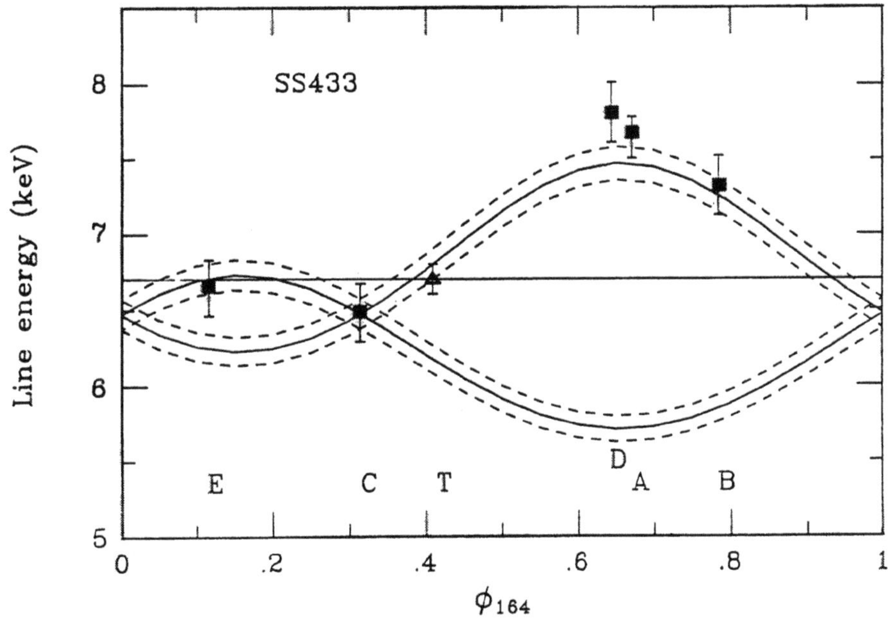

Figure 5: Energy of X-ray emission in SS433 and the predicted kinematic model position of 6.7 keV line (solid line) versus 162 day phase (from Watson et al. 1986)

which gives $V \sim 3 \times 10^{32}$ cm^3. One can also estimate the base radius r_x of a cyclindrical jet of height 10^{12} cm that will give this volume as $\sim 10^{10}$ cm. The mass flow rate in the jet is then:

$$\dot{M} \sim \beta_j c n m_p \pi r_x^2 \sim 10^{-6} M_\odot \; yr^{-1}$$

and the kinetic luminosity is:

$$L_j \sim \frac{1}{2}\dot{M}\beta_j^2 c^2 \sim 2 \times 10^{39} \text{ erg s}^{-1} \quad .$$

Thus, the X-ray data, in a 'self consistent' manner, reveals important parameters of the jets. If some mechanism that can reheat the jets existed, then n can be higher which will make V and hence $\dot{M}$ smaller. The fact that the jet material is at X-ray temperatures close to the neutron star, renders the H Lyman α line-locking mechanism proposed by Milgrom (1979) unlikely, since the material would already be highly ionized. Coordinated observations of the X-ray and optical eclipse of SS433 by Stewart et al. (1986), over one binary cycle, show that the X-ray eclipse is coincident with the deeper optical minimum. This is interpreted as the eclipse of the X-ray emitting material by the companion star. The characteristics of the X-ray eclipse, and the fact that X-rays come from the jets, constrain the relative sizes of the system as shown in Figure 6.

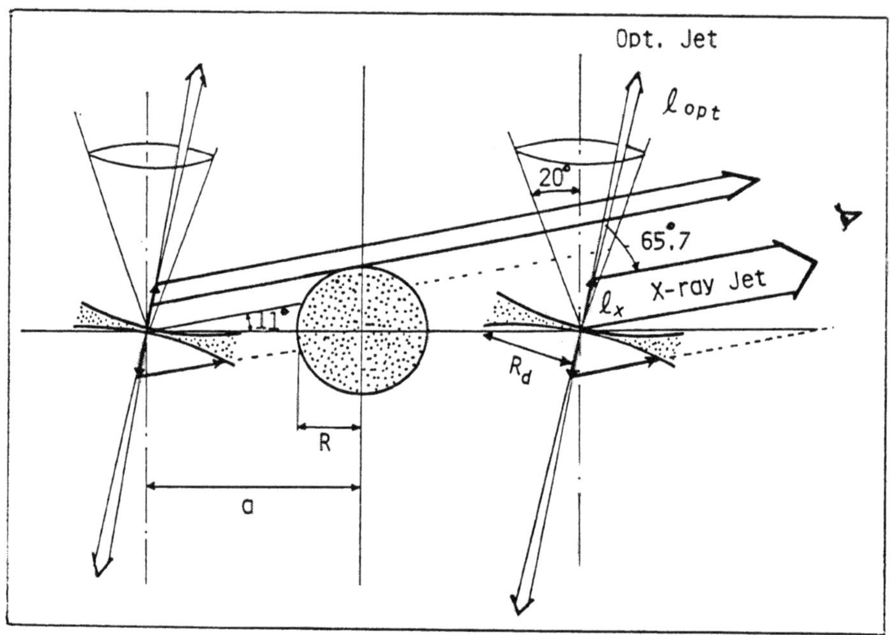

Figure 6: A schematic of the SS433 geometry as inferred from X-ray and optical data (from Matsuoka 1986)

Assuming that the companion star fills its Roche-lobe, Stewart et al. (1986) get an allowable size vs mass ratio q region shown in Figure 7.

The absolute value of the dimension and masses can then be obtained from the radial velocity amplitude $K_* \sim 195$ km/s of the HeII lines, which yields:

$$M_x \sim 11q(1+q)^2 \, M_\odot$$

$$M_* \sim 11(1+q)^2 \, M_\odot$$

$$a \sim 3.5 \times 10^{12}(1+q) \; cm \quad .$$

For $q = 1$, both stars should be $\sim 44 \, M_\odot$, implying that the compact object is a black hole. To have a neutron star of 1.4 $M_\odot$ as the compact object, q has to be 0.1, and the companion star has to be $\sim 11 \, M_\odot$ B star; this possibility is allowed by the X-ray data. Leibowitz's (1984) self consistent fit of the photometric data favors $q \sim 4$ but allows the possibility of q values as low as 0.8. On the other hand $q = 4$ is excluded by the X-ray data. Due to the fact that Leibowitz's model is not unique, and the fact that it is difficult to understand consistently the various components of the radial velocity curve, the possibility of having a neutron star in SS433 is still viable.

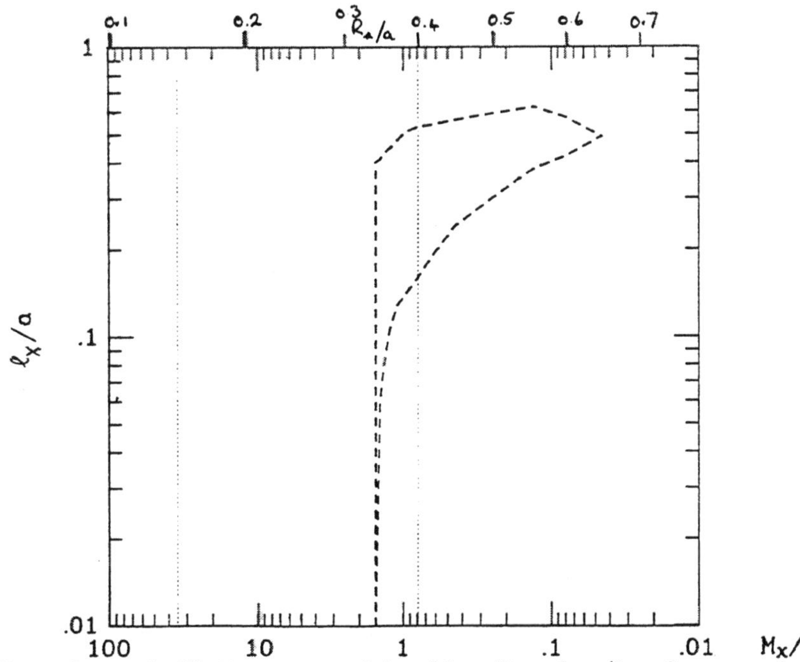

Figure 7: Constraints on the SS433 geometry as inferred from X-ray data (from Stewart et al. 1986)

Another interesting aspect of SS433 is the close correlation of the optical and X-ray fluxes (Matsuoka 1986), which suggests that the optical emission is may be due to reprocessing of the X-rays. In other X-ray sources, such as Sco X-1, and Her X-1, where similar reprocessing occurs, the ratio L_x/L_{opt} is of the order of 10^2 to 10^4. Whereas, even before the $A_v \sim 8^m$ of interstellar extinction is taken into account, $L_x/L_{opt} \sim 1$ for SS433. However, it is quite likely that we never observe the X-ray source directly. For example, in Her X-1, at certain times the disk can block the central source. In Her X-1 the inclination $i \sim 85^o$. In SS433, the inclination of the precession axis is about 79°; however, the precession axis may not be perpendicular to the orbital plane. We may actually be observing SS433 at $i \sim 90^o$. Furthermore, the large implicit mass transfer rates ($\dot{M} \sim 10^{-6}$ $M_\odot$ yr^{-1}), can make the disk considerably thick. The residual flux of Her X-1, during off states is $\sim 10^{-2}$ of the normal flux. Similar scaling to SS433 would imply that the central source can easily have an X-ray luminosity in excess of 10^{37} erg s^{-1} .

To top the spectrum, let me say few words about the intense gamma ray line emission observed from SS433 around 1 MeV (Lamb et al. 1983). The energy of the observed line emission has been interpreted as due to Doppler-shifted ^{24}Mg nuclear transition. The observed luminosity of 2×10^{37} erg s^{-1} raises the energy flow in ^{24}Mg nuclei alone to $\gtrsim 10^{42}$ erg s^{-1} , which is inconsistent with the magnitude of energy flow estimated from other channels. A subsequent, independent experiment does not confirm this level of luminosity (MacCallum et al. 1985).

6. Comparison of the Three Sources

The huge diversity in the category of objects with jets that range from stellar to galactic, appears to apply to the characteristics of this very small subclass. The range of various parameters that describe these systems, make it difficult to point out converging aspects.

As far as *the masses* of the three systems are concerned, for Sco X-1 and Cyg X-3 we must be dealing with low-mass systems, where the compact star is a neutron star of $\sim 1.4\ M_\odot$, and the companion star is around $1\ M_\odot$. For SS433, the compact star can be a black hole up to $100\ M_\odot$, with a similar mass component. If we insist on having a neutron star, the companion mass should be around $10\ M_\odot$. If we are willing to give up the interpretation of the radial velocity measurements from HeII lines, we can even have an evolved solar mass companion with a neutron star, thereby having similar binary systems in all three objects.

As far as the jet phenomena are concerned, the central engine should not care about the companion as much as it would care about the *available* $\dot{M}$, and possibly its history. $\dot{M}$ should be sufficient to power the central source and to provide the mass flow in the jets if the companion star fills its Roche-lobe. This can happen during the main-sequence life of the companion if the separation is small or the companion mass is large. Otherwise, large mass transfer can take place after the companion evolves away from the main-sequence (van den Heuvel 1983). The *X-ray luminosity* of the compact object indicates the accreted mass rate. Again, both for Sco X-1, and Cyg X-3, the luminosities $L_x \sim 10^{37}$ to 10^{38} erg s^{-1} indicate that we are dealing with accretion rates of the order of 10^{-9} to $10^{-8}\ M_\odot$ per year. For SS433 we see only 10^{35} erg s^{-1}. However, as pointed out in the preceding section, the direct view of the source may be blocked by the disk; the true X-ray luminosity may be larger than 10^{37} erg s^{-1}, similar to the other two. Nevertheless, this is still 2 orders of magnitude less than the inferred kinetic luminosity of the jets. Similarly, the mass flow rate of $10^{-6}\ M_\odot$ per year in the jets, is about three orders of magnitude larger than the $\lesssim 10^{-9}\ M_\odot$ per year, that is inferred from the X-ray luminosity.

The orbital periods, together with the mass estimates, give the *binary separation* a of the system:

$$a = \left[\frac{G(M_* + M_x)P_{orb}^2}{4\pi^2}\right]^{\frac{1}{3}} = 2.9 \times 10^{11}\left(\frac{M_* + M_x}{M_\odot}\right)^{\frac{1}{3}}\left(\frac{P_{orb}}{day}\right)^{\frac{2}{3}} .$$

For Sco X-1 ($P \sim 0.79$ day), and Cyg X-3 ($P \sim 0.2$ day) we must be dealing with 1 to 2 $\times 10^{11}$ cm separations (1.5 to 3 $R_\odot$). For SS433 ($P \sim 13$ day), the separation must be of the order of 2 to 8 $\times 10^{12}$ cm. Notice that higher mass estimates with black holes only makes a difference of a factor of ~ 5 in the separation because of the $M^{\frac{1}{3}}$ dependence.

What sort of *companion stars* are we dealing with? Considering that Sco X-1 has a $\sim 1\ M_\odot$ companion, its period of 0.79 days implies that it has to be an evolved star in order to fill its Roche-lobe. For Cyg X-3, the short orbital period allows the existence of a $\sim 1\ M_\odot$ main-sequence star. Presumably we never resolve the light from the companion under the overpowering presence of the reprocessed light from the accreting material. In SS433, if we interpret the "dark events" during the eclipses of the X-ray source, where the optical intensity drops down by a factor of ~ 10 (Kemp et al. 1986), as the remaining optical light from the companion, then this star should have a luminosity of $\sim 10^{37}$ erg s^{-1} (with $A_v \sim 8^m$), and it should correspond to a B star with a mass of $\sim 10\ M_\odot$. In this case it would be slightly evolved from the main-sequence to meet the Roche-lobe filling criterion. A more massive O type companion can still be on the main-sequence.

The *jet features* of the three sources can only be compared in the radio region since for Sco X-1 and Cyg X-3, jets are only apparent in these wavelengths. The radio luminosities L_R of the central region and the jets in SS433 and Cyg X-3 appear to be similar; both sources vary in the 5×10^{32} to 5×10^{33} erg s^{-1} range. Cyg X-3 can go up a factor of 10 more during giant radio flares. In contrast, Sco X-1 seems 4 orders of magnitude weaker; the central source fluctuates between 3×10^{28} to 3×10^{30} erg s^{-1} . The radio lobes also have $\sim 10^{29}$ erg s^{-1} luminosity. These extended components vary within 10 to 30% on timescales $\lesssim$1 year, and appear to be correlated (Geldzahler, and Fomalont 1985).

The *extent and speed of the jets* are harder to compare since their resolution depends on their luminosity, distance, and the background flux. Nevertheless, the expanding features of Cyg X-3 can be followed up to 10^{16} cm; SS433 to 3×10^{17} cm. The projected expansion speed of the jets in Cyg X-3 is about .35 c (at 10 kpc), and in SS433 it is well established as .26 c. In the case of Sco X-1, there appears to be no bridging between the slow moving (35 km/s) lobes at $\sim 10^{18}$ cm and the central source. If we assume that the intensity correlations of the lobes on timescales of $\lesssim$1 year are causally related, the speed of the flowing material may be ~ 0.5 c.

Finally, we can make few comments on the *long term periodicities* of these three objects. The 162 day precession period of SS433 is well established. Some long timescale X-ray variations of 17, 20, and 34 days have been claimed for Cyg X-3. However, Priedhorsky and Terrell (1986) see no periodic behavior with the Vela 5B satellite; they suggest that the nonperiodic X-ray fluctuations are better modelled with random, ~ 100 day long shot-noise pulses. There is no long term periodicity observed for Sco X-1.

The comparative features of the three sources are summarized in Table 1.

7. THE CENTRAL ENGINE

Based on the discussion of the previous sections, I attempt to outline a crude picture for the central engine and jet production that may be operating in all three sources. Similar models have been previously discussed in the literature (Davidson and McCray 1980; Kundt 1985; Priedhorsky 1986).

Table 1. Summary of the properties of the three X-ray sources with jets.

	Sco X-1	**Cyg X-3**	**SS433**
M_x ($M_\odot$)	~ 1.4	~ 1.4	~ 1.4 (to 50?)
M_* ($M_\odot$)	~ 1	$\sim 1 - 4$	~ 10 (to 50?)
$\dot{M}_*$ ($M_\odot$ yr^{-1})	10^{-8}	10^{-6} ?	$10^{-6} - 10^{-11}$
P_{orb} (d)	0.78	0.20	13.09
a (cm)	$\sim 3 \times 10^{11}$	$\sim 3 \times 10^{11}$	$\sim 3 \times 10^{12}$
L_{opt} (erg s^{-1})	$10^{35} - 10^{36}$	?	10^{37}?
L_x (erg s^{-1})	$10^{37} - 10^{38}$	10^{38}	10^{35} (to 10^{38})
L_{rad} (erg s^{-1})	10^{30}	10^{33}	10^{33}
l_{jet} (cm)	10^{18}	10^{16}	10^{17}
v_{jet} (c)	~ 0.5 ?	0.35	0.26
P_{QPO} (s)	.05 – .10	100 – 1000	?
P_{prec} (d)	*none*	10 – 100 ?	162
d (kpc)	0.5 – 1.0	10 – 12	5.5 – 3

The heart of the central engine must be a neutron star. Experimentally this is a majority decision with two for (Sco X-1, and Cyg X-3), and one undecided (SS433).

The neutron star must be accreting near its Eddington limit, $\dot{M}_x \sim 10^{-8}$ $M_\odot$ yr^{-1}, and $L_x \sim 10^{38}$ erg s^{-1}. These luminosities are observed for Sco X-1 and Cyg X-3; in SS433, I assume that we see less owing to the fact that the central source is blocked by the disk.

The mass transfer $\dot{M}_*$ from the companion star must be through Roche-lobe overflow plus an intermediary disk. $\dot{M}_*$ must be larger than that which is accreted by the neutron star. In SS433 there is direct evidence for this from the inferred $\dot{M}_{jet} \gtrsim 10^{-6}$ $M_\odot$ yr^{-1}. In Sco X-1, $\dot{M}_*$ is probably slightly higher than the accreted $\dot{M}_x$. In Cyg X-3 it could also be considerably higher. Jets must be the means of getting rid of the material that can not be accreted.

The ejected plasma in the jets has relativistic speeds with $v \sim 0.3$ to 0.5 c. These velocities have been clearly observed in SS433 and Cyg X-3; in Sco X-1 there is some evidence for it. The magnitude of these velocities are similar to that expected for free-fall or Keplerian rotation near the surface of a neutron star. They are also appropriate for the rotation velocity near the equator of a neutron star with weak surface field, $B_s \lesssim 10^9$ gauss, that has been spun-up by accretion torques corresponding to $\dot{M}_x \sim 10^{-8}$ $M_\odot$ yr^{-1}. A line-locking mechanism would be an unlikely alternative to providing the observed velocities, since 10^8K temperatures of the X-ray jets would exclude such a process in SS433. Furthermore, if SS433 is a neutron star (as assumed), the momentum flux at the Eddington limit would fail by two orders of magnitude to drive the momentum flux of the jets.

If we thus associate the relativistic speeds with the free-fall or Keplerian velocities near the neutron star, we imply that the material that is not accreted must get very close to the neutron star before being ejected as jets. Let us assume that the accretion disk comes up to the Alfven radius r_A where the Keplerian velocities v_K are:

$$v_K \sim \frac{1}{3} c \sim \left(\frac{GM_x}{r_A}\right)^{\frac{1}{2}}$$

which, for a 1.4 $M_\odot$ neutron star gives $r_A \sim 3 \times 10^6$ cm, i.e. 2 to 3 times the neutron star radius. For the disk to come this close to the neutron star, even with the very high mass flow rates of $\dot{M}_* \sim 10^{-6}$ $M_\odot$ yr^{-1} (as in SS433) , the surface magnetic field has to be $\lesssim 10^9$ gauss. This is considerably weaker than the 10^{12} gauss field nominally assumed for neutron stars. This argument implies that neutron stars with jets should have weak magnetic fields, may it be due to decay or low value at birth. However, if , as suggested by Kundt (1986), the infalling material comes down as a spray of massive clumps on to the surface of the neutron star, the surface magnetic field can still be of the order of 10^{12} gauss.

The angular momentum flow $\dot{J}$ at r_A is of the order of:

$$\dot{J}(r_A) \sim \dot{M} r_A \frac{1}{3} c \sim 2 \times 10^{36} \left(\frac{\dot{M}}{10^{-6} M_\odot \; yr^{-1}}\right) g \; cm^2 \; s^{-2}$$

The neutron star may gain some fraction of this angular momentum flow until it comes to an equilibrium period in time scales $\sim 10^4$ to 10^5 years. The equilibrium periods expected are in the millisecond range. Once the neutron star is at this equilibrium period, the incoming angular momentum must be carried back out by the jets. Similar to a picture of Sco X-1 suggested by Priedhorsky (1986), the neutron star may then start acting like a flywheel. At times when the $\dot{M}_*$ of the disk fluctuates positively, r_A is pushed in, and the neutron star gains angular momentum. At times when the mass flow rate fluctuates negatively, the spinning neutron star can transfer angular momentum and energy to the material of the disk. Such a condition is also supported by the large variability observed in these sources. Effectively, the gravitational energy release by the matter flowing toward the compact object is the ultimate energy source of the jets.

Under this scheme, the power advected by the jets should be proportional to $\dot{M}_*$. If we assume that the radio luminosity is a good indicator of the jet power, then SS433 and Cyg X-3 should have similar $\dot{M}_*$, since their radio luminosities are about the same, 5×10^{32} to 5×10^{33} erg s^{-1} . Normalizing this luminosity to $\dot{M}_* \sim 10^{-6}$ $M_\odot$ yr^{-1} of SS433, we can estimate that Sco X-1, with radio luminosity of 3×10^{28} to 3×10^{30} erg s^{-1} should have $\dot{M}_*$ in the range 10^{-10} to 10^{-8} $M_\odot$ yr^{-1}; a reasonable value considering roughness of the estimations.

It is not clear how the rotational symmetry of the flow can be diverted into linear motion of the jets. One possibility for breaking the rotational symmetry is to have the rotation axis of the neutron star not perpendicular to the disk plane. The jets may flow along the lines where the equatorial plane of the neutron star intersects the disk plane. The additional freedom of having a non aligned dipole axis may further alter the jet geometry. Unfortunately, the details of mass transfer around the neutron star magnetospheric boundary are still not well understood (Davidson and Ostriker 1973; Lamb et al. 1973; Ghosh and Lamb 1979a,b; Arons and Lea 1980; Anzer and Börner 1983). Hence, it is very difficult to give a quantiative description of how jets may emerge near the magnetosphere.

8. Precession

It is interesting to consider the precession of the jets of SS433 as an additional constraint on the model. We can immediately reject the possibility of forced precession of the neutron star due to the angular momentum carried away by the jets. In such a case the period of precesssion would be of the order of the spin-up timescale, i.e.:

$$P_{prec} \sim 2\pi \frac{J}{\dot{J}} \sim 10^5 \ yr$$

and not 162 days. The other possibilty, which has been applied as well to a number of other sources showing long periods such as the 35 day cycle of Her X-1, is the slaved disk precession (Roberts 1974; Petterson 1975, 1977). In these models, it is the precession of the normal star that the disk is slaved to. There exist some serious problems with this model. One problem is the decay time of wobble amplitude of a non-rigid object such as a normal star (Chevalier 1976). Another problem is to transfer a warped structure that gets started on the outside, to the center of the disk, where the engine is. Lipunov, Semenov, and Shakura (1981) have shown that the interaction between the dipole field of the neutron star and a diamagnetic disk, in conjunction with the viscous forces, will force the inner parts of the disk to be coplanar with the rotation equator of the neutron star. The Lense-Thirring effect (Bardeen and Petterson 1975) will act in the same way. It can be estimated that at distances $\lesssim 10^8$ cm from the neutron star, the disk should become coplanar with the rotation plane and not influenced by twists in the outer part. Effectively, near the central engine , the disk should not wobble, and modulate the beam direction. Furthermore, *EXOSAT* observations of Her X-1, the best studied long period variable, have shown that the large changes in 1.24 s pulse profile during the 35 day cycle, can be best explained by the free precession of the dipole axis of the neutron star (Trümper et al. 1986). It is attractive to apply a similar free precession model to SS433.

A freely precessing neutron star has a fixed angular momentum vector in space; only the figure axis can precess at large angles with respect to the rotation axis. The rotation vector also precesses around the angular momentum vector, but the amplitude of this precession is extremely small, i.e. of the order of the oblateness

(Lamb et al. 1975). The magnetic dipole axis, which can be considered fixed on the surface, can also precess with large amplitudes. The precession frequency ω_{pr} is related to the rotation frequency ω_r as (Pines 1985):

$$\omega_{pr} \simeq \frac{3}{2}\frac{B}{A}\epsilon_0\omega_r$$

where B is the elastic energy content of the crust, A is the gravitational energy content $3GM^2/25R$, and ϵ_0 is the reference oblateness of a strain free crust. For SS433, we have already remarked that the rotation frequencies ω_r must be $\sim 10^3$ rad s^{-1} ($P \sim 6$ ms). This implies that, to give a free precession period of 162 days ($\omega_{pr} \sim 5 \times 10^{-7}$ rad s^{-1}), the present oblateness ϵ has to be:

$$\epsilon \simeq \frac{3}{2}\frac{B}{A}\epsilon_0 \sim 3 \times 10^{-10} \ .$$

Theoretical values for $\frac{B}{A}$ range from 5×10^{-6} for a soft equation of state to 3×10^{-4} for a stiff equation of state, implying that ϵ_0 must be (1-60)$\times 10^{-6}$. These values are a factor of 10^3 smaller than those derived for Her X-1 (Pines 1985). The present reference oblateness ϵ_0 is either the frozen-in reference oblateness at birth, $\epsilon_{birth} \sim I\omega_{birth}^2/4A$, or it must have evolved through plastic flow or star quakes to a critical value which is called the strain angle of the crust. If the present reference oblateness of SS433 is the value at birth, or the value after the last plastic flow adjustment, it would imply that the neutron star had, sometime during its history, periods of 40 to 600 milliseconds corresponding to soft and hard equation of state, respectively. Being longer than the assumed present period, this range of past reference oblateness periods support the initial hypothesis that we are dealing with spun-up neutron stars.

With the free precession model, there is the additional problem of making jets that precess according to the kinematic model of SS433. Even though the dipole axis may precess with respect to the rotation axis with the observed amplitudes of $\sim \pm 20^0$, it should form a cone about the rotation axis and thus have conical symmetry. It would be a challenging problem to create a linear jet from this rapidly rotating dipole cone.

Precessing neutron star dipole axis and the resulting changes in the X-ray beam pattern will produce illumination changes over the disk and the Lagrangian point and may lead to a kind of "slaved" disk precession situation, as is suspected for Her X-1/Hz Her system. However, the clock of the system would still be associated with the free precession of the neutron star.

9. Summary and Conclusions

The three galactic X-ray binary sources that show definite evidence for jets , Sco X-1, Cyg X-3, and SS433, have been reviewed. The range of various parameters that describe these systems make it difficult to point out their converging aspects.

A rough picture for the central engine has been presented which may be operating in all three sources; it is a low surface field neutron star ($B_s \lesssim 10^9$ gauss) that has been spun-up by the accretion torques to a period of few milliseconds. To make the jets, the companion has to provide super-Eddington mass flow through a disk, $\dot{M} \gtrsim 10^{-8}$ $M_\odot$ yr^{-1}. The material that cannot be accreted to the surface of the neutron star is ejected as jets near the Alfven surface, with the help of centrifugal forces and radiation pressure provided by the fast spinning neutron star. Similar models have been previously discussed in the literature (Davidson and McCray 1980; Priedhorsky 1986).

Millisecond pulsars with low magnetic fields are objects that have already been observed (PSR: 1937+21, 1855+09, 1953+29), and interpreted as accretion spun-up neutron stars in binaries (Alpar et al. 1982). These objects are generally thought to occur in low mass X-ray binaries since their longer lifetime as X-ray sources ($\sim 10^8$ to 10^9 yr) may be long enough to allow time for magnetic field decay and spin-up. The fact that SS433 is in a more massive system would imply that it has a youngish neutron star ($\lesssim$few$\times 10^6$ yr) and hence it must be in its early stages of spin-up.

Although SS433 may be a rare object, we should see more sources with jet signatures similar to Sco X-1. Geldzahler (1983) has surveyed 14 X-ray sources with the VLA, which showed Sco X-1 like characteristics at other wavelengths: ultraviolet excess, rapid and sporadic X-ray variability or high X-ray luminosity. 5 of these objects were detected as unresolved radio sources although none exhibited extended structure. Non-detection of extended lobes can also be due to their low surface brightness. Another feature of Sco X-1 that maybe related to the production of jets is the quasi-periodic oscillations (QPO) at 6 to 20 Hz range of the X-ray intensity (Middleditch and Priedhorsky 1985). Of the 12 known QPO sources (van der Klis 1986), 8 have been observed in the radio, and 7 of them are positive detections. I am convinced that, further observational work in the X-ray and radio regions would reveal other galactic X-ray binary jet sources similar to Sco X-1. Increasing the sample size from a mere 3 to some statistically more meaningful numbers, would be of considerable help in understanding the nature of these objects.

ACKNOWLEDEMENTS: I would like to thank A. Alpar, W. Brinkmann, W. Kundt, and F. Verbunt for helpful comments and criticisms. This work was supported by the Deutsche Forschung Gemeinschaft.

REFERENCES:

Abell, G.O., and Margon, B.: 1979, *Nature,* **279**, 701,

Ables, J.G.: 1969, *Proc. Astr. Soc. Australia*,**1**,237.

Achterberg, A., Blandford, R.D., and Goldreich, P.: 1983, *Nature,* **304**, 607.

Alpar, M.A., Cheng, A.F., Ruderman, M.A., and Shaham, J.: 1982, *Nature,* **300**, 728.

Anderson, S.F., Margon, B., and Grandi, S.A.: 1983, *Astrophys. J.* **269**, 605.

Andrew, B., and Purton, C.: 1968, *Nature,* **218**, 855

Anzer, U., and Börner, G.: 1983, *Astron. Astrophys.* **122**, 73.

Arons, J., and Lea, S.: 1980, *Astrophys. J.* **210**, 792.

Bardeen, J.M., and Petterson, J.A.: 1975, *Astrophys. J.* **195**, L65.

Becklin, E.E. et al.: 1974, *Astrophys. J. (Letters)* **192**, L113.

Blandford, R.J., and Rees, M.J.: 1974, *Mon. Not. R. astr. Soc.* **169**, 395.

Bonnet-Bidaud, J.M., and van der Klis, M.: 1981, *Astron. Astrophys.* **101**, 299.

Bonsignori-Facondi, S.R., and Braccesi, A.: 1986, *Astron. Astrophys.* **163**, 77.

Boynton, P.E., Crosa, L.M., and Deeter, J.E.: 1980, *Astrophys. J.* **237**, 169.

Bradt, H.V. et al.: 1975, *Astrophys. J.* **197**, 443.

Braes, L.L.E., and Miley, G.K.: 1972, *Nature,* **237**, 506.

Chadwick, P.M. et al.: 1985, *Nature,* **318**, 642.

Chevalier, R.A.: 1976, *Astrophys. J. (Letters)* **18**, 35.

Clark, D.H., and Murdin, P.: 1978, *Nature,* **276**, 45.

Cowley, A.P., ans Crampton, D.: 1975, *Astrophys. J. (Letters)* **201**, L65.

Crampton, D., Cowley, A.P., Hutchings, J.B., and Kaat, C.: 1976, *Astrophys. J.* **207**, 907.

Crampton, D., Cowley, A.P., Hutchings, J.B.: 1980, *Astrophys. J. (Letters)* **235**, L131.

Crampton, D., and Hutchings, J.B.: 1981, *Astrophys. J.* **251**, 604.

Crosa, L., and Boynton, P.E.: 1980, *Astrophys. J.* **235**, 999.

Davidson, K., and Ostriker, J.P.: 1973, *Astrophys. J.* **179**, 585.

Davidson, K., and McCray, R.: 1980, *Astrophys. J.* **241**, 1082.

Fabian, A.C., and Rees, M.J.: 1979, *Mon. Not. R. astr. Soc.* **187**, 13.

Fomalont, E.B., Geldzahler, B.J., Hjellming, R.M., and Wade, C.M.: 1983, *Astrophys. J.* **275**, 802.

Geldzahler, B.J.: 1983, *Astrophys. J. (Letters)* **264**, L49.

Geldzahler, B.J. et al.: 1983, *Astrophys. J. (Letters)* **273**, L65.

Geldzahler, B.J., and Fomalont, E.B.:, 1985, *Bull. Am. Astr. Soc.*, **17**, 854

Ghosh, P., and Lamb, F.K.: 1979a, *Astrophys. J.* **232**, 259.

Ghosh, P., and Lamb, F.K.: 1979b, *Astrophys. J.* **234**, 296.

Giacconi, R., Gursky, H., Paolini, F.R., and Rossi, B.B.: 1962, *Phys. Rev. Letters*, **9**, 439.

Gottlieb, E.W., Wright, E.L., and Liller, W.: 1975, *Astrophys. J. (Letters)* **195**, L33.

Gregory, P.C.: 1972, *Nature,* **239**, 439.

Gregory, P.C., and Fahlman, G.G.: 1980,*Nature,* **287**, 805.

Grindlay, J.E., Band, D., Seward, F., Leahy, D., Weisskopf, M.G., and Marshalll, F.E.: 1984, *Astrophys. J.* **277**, 286.

Henrichs, H.F.: 1983, in *Accretion Driven Stellar X-Ray Sources*, eds. W.H.G. Lewin and E.P.J. van den Heuvel (Cambridge University Press, Cambridge), p. 393.

Hjellming, R.M., and Wade, C.M.: 1971, *Astrophys. J. (Letters)* **158**, L155.

Hjellming, R.M., and Johnston, K.J.: 1981, *Astrophys. J. (Letters)* **246**, L141.

Hjellming, R.M., and Johnston, K.J.: 1985, in *Radio Stars*, eds. R.M. Hjellming, and D.M. Gibson (Reidel, Dordrecht), p. 309.

Holt, S.S., Boldt, E.A., and Serlemitsos, P.J.: 1976, *Astrophys. J. (Letters)* **205**, L27.

Katz, J.I., Anderson, S.F., Margon, B., and Grandi, S.A.: 1982, *Astrophys. J.* **260**, 780.

Kemp, J.C. et al.: 1986, *Astrophys. J.* **305**, 805.

Kundt, W., and Gopal-Krishna:1984, *Astron. Astrophys.* **136**, 167.

Kundt, W.: 1985, *Astron. Astrophys.* **150**, 276.

Kundt,W.: 1986, in *The Evolution of Galactic X-Ray Binaries*, eds. J. Trümper, W.H.G. Lewin, and W.Brinkmann, (Reidel, Dordrecht), p. 263.

Lamb, F.K, Pethick, C.J., and Pines, D.: 1973, *Astrophys. J.* **184**, 271.

Lamb, D.Q, Lamb, F.K., Pines, D., and Shaham, J.: 1975, *Astrophys. J. (Letters)* **198**, L21.

Lamb, R.C., Fichtel, C.E., Hartman, R.C., Kniffen, D.A., and Thompson, D.J.: 1977, *Astrophys. J. (Letters)* **212**, L63.

Lamb, R.C., Ling, J.C., Mahoney, W.A., Riegler, J.R., Wheaton, W.A., and Jacobson, A.S.: 1983, *Nature,* **305**, 37.

Leibowitz, E.M.: 1984, *Mon. Not. R. astr. Soc.* **210**, 279.

Liebert, J., Angel, J.R.P., Hege, E.K., Martin, P.G., and Blair, W.P.: 1979, *Nature,* **279**, 384.

Lipunov, V.M., Semenov, E.S., and Shakura, E.S.: 1981, *Sov. Astr.*, **25**, 439.

Lloyd-Evans,J., Coy, R.N., Lambert, A., Lapikens, J., Patel M., Reid, R.J.O., and Watson, A.A.: 1983, *Nature,* **305**, 784.

MacCallum, C.J., Huters, A.F., Stang, P.D., and Leventhal, M.: 1985, *Astrophys. J.* **291**, 486.

Margon, B.: 1984, *Ann. Rev. Astron. Astrophys.*, **22**, 507.

Margon, B.: 1985, in *Japan-U.S. Seminar on Galactic and Extragalactic Compact X-ray Sources*, eds. Y. Tanaka, and W.H.G. Lewin, (Institute of Space and Astronautical Science Report, Tokyo), p. 137.

Margon, B., Ford, H.C., Grandi, S.A., and Stone, R.P.S.: 1979, *Astrophys. J. (Letters)* **233**,L63.

Matsuoka, M.: 1986, *Proceedings of the ESA Conference on the Physics of Accretion onto Compact Objects*, Tenerife, Spain, April 1986.

McClintock, J.E.: 1986, *Proceedings of the ESA Conference on the Physics of Accretion onto Compact Objects*, Tenerife, Spain, April 1986.

Middleditch, J., and Priedhorsky, W.: 1986, *Astrophys. J.* **306**, 230.

Milgrom, M., and Pines, D.: 1978, *Astrophys. J.* **220**, 272

Milgrom, M.: 1979, *Astron. Astrophys.* **78**, L9.

Miyamoto, S., and Matsuoka, M.: 1977, *Sp. Sci. Rev.*, **20**, 687.

Molnar, L.A.: 1986, in *Proceedings of the Conference on the Physics of Accretion onto Compact Objects, April 1966, Tenerife, Spain* eds. M. Watson, K. Mason, and N. White.

Ögelman, H.: 1986, *Astron. Astrophys.* , in press.

Petterson, J.A.: 1975, *Astrophys. J. (Letters)* **201**, L61

Petterson, J.A.: 1977, *Astrophys. J.* **218**, 783

Pines, D.: 1985, in Lectures Presented at the NATO Inst. *High Energy Phenomena and Collapsed Stars*, Sept 1985

Priedhorsky, W., and Terrell, J.: 1986, *Astrophys. J.* **301**, 886.

Priedhorsky, W.: 1986, *Astrophys. J. (Letters)* **306**, L97.

Roberts, J.W.: 1974, *Astrophys. J.* **187**, 575.

Ryle, M., Caswell, J.L., Hine, G., and Shakeshaft, J.: 1978, *Nature,* **276**, 571.

Sandage, A.R. et al.: 1966, *Astrophys. J.* **146**, 316.

Samorski, M., and Stamm, W.: 1983, *Astrophys. J. (Letters)* **268**, L17.

Savonije, G.J.: 1983, in *Accretion Driven Stellar X-Ray Sources*, eds. W.H.G. Lewin and E.P.J. van den Heuvel (Cambridege University Press), p. 343.

Spencer, R.E., and Johnston, K.J.: 1986, in *Manchester Conference on RS Ophuici, December 1985*, ed. M. Bode (VNU Press)

Stepanian, A.A.: 1984, *Adv. Space Res.*, **3**, 123.

Stewart, G.C. et al.: 1986, preprint

Swanenburg, B.N. et al.: 1981, *Astrophys. J. (Letters)* **243**, L69.

Trümper, J., Kahabka, P., Ögelman, H., Pietsch, W., and Voges, W.: 1986, *Astrophys. J. (Letters)* **300**, L63.

van den Heuvel, E.P.J., and De Loore, C.: 1973, *Astron. Astrophys.* **25**, 387.

van den Heuvel, E.P.J.: 1983, in *Accretion Driven Stellar X-Ray Sources*, eds. W.H.G. Lewin and E.P.J. van den Heuvel (Cambridge University Press, Cambridge), p. 303.

van der Klis, M.: 1986, in *Proceedings of the Conference on the Physics of Accretion onto Compact Objects, April 1966, Tenerife, Spain* eds. M. Watson, K. Mason, and N. White.

Watson, M.G., Willingdale, R., Grindlay, J.E., and Seward, F.D.: 1983, *Astrophys. J.* **273**, 688.

Watson, M.G., Stewart, G.C., Brinkmann, W., and King, A.R.: 1986, *Mon. Not. R. astr. Soc.* **222**, 261.

Weekes, T.C., Murphy, H.P., Schild, R., Gursky, H., Geary, J., and Stephenson, T.: 1981, *Pub.A.S.P.*, **93**, 474.

White, N.E., and Holt, S.S.: 1982, *Astrophys. J.* **257**, 318.

Willingale, R., King, A.R., and Pounds, K.A.: 1985, *Mon. Not. R. astr. Soc.* **215**, 295.

T TAURI STARS, PRE-T TAURI STARS, AND STELLAR JETS

Martin Cohen
Radio Astronomy Laboratory
University of California
Berkeley, CA 94720
U.S.A.

ABSTRACT. The properties of the pre-main-sequence T Tauri stars are described in relation to the likely properties of their predecessors. These predecessors, the "pre-T Tauri stars", are thought to be the stars responsible for exciting Herbig-Haro objects. Pre-T Tauri stars are capable of generating "stellar jets" by which bipolar flows carry material out into the ambient medium out of which low-mass stars form. Manifestations of these bipolar flows are: strings of optical Herbig-Haro objects; radio jets; large flattened dusty structures orthogonal to the flows. We describe the observations (optical, infrared, far-infrared and radio continuum) that provide evidence for these components in order to establish a framework for other authors in this volume to attempt to explain stellar and extragalactic jets by some unified model.

1. T TAURI STARS

1.1 Spectroscopic Characteristics: Chromospheres

Several recent reviews have treated the topic of the T Tauri stars (hereafter TTS: see, for example, Bertout 1984, Cohen 1984). From their locations in dark clouds and places in the Hertzsprung-Russell diagram one can infer that TTS represent the pre-main-sequence phases of roughly solar mass stars (Cohen and Kuhi 1979). Optically they are characterized by photospheres (when these are recognizable) between early K and lateish M overlaid by bright, rich emission-line spectra. These spectra are attributed to chromospheres. One can deduce this from the difference between a TTS spectrum and that of a control star of identical photospheric type: all the discrete lines of FeII, MgI, etc. are plainly in emission in TTS. Similarly, there are strong correlations between the intensities of the emission lines of FeII and HeI in TTS, supporting the notion of chromospheres in these stars. So strong are these lines in a small subset of TTS (perhaps 5-8% in different stellar nurseries) that these few stars cannot be located in

W. Kundt (ed.), Astrophysical Jets and Their Engines, 91–102.

the Hertzsprung-Russell diagram for their photospheric absorptions are totally obliterated by emission lines.

1.2 Infrared Observations: Circumstellar Dust Emission

To determine bolometric luminosities requires the use of infrared photometric techniques and the recent IRAS mission clearly demonstrates the merit of defining these luminosities from data that are extended at least as far as 12/25 μm (the two shorter IRAS bands). The excess emission in the infrared, over and above photospheric radiation in these stars, is attributed to thermal emission by circumstellar dust grains (Cohen and Kuhi 1979). The grains must clearly exist across a wide range of temperature because one sees excess radiation at wavelengths as short as 2.2 μm and as long as 60 or even 100 μm with IRAS in some stars. The circumstellar emission therefore covers temperatures between the vaporization zone (at about 1300-1500 K for silicates) and the edges of these pre-main-sequence stars' dust shells ($\sim$ 50K).

The chemical character of the dust grains can be inferred from the suggestion of a maximum temperature close to 1300K in a near-infrared color-color plane (e.g. Rydgren, Schmelz and Vrba 1982) and from the direct observation of broad emission features that cover the entire 8-13 μm window (Cohen 1980; Cohen and Witteborn 1985). Silicates are therefore thought to be present.

It is also possible to infer something of the geometry of this circumstellar material. From the greater likelihood of finding silicate emission features than silicate absorption features (4 to 1), and from correlations between location in the near-infrared color-color plane and presence of emission or absorption features, Cohen and Witteborn (1985) concluded in favor of a flattened geometry for these circumstellar dust envelopes in TTS. Rydgren and Cohen (1985) have also listed various clues to a disklike dust geometry in the TTS. The uniqueness of the star HL Tau, that shows a 3.1 μm ice absorption (Cohen 1975) as well as the deepest silicate absorption at 10 μm (Cohen and Witteborn 1985) and the highest optical linear polarization among TTS (Bastien 1982a,b), suggests that we only rarely see TTS through the planes of their proto-solar nebulae and dusty disks. The rarity of sizable (above 1- 2%) optical degrees of polarization also supports this conclusion. The spectral character of polarization is of interest for it shows that the TTS have very flat polarization curves with wavelength and that even Hα, which is thought to come from a region much more extensive than the normal chromosphere (Heidmann and Thomas 1980), is polarized exactly as is the adjacent continuum. Therefore, the polarizing dust screen is very distant from the star and presumably is associated with the circumstellar dust seen radiating at 2 μm and beyond.

1.3 High Energy Radiation: Flares and Variability

X-ray emission and ultraviolet data of TTS paint a picture of stars with active outer atmospheres although their coronae do not attain temperatures as high as the sun's (e.g. Walter and Kuhi 1981). Flaring activity characterizes the X-ray radiation (e.g. Walter and Kuhi 1984), and ultraviolet spectral and photometric variability are well-established. An ensemble of flares, like the sun's, but much more energetic, satisfactorily describes these variations (Worden et al. 1981; Montmerle et al. 1983).

1.4 Radio Continuum Emission: Flares

In the radio continuum it is difficult to detect most TTS (Bieging, Cohen and Schwartz 1984). The few that have been seen with the VLA are either unusually active and really represent the pre-TTS rather than the TTS phase, or are detected during substantial flares. For example, V410 Tau is optically one of the weakest of TTS in terms of its emission-line activity yet it has been seen frequently in the radio continuum (Cohen, Bieging and Schwartz 1982; Cohen and Bieging 1986). It displays two very distinct states of radio emission. In its "on" state it attains a 15 GHz flux as high as 60 mJy with a spectral index (between 1.4 and 22 GHz) of $\alpha = +1$ (i.e. $S_{\nu} \sim \nu^{\alpha}$). By contrast its more recent behavior has shown a flat index ($\alpha \sim 0$) at a level just below 1 mJy between 5 and 15 GHz (Cohen and Bieging 1986). Such data clearly indicate the importance of variability to the T Tauri phenomenon in general.

1.5 Stellar Rotational Velocities: Low v sin i s

Stellar rotation is clearly of importance to our understanding of the TTS. Vogel and Kuhi (1981) found that the vast majority of TTS that lie on their convective tracks were not perceptibly rotating (limits of order 25 km/s in v sin i). More recent measurements by Bouvier et al. (1986) and by Hartmann et al. (1986) have succeeded in determining velocities for many of these slowly rotating stars: their v sin i values are of order 10-15 km/s typically. V410 Tau is an outstanding exception on the convective track (based on its bolometric luminosity and its inferred photospheric temperature), perhaps the only one among the Taurus-Auriga stars, since it shows rotational broadening of order 75 km/s. The other fast rotators are rather more luminous TTS on their radiative tracks. One has the impression that the length of convective track available to a TTS determines the degree to which it has spun down. Certainly there is no abrupt change in rotational properties along the convective track that could be attributed to the formation of preplanetary systems. Both this qualitative conclusion and the more quantitative information that Cohen (1983) has derived for HL Tau lead to the feeling that the onset of planet-building must have started at a very early stage, perhaps before a million years have elapsed for a 1 $M_{\odot}$ TTS.

1.6 Typical Parameters

The typical TTS in the well-studied Taurus-Auriga dark cloud complex, for example, would be characterized by the following physical parameters: effective temperature 4000K; radius $\sim 3\ R_{\odot}$; age 10^5 to 10^6 yr; $A_V \sim 1$ mag, mostly from intracluster material; optical linear polarization 1%; coronal temperature $\sim$ a few x 10^5K; transition zone $\sim 10^5$K; velocity scales for stellar material from

50-300 km/s; radio continuum emission <1 mJy at 5 GHz; v sin i ~ 15 km/s; L(bol) ~ a few $L_{\odot}$.

2. HERBIG-HARO OBJECTS AND THEIR EXCITING STARS

2.1 Herbig-Haro Objects: Shocks

For a detailed review of the properties of the Herbig-Haro (hereafter HH) Objects the reader is referred to Schwartz (1983). For our purposes it will suffice to characterize them as dominantly line-emitting clouds cooling after suffering shock heating. Optically and in the ultraviolet most HHs seem to have encountered shocks with relative velocities ~ 100 km/s. However, their true structure is rather more complicated than this single parameter would suggest. In the near-infrared, near 2 μm, one sees vibration-rotation emission lines from molecular hydrogen typical, it is suggested, of shocks with only 10-15 km/s velocities (Elias 1980). [The early detection of molecular hydrogen emission from T Tau itself (Beckwith et al. 1978) is now recognized to arise because T Tau (or its southern radio/infrared companion) excites a nearby HH object.] At far-infrared wavelengths the [OI] emission line at 63 μm is also detected predominantly from a rather special subset of HH nebulae, namely those whose optical spectra are describable as of "very-low excitation" (hereafter VLE: Cohen et al. 1987). Their optical spectra are dominated by the red [SII] and [OI] lines which exceed the strength of even Hα by factors of order at least 2- 3 (e.g. Dopita et al. 1987). The 63 μm line could come from regions heated by even lower velocity shocks than the molecular hydrogen emitting zones. Consequently it may be more fruitful to think of HH objects as having inner optical/UV-emitting cores from high-velocity shocks but embedded in "cocoons" of much lower-velocity material shocked by the sidewise expansion of the cocoon material.

2.2 Proper Motions

High proper motions are known for a modest number of knots in different HH families (e.g. Herbig and Jones 1981; Jones et al. 1984). These, combined with radial motions, suggest total space velocities that can attain several hundred km/s. The observed shock spectra of fast-moving HH knots can be matched by models with rather lower velocities than these, indicating that the apparent tangential motions truly represent physical bulk motions. A remarkable fact about these "interstellar bullets" is that their motions can even be used to indicate the directions from which they originally came and even, in the case of two-sided families, to bound the location of their responsible stars. This clearly represents a purely empirical approach but there is definite predictive power in this method to identify the locations of heavily embedded infrared sources that might otherwise be undiscovered (e.g. the exciting stars of RNO 40 and RNO 43a: Jones et al. 1984).

2.3 The Relationship between T Tauri Stars and Herbig-Haro Objects

Some 700 TTS are known in different dark clouds and perhaps 100 HH objects (some simple knots, others more complex families with 10 or more brightish knots and bridges of interconnecting material). However, the overlap between these two sizeble classes of object in dark clouds is curiously small. In a very few cases one sees an obviously intimate association between a small HH knot and a visible star, identifiable as a TTS usually from at least its chromospheric (typically FeII) emission-line spectrum. More commonly one sees a red continuum scattered within an HH nebula from the responsible exciting star (not to be confused with the enhanced hydrogenic two-photon continuum that arises purely from nebular processes in the HH itself). A very few of these reflected continua lend themselves to spectral classification in which cases, again, one is likely to recognize a TTS spectrum of emission lines (Fig. 1), even in one case (HH55: Cohen, Dopita and Schwartz 1986a) a photosphere of M3.5 stellar absorption features. From these statistics and from the most frequently encountered situation in which no visible object seems associated with an HH nebula and no stellar continuum is scattered, one concludes that it is likely not the TTS but *their predecessors* that are the true exciting stars of the HH objects.

2.4 CO Bipolar Flows

This impression is also conveyed by the number of bipolar CO flows said to be associated with TTS (Edwards and Snell 1982). If a readily discernible bipolar flow is found from a TTS then it will be found that the star is actually, *a fortiori*, an HH-exciting star (hereafter HHES). More direct searches for high-velocity molecular gas associated with the environments have been quite successful (e.g. Snell and Edwards 1981; Edwards and Snell 1984) while the work of Bally and Lada (1983) suggests that bipolar mass flows may characterize the early phases of evolution of stars over a broad range of mass.

One is much more likely to detect radio continuum emission with the VLA from an HHES than from a random TTS. The activity that spawns the fast-moving HHs, drives the large-scale bipolar flows in dark clouds, and manifests itself as radio emission has already decayed in the average TTS. We need to observe the phase that precedes the pre-main-sequence TTS in order to note these symptoms of peculiarity.

2.5 The Connection with FU Orionis Stars

There is, perhaps, another class of stellar objects associated with the excitation of HH objects, namely potential FU Ori stars. The FU Ori stars themselves (cf. Herbig 1977) are somewhat mysterious objects and it is difficult to decide what stage of early stellar evolution they represent. They are recognized by their dramatic optical brightenings, typically by 5 mags, although it is not known whether these are accompanied by corresponding increases in bolometric luminosity. Nor is it clear whether there is intrinsic stellar change, as Herbig (1977) has argued in some cases, or merely a "clearing" of the dusty, immediately circumstellar, environment.

Both the star within HH57, which has appeared in the last decade and can be spectroscopically studied directly (Cohen, Dopita and

Schwartz 1986b), and the reflected spectrum in HH102, presumed to be that of L1551 IRS5 (Mundt et al. 1985), reveal characteristics of FU Ori stars in their optical spectra. However, it is certainly intriguing that there is now a definite connection between one of these stars and an HH nebula (in HH57). [Such a relationship was suggested much earlier by Gyul'budagyan (1975) and by Dopita (1978).] This is thought by some (e.g. Cohen, Dopita and Schwartz 1986a) to indicate the probable non-recurrent nature of FU Ori flares or, at least, the fact that they are allied only with the very earliest phases of protostellar evolution rather than with the pre-main-sequence stage.

2.6 "Jets"

Among the HH objects there are some that consist of strings of aligned knots or linear wisps of nebulosity. The term "jet" is now bandied about the literature without any formal definition yet it is not a well-specified description of an HH. To eliminate confusion by random linear streaks of nebulosity we shall use the term "jet" to signify a roughly linear, probably knotty, HH nebula whose exciting star is either contained within the body of the nebula or lies at one extremity. We specifically exclude systems of HH knots that are very distant from their exciting star (e.g. HH43 cf. HH43 IRS1: Cohen and Schwartz 1983). It is the quasi-periodicity of some HH jets (e.g. HH7-11; "DG Tau B") that has led to the attempt to understand at least some of these in terms of intrinsic shock structures (e.g. Falle, Innes and Wilson 1986).

With this strict definition we can make several general statements based on detailed studies of a handful of jets. As we shall see below, the exciting stars of HH jets are by no means luminous objects. The jets may be one- or two-sided and it is not always the blue-shifted flow that is best-developed, in spite of the valid argument that, in dark clouds, the red-shifted flows should be more deeply buried in dust than the blue. A good example of this is "DG Tau B" (Jones and Cohen 1986) whose red lobe shows much more extended activity than its blue. Typical flow speeds (based mainly on radial velocities) are of order 200 km/s. VLE emission characterizes if not all than at least most HH jets (Cohen and Jones 1987) and the degree to which [SII] exceeds the strength of Hα is quite unprecedented in some: one finds I(6717+6731)/I(Hα) can attain values as high as 12 (Fig. 2)! There seem to be two patterns that emerge from the study of jets (though not all jets obey both) namely a monotonic decrease in electron density with increasing distance from the HHES, and likewise a systematic lowering of the excitation of the spectrum. Jones and Cohen (1986) and Cohen and Jones (1987) report that $N_e \sim r^{-1}$ in 3 of the 5 jets they have so far studied.

One might account for this decrease in excitation by: (i) a steady acceleration of the knots within the driving flow, lowering the relative shock velocity; or (ii) a "flaring out" of the jet so that it is increasingly obliquely shocked as it moves away from the star; or (iii) perhaps a steady change from atomic to molecular conditions thereby strengthening molecular hydrogen emission at the expense of the Balmer lines.

2.7 Radio "jets"

Cohen, Bieging and Schwartz (1982) spatially resolved a radio continuum emission region associated with the HHES, L1551 IRS5 (Fig. 3). Subsequent studies (Bieging, Cohen and Schwartz 1984; Bieging and Cohen 1985) indicate that this region is resolved in the direction of flow from IRS5 but not perpendicular to it. The observational limit on length-to- thickness ratio is 10:1. Simple modelling as an optically somewhat thick thermal source indicates a ratio that could attain several hundred to 1! But even 10:1 justifies the use of the term "jet".

The second HHES associated with spatially resolvable radio emission is DG Tau whose VLA maps show a time-variable 5 GHz structure (Cohen and Bieging 1986), sometimes pointlike, more commonly biconical with flow axis towards a nearby HH knot (Bieging, Cohen and Schwartz 1984). New data at 15 GHz (Cohen and Bieging 1986) clearly show a bipolar morphology, aligned with that previously seen at 5 GHz, and indicate that the deviation from spherical flow has already occurred within only 16 A.U. of this strong emission-line TTS. The pinched "waist" of the radio maps lies in the same direction as the electric vector of optical polarization and the sense of possible elongation at 100 μm (Cohen, Harvey and Schwartz 1985).

It is possible to derive a radio spectral index between 5 and 15 GHz from the maps of L1551 IRS5 (Bieging and Cohen 1985). The average from eight points within the elongated jet yields an index of $\alpha \sim +0.3$, interestingly the same as that of some of the radio components at the centre of our own galaxy and as the sources allied with several active galactic nuclei.

Another possible structural parallel with extragalactic jets is that some HHES seem to have steadily and systematically altered their directions of flow. For example, in some families (e.g. HH1/2/3 and HH38/43), the vector connecting the centroid of each knot to the location of the HHES appears to have rotated between the events responsible for the creation of successive knots. In L1551 IRS5 too, the central 5 GHz contours of the partially-resolved jet rotate from an almost east-west alignment at the core to an axis much closer to NE-SW, in keeping with the axis of the much larger-scale CO flow. This apparent "precession" in young stellar jets may not correspond to true physical precession as in the extragalactic context, however.

2.8 Typical Parameters of HH Objects

From the forbidden emission lines in the optical spectrum we can estimate the physical state of the gas involved in the radiating HH shocks: relative shock velocity $\sim$ 100 km/s; electron density from 100 -1000 cm^{-3}; mass in the emitting surface of the knot very small (only a few earth masses!); space velocity $\sim$ 200 km/s; molecular hydrogen emission seen in the ultraviolet (Lyman bands) and the near-infrared (typically observed in the S(1) 1-0 vibration-rotation line); 2 photon continuum seen strongly in the ultraviolet spectrum and into the blue in the VLE nebulae. For those HH knots that constitute part of the body of a jet there is a hint that the knots closest to the exciting star have unusually high densities for the class (e.g. Cohen and Fuller 1985), of order 10,000 cm^{-3}, close to the limit of saturation for the ratio of red sulphur line intensities. The typical length of the confined portion of HH jets is $\sim 10^{17}$ cm.

3. THE EVOLUTIONARY STATUS OF HH-EXCITING STARS

3.1 Bolometric Luminosities

One way of learning something about the evolutionary status of the HHES is to determine their bolometric luminosities (e.g. Cohen and Schwartz 1983). This can be done whether the star is optically known or not, via near-, mid-, and far-infrared observations since most of the radiation emitted by HHES is at infrared wavelengths. There are some technical aspects to this scheme which bear consideration. First, one should now utilize the IRAS databases to provide good estimates of the far-infrared luminosity even though there already exist airborne data of some HHES, with higher spatial resolution than IRAS. This situation reflects the spatially extended nature of the far-infrared sources associated with HHES. In a few cases one definitely resolves a non-stellar far-infrared emission zone around the HHES, elongated in the direction of flow from the exciting star (e.g. that around L1551 IRS5 detected with the KAO: Cohen, Harvey and Schwartz (1985) and with IRAS: Edwards et al. 1986). Airborne methods usually do not detect these very extended low surface brightness zones and therefore underestimate the true brightness of the sources by as much as a factor of 2 at 100 μm (Cohen and Schwartz 1987). Further, the color temperatures of the coolest matter detected by airborne measurements are only 75% of those defined by IRAS observations of the same sources, confirming this impression of spatial extension of the far-infrared emission into regions with cooler dust temperatures (Cohen and Schwartz 1987). Second, the spectra of HHES are often so steep that ground-based photometry, even at 10 μm, is incapable of recognising the source. Again, IRAS may be able to detect weak mid-infrared emission, if not in the Point Source Catalogue then at least upon coaddition of the several passes through the relevant piece of the sky.

By these means, Cohen and Schwartz (1987) have recently determined bolometric luminosities for 25 objects that all have a very high probability of truly being HHES. The histogram of L(bol) is highly revealing. While it is still true that the upper end of the HHES is represented by R Mon (at almost 800 $L_\odot$) the main contribution of IRAS has been to demonstrate the extremely low luminosities of some HHES. The range now goes down to only 0.06 $L_\odot$ and still these HHES are associated with well-defined HH jets! This lower bound should certainly be a factor in assessing the validity of specific theoretical models of jet production.

3.2 Spatially Resolved Far-Infrared Emission

It has been possible to obtain airborne far-infrared maps of several HHES. In a number of cases there are definite indications of spatial resolution of a far-infrared emitting structure in the direction orthogonal to the flow from the stars, but not parallel to the flow direction (Fig. 4). These are clearly flattened structures, with radii on the order of a few thousand A.U. (Cohen, Harvey and Schwartz 1985) but it is not yet possible to determine the observed axial ratios with either IRAS or the Kuiper Airborne Observatory. The empirical limits are only >3 - 4:1. Presumably these dusty structures might represent the extensive, somewhat flattened, still-infalling regions associated with these HHES.

A disklike or at least flattened morphology would not be surprising. Indeed it is striking that there are great discrepancies between the visual extinctions along the direct paths to the infrared sources that represent the sites of heavily-embedded HHES and the extinctions towards those HH objects seen closest to the stars. Within only 2" the values of A_V can plummet from $\sim$ 10 mag to only $\sim$ 1 [e.g. for "DG Tau B" (Jones and Cohen 1986) and L1551 IRS5 (Cohen and Fuller 1985)]. This too is suggestive of the presence of dusty disks which eventually turn into the vestigial remnants recognizable around the pre-main-sequence TTS themselves.

3.3 The Potential Role of Binarity

There are definite suggestions that binarity is associated with systems that produce stellar radio jets and/or aligned HH nebulae. Even in an evolved stellar system binarity alone seems capable of the high-speed ejection of HH knots (cf. OH0739-14, a bipolar nebula containing an M9III star with a suspected faint blue companion: Cohen et al. 1985) unless it can be proven that a fast rotating compact blue star is also present. For L1551 IRS5 Bieging and Cohen (1985) show an apparent binary at 15 GHz, one component of which (the brighter at 15 GHz) coincides with the central source and the core of the 5 GHz jet from IRS5 (Fig. 3). The direction in which the 5 GHz jet core emerges is also perpendicular to the line joining the two high-frequency components. Both radio components of this suggested binary have steep radio spectra ($\alpha \sim +1$). For the T Tau system too, it is interesting that the isolated HH knot to the west of this close potential binary system lies in a direction orthogonal to the inferred orbital plane of the binary, and even, perhaps, has a proper motion in this orthogonal direction (Bertout, cited by Cohen 1984). T Tau and L1551 IRS5 are currently the only young, low-mass HHES associated with potential binaries. [A claim by Mundt and Fried (1983) that a visible binary might lie at one end of the "DG Tau B" jet has recently been refuted by Jones and Cohen (1986).]

4. REFERENCES

Bastien, P. 1982a, A.A.Suppl., 48, 153.
Bastien, P. 1982b, A.A.Suppl., 48, 513.
Bally, J. and Lada, C. J. 1983, Ap.J., 265, 824.
Beckwith, S., Gatley, I., Matthews, K. and Neugebauer, G. 1978, Ap.J. (Letters), 223, L41.
Bertout, C. 1984, Rep. Prog. Phys.47, No. 2.

Bieging, J. H. and Cohen, M. 1985, Ap.J. (Letters), 289, L5.
Bieging, J. H., Cohen, M. and Schwartz, P. R. 1984, Ap.J., 282, 699.
Bouvier, J., Bertout, C., Benz, W. and Mayor, M. 1986, A.A., in press.
Cohen, M. 1975, M.N.R.A.S., 173, 279.
Cohen, M. 1980, M.N.R.A.S., 191, 499.
Cohen, M. 1983, Ap.J. (Letters), 269, L70.
Cohen, M. 1984, Physics Reports, 116, 173.
Cohen, M. and Bieging, J. H. 1986, A.J., in press.
Cohen, M., Bieging, J. H. and Schwartz, P. R. 1982, Ap.J., 253, 707.
Cohen, M., Dopita. M. A. and Schwartz, R. D. 1986a, Ap.J. (Letters), 307, L21.
Cohen, M., Dopita, M. A. and Schwartz, R. D. 1986b, Ap.J. (Letters), 302, L55.
Cohen, M., Dopita, M. A., Schwartz, R. D. and Tielens, A. G. G. M. 1985, Ap.J., 297, 702.
Cohen, M., Erickson, E. F., Haas, M., Hollenbach, D. J. and Wolf, J. 1987, in preparation.
Cohen, M. and Fuller, G. A. 1985, Ap.J., 296, 620.
Cohen, M., Harvey P. M. and Schwartz, R. D. 1985, Ap.J., 296, 633.
Cohen, M. and Jones, B. F. 1987, in preparation.
Cohen, M. and Kuhi, L. V. 1979, Ap.J.Suppl., 41, 743.
Cohen, M. and Schwartz, R. D. 1983, Ap.J., 265, 877.
Cohen, M. and Schwartz, R. D. 1987, Ap.J., in press.
Cohen, M. and Witteborn, F. C. 1985, Ap.J., 294, 345.
Dopita, M. A. 1978, Ap.J.Suppl., 37, 117.
Dopita, M. A., Caganoff, S., Schwartz, R. D. and Cohen, M. 1987, in Proceedings of IAU Symposium No. 115, "Star Forming Regions", eds. J. Jugaku and M. Peimbert (Dordrecht: Reidel).
Edwards, S. and Snell, R. L. 1982, Ap.J., 261, 151.
Edwards, S. and Snell, R. L. 1984, Ap.J., 281, 237.
Edwards, S., Strom, S. E., Snell, R. L., Jarrett, T. H., Beichman, C. A. and Strom, K. M. 1986, Ap.J. (Letters), 307, L65.
Elias, J. H. 1980, Ap.J., 241, 728.
Falle, S. A. E. G., Innes, D. E. and Wilson, M. J. 1986, submitted to M.N.R.A.S.
Gyul'budagyan, A. L. 1975, Astrofizika, 11, 531.
Hartmann, L., Hewett, R., Stahler, S. and Mathieu, R. 1986, Ap.J., in press.
Heidmann, N. and Thomas, R. N. 1980, A.A., 87, 36.
Herbig, G. H. 1977, Ap.J., 217, 693.
Herbig, G. H. and Jones, B. F. 1981, A.J., 86, 1232.
Jones, B. F. and Cohen, M. 1986, A.J., in press.
Jones, B. F., Cohen, M., Sirk, M. and Jarrett, R. 1984, A.J., 89, 1404.
Montmerle, T., Koch-Miramond, L., Falgarone, E. and Grindlay, J. E. 1983, Ap.J., 269, 182.
Mundt, R. and Fried, J. W. 1983, Ap.J. (Letters), 274, L83.
Mundt, R., Stocke, J., Strom, S. E., Strom, K. M. and Anderson, E. R. 1985, Ap.J. (Letters), 297, L41.
Rydgren, A. E. and Cohen, M. 1985, in Protostars and Planets II, eds. D. C. Black and M. S. Matthews (Univ. Ariz. press), p. 371.
Rydgren, A. E., Schmelz, J. T. and Vrba, F. J. 1982, Ap.J., 256, 168.

Schwartz, R. D. 1983, Ann. Rev. A. A., 21, 209.
Snell, R. L. and Edwards, S. 1981, Ap.J., 251, 103.
Vogel, S. and Kuhi. L. V. 1981, Ap.J., 245, 960.
Walter, F. and Kuhi, L. V. 1981, Ap.J., 250, 254.
Walter, F. and Kuhi, L. V. 1984, Ap.J., 284, 194.
Worden. S. P., Schneeberger, T. J., Kuhn, J. R. and Africano, J. L. 1981, Ap.J., 244, 520.

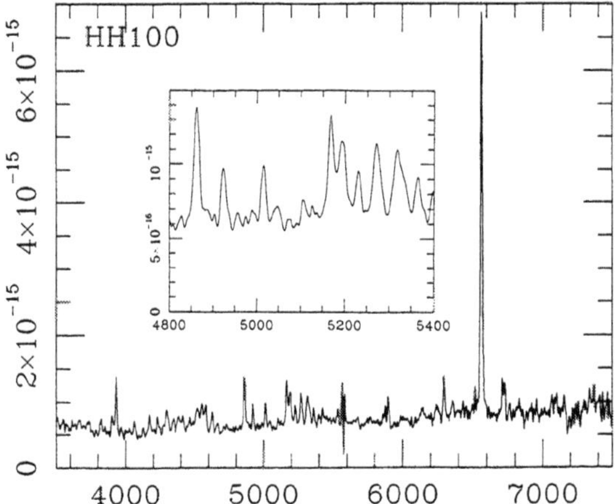

Fig. 1: Spectrum of HH100 showing the clear reflection of the powerful chromospheric emission lines of an active T Tauri star. Inset gives details near H-beta where many FeII lines are seen.

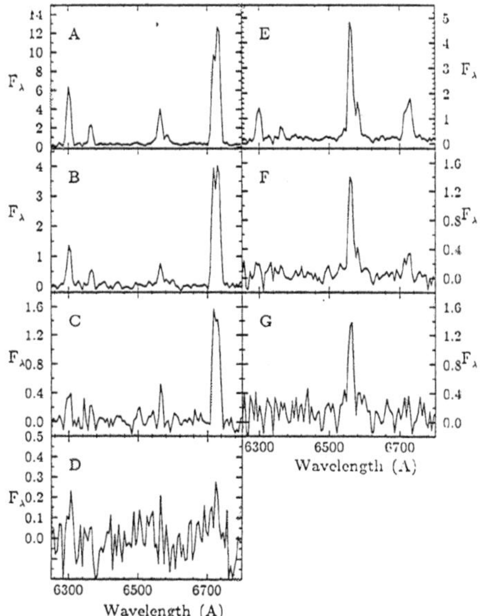

Fig. 2: Red spectra of knots along the jet, "DG Tau B", from star (between A and E) along the red-shifted lobe (ABCD), and also along the blue-shifted lobe (knots EFG). Note the steady decrease in excitation along the red lobe from A to D, and the extremely strong [SII] compared with H-alpha in knot D.

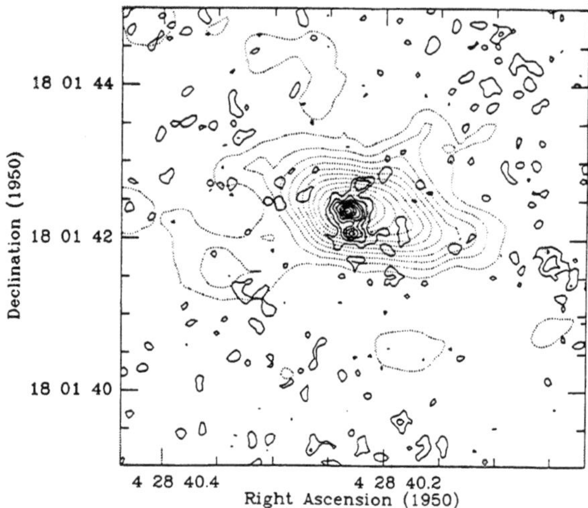

Fig. 3: VLA radio continuum images of L1551 IRS5: solid lines are the 15 GHz image (the apparent binary); dotted lines are the 5 GHz image, showing the radio "jet" emerging from the northern 15 GHz point source and stretching to NE and SW from it.

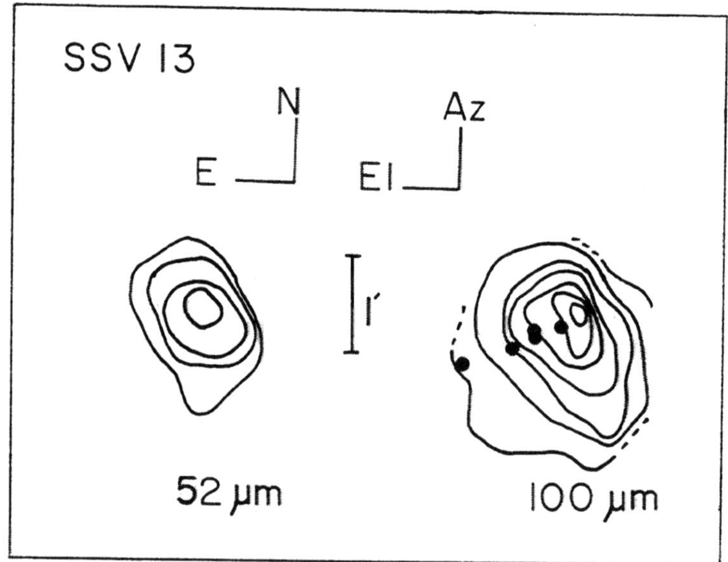

Fig. 4: 50 and 100 micron airborne maps made of the exciting star of the HH 7-11 chain of objects. HHs are represented by filled circles in the 100 micron image. The far-infrared peak coincides with the near-infrared source, SSV13, thought to be the exciting star for all these aligned HHs. Spatial extension is clear, perpendicular to the flow (HH line) but not parallel.

THE PHYSICS AND THE STRUCTURE OF AGN

Hagai Netzer
School of Physics and Astronomy, Tel Aviv University
Tel Aviv 69978
Israel

1. DEFINITIONS AND FUNDAMENTAL QUESTIONS

Active Galactic Nuclei (AGN) is a common name for quasars, Seyfert galaxies, liners, BL Lac objects and some narrow emission line galaxies. They show energy output not usually associated with stellar processes which is centered in a small nuclear region. An extragalactic object is classified as AGN if at least one of the following criteria is fulfilled:

a. Compact nuclear region brighter than the corresponding region in galaxies of similar type.
b. Nonstellar (or nonthermal) nuclear continuum emission.
c. Nuclear emission lines indicating non-stellar excitation mechanism.
d. Variable continuum and/or emission lines.

The main emphasis in this review is on the more luminous AGN, where all four criteria are fulfilled. They cover the luminosity range of $10^{43}<L_{bol}<10^{48}$ erg/sec and a redshift range of $0<z<4.01$ (correct for September 1986). Section 3 gives a brief description of the less luminous AGN, with luminosities as low as 10^{41} erg/sec.

Most of our information on AGN comes from spectral analysis. Some examples are shown in Fig. 1. The more luminous AGN show a combination of broad ($\sim$ 5000 km/sec FWHM) strong permitted lines and narrow ($\sim$ 500 km/sec) permitted and forbidden lines. The most prominent broad lines are L_α, CIV 1549, OVI 1035, CIII]1909, MgII 2798, the hydrogen Balmer lines and FeII lines. The strongest forbidden lines are due to [OIII], [OII], [NII], [NeIII] and [NeV]. The excitation covers a very large range; from [FeII] to [Fe XI] and from OI to OVI.

The infrared to X-ray continuum is definitely nonstellar in origin. It can be described, over a limited range in frequency, by a power-law, $F_\nu = c\nu^{-\alpha}$, with a spectral index α. An example is shown in Fig. 2. In almost all objects the infrared continuum is steep, with $\alpha>1$ and there is a noticable continuum flattening at $\lambda<7000$Å where $\alpha\sim0.5$. Further steepening below 912Å is another common feature.

AGN are powerful X-ray sources and $\alpha=0.7$ gives an extremely good fit to their 3-70 KeV continuum. The soft X-ray (0.5-3 KeV) differs in

W. Kundt (ed.), Astrophysical Jets and Their Engines, 103–124.

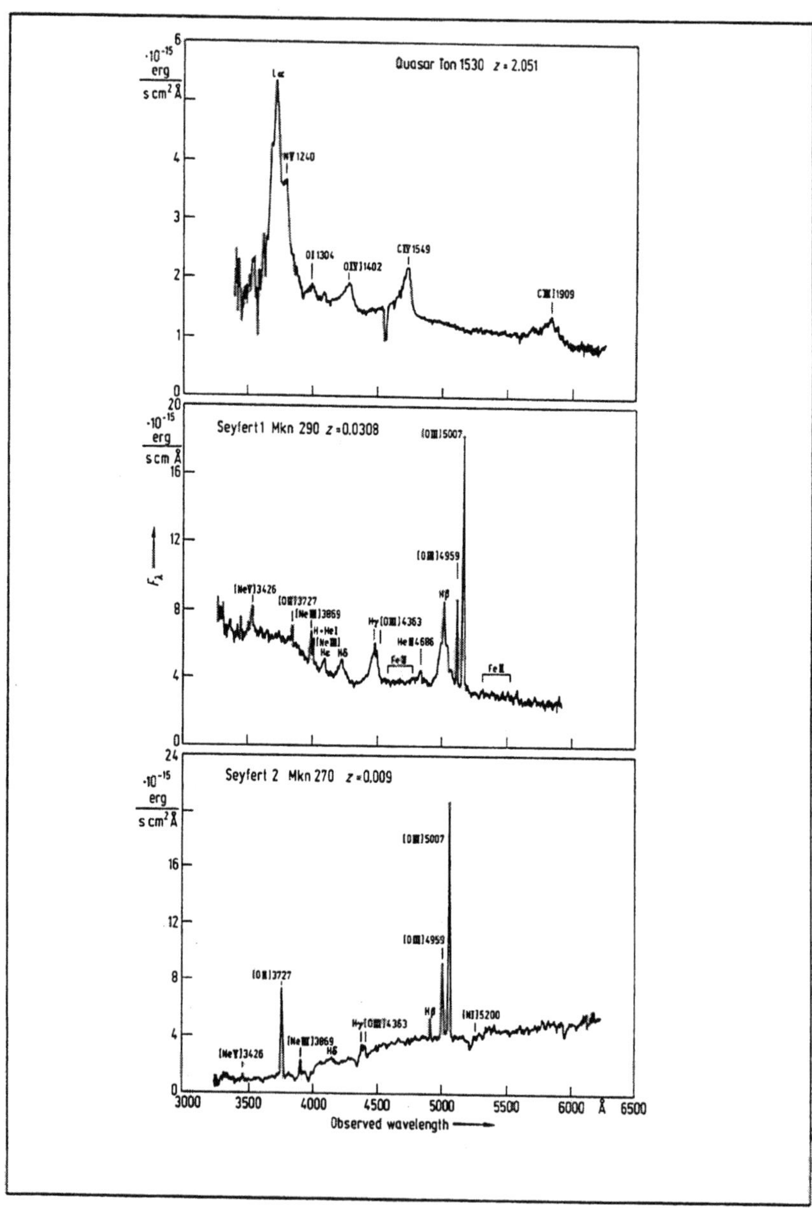

Fig. 1: Spectra of three AGN.

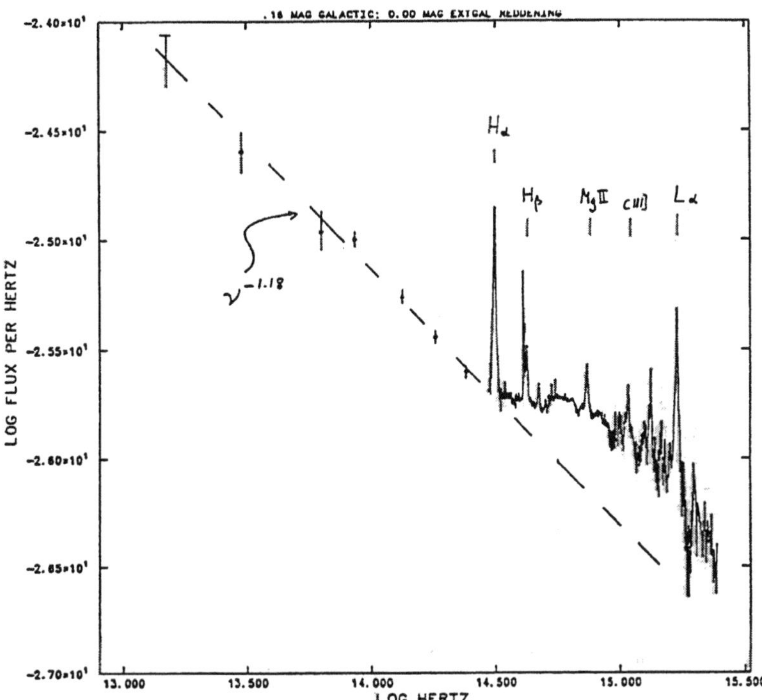

Fig. 2: The infrared to ultraviolet spectrum of the quasar 0742+318 obtained, simultaneously at Hawaii (IRTF), Texas (McDonald Observatory) and from space (IUE). Note the continuum flattening at $\lambda < 6500$ Å.

slope, from object to object with $0.5<\alpha<2$. (In fact it may not be fitted by a single power-law, at all). Extrapolation of the observed 2 KeV flux, with $\alpha=0.7$, to longer wavelength is 100-300 times fainter than the observed 1000Å continuum. The shape and flux of the unobservable part of the continuum ($50<\lambda<900$Å) is crucial to the understanding of the physics of AGN and is an important area of research.

The main objectives in studying AGN is to understand their nature and the physical processes governing their behaviour. The following sections give a summary of our present understanding and address, more specifically, three basic questions that are at the frontier of our knowledge:

a. What is the mass of the central object in AGN?
b. Is there a mass-luminosity relationship in AGN?
c. What are the evidences for massive accretion disks in AGN?

Section 2 of this review describes the geometry and structures of the brightest AGN, and the basic assumptions of photoionization as applicable to such objects. Section 3 extends and generalizes this theory to the fainter objects of this class and Section 4 addresses the question of AGN continuua and accretion disks. Partial answers to questions a-c are given at the end of Section 2, in Section 4 and in a conclusion section.

2. MODELING AGN

2.1. Geometry

The innermost part of AGN contains a massive object, probably a black hole. A small region around it is the location of the optical-UV-X-ray emitting continuum. The overall dimension, as deduced from light variability arguments, is of the order of $10^{16}L_{46}$cm where L_{46} is the optical-UV luminosity in units of 10^{46} ergs/sec. There are clear indications that the X-ray source is smaller than the optical source and that there is some correlation between luminosity and dimension (e.g. Barr and Mushotzky 1986).

The line emitting region can be divided into the Broad Line Region (BLR) with dimensions of $\sim 10^{18.5}L_{46}$ cm, and the Narrow Line Region (NLR) with dimension $\sim 10^{21}L_{46}$ cm. The gas in the BLR is probably in small clouds, or filaments, with typical density of $\sim 10^{10}$ cm^{-3} and column density of 10^{22-23} atoms/cm^2. The motion of these clouds, with relative velocities of 3000-10000 km/sec, gives rise to the broad line profiles. The NLR gas is of much smaller densities ($\sim 10^{4-6}$ cm^{-3}) as can be inferred from the presence of strong narrow forbidden lines. It is not clear whether there is a smooth transition zone, of intermediate densities and velocities, between these two regions.

Simple consideration indicates that BLR clouds cover only a small part of the continuum and most of its radiation can freely escape. A useful quantity is "the covering factor", $\Omega/4\pi$, where Ω is the solid angle sustended by all clouds. The associated filling factor (fraction by volume) is extremely small, $\sim 10^{-5-6}$.

The powerful ultraviolet continuum suggests photoionization as the dominant heating and excitation process in the BLR and NLR. This can be seen by analysing observed line ratios such as CIII] 1909/CIII 977 that indicate $T_e \sim 20,000$K. This temperature is too low to collisionally ionize carbon to C^{+2} and definitely not enough to explain the strong lines of N^{+4} and O^{+5}. The co-existence of low and high ionization species, at relatively low temperatures, indicate that collisional processes (shocks, fast particles, ...) cannot be the dominant energy input sources.

2.2. Photoionization Models

2.2.1. General principles: Consider a gas cloud at a distance R from a continuum source whose ionizing flux per unit frequency is $F\nu$. The steady state ionization of element X is given by solving equations of the form:

$$\frac{X^{i+1}}{X^{i}} = \int_{\nu_{min}}^{\infty} \frac{F_\nu \, e^{-\tau_\nu} \sigma_\nu \, d\nu}{(4\pi R^2 h\nu)(\beta N_e)} \tag{1}$$

where σ_ν is the photoabsorption cross section, ν_{min} the minimum ionizing frequency, τ_ν the optical depth to the illuminated surface of the cloud and β the recombination coefficient.

It is convenient to define an ionization parameter, U, which is related to the total ionizing photon flux Q(H), where:

$$Q(H) = \int_{\nu_H}^{\infty} \frac{F_\nu d\nu}{h\nu} \tag{2}$$

and

$$U = \frac{1}{4\pi R^2} \frac{Q(H)}{N_e} \quad \text{cm s}^{-1} \quad . \tag{3}$$

The degree of ionization of most elements, at the ionized face, is proportional to U. For example, $n(H^+)/n(H^o) \simeq U/10^{5.3}$cm s^{-1} and $n(He^{++})/n(He^+) \simeq U/10^{7.2}$ cm s^{-1}. The emitted spectrum is defined by U and the continuum slope α. $10^8<U<10^9$ cm s^{-1} gives a good fit to the observed spectrum of many AGN if $\alpha\sim1$.

A second set of equations is needed for solving the temperature. Most heating, in the ionized part, is due to ionization and most cooling due to free-free, bound free and collisional excitation processes.

2.2.2. Photoionization of the BLR. Photoionization models for AGN's high-density clouds are rather complicated (see Davidson and Netzer 1979). The nonstellar continuum produces large flux of ionizing photons at different frequencies. This results in overlapping of different stages of ionization in comparison with ionization by thermal sources, where the ionization fronts are well defined. The most extreme case is that of a hard ($\alpha<1$) low ionization parameter continuum which produces an extended "warm" (6000<T<10,000K) partly neutral ($n(H^+)/n(H)\sim0.05$) zone where cooling is by the low excitation lines of CII], MgII H and FeII.

Of great importance in the "warm" region are processes related to the X-ray continuum. They include the Auger process and inner-shell ionization, the heating and ionization by fast secondary electrons and Compton heating. Non standard techniques are needed to calculate these processes. Charge exchange reactions, with neutral hydrogen, are also very important, due to the large fraction of neutral gas at the relatively high temperatures.

Perhaps the most important difference between the BLR and low density nebulae is the large optical depth in many lines, especially the hydrogen Balmer lines. It can be shown (e.g. Ferland and Netzer 1979) that the Balmer optical depth is proportional to $Q(H)/R^2$ and that $\tau(H_\alpha)>10^3$ under most conditions. The Balmer optical depth is thus a sensitive indicator of the cloud-central source distance. Other lines that are thin in low density nebulae and thick in the BLR are HeI 5876 and perhaps also HeII 1640 (MacAlpine, 1976, Netzer et al. 1985). This situation requires the introduction of special transfer techniques in order to calculate the emergent spectrum.

Most recent models treat line transfer with the "escape probability" method. This involves a location dependent function, $\varepsilon(\tau)$, which multiplies the transition probability of the lines. The functional form of $\varepsilon(\tau)$ is known to a good approximation, but there are complications due to the different forms for resonance and interlocking lines (e.g. Collin-Souffrin and Dumont 1986).

The main advantage of the method is that a large number of optically thick lines can be treated simultaneously. There are also ways to treat line fluorescence and line-continuum interaction within this formalism (Netzer et al. 1985). The disadvantage is that all line transfer is treated locally, without allowing photon diffusion in space. It is not yet clear what is the level of uncertainty introduced by this simplification.

2.2.3. Results of photoionization calculation. A photoionization model produces a complete line list for one cloud with specified U,N and abundances. Fig. 3 is an example showing relative line ratios as a function of the ionization parameter for typical BLR and NLR clouds. A complete model includes the weighted mean from a large number of clouds, over the expected range of U and N_e.

The physical parameters that are obtained from such models are the chemical abundances, the densities, the covering factor and the column density of the clouds. Abundances are rather difficult to get since strong lines like CIV 1549, are important coolants and the CIV 1549/L_α ratio cannot be used to deduce C/H. Best abundance indicators are some weak lines that are insensitive to temperature variations (Shields 1976). These are NIV]1486/CIV 1549, NIII]1750/CIII] 1909 and OIII]1663/CIV 1549+CIII]1909. Unfortunately, there are no good line ratios to deduce C/H, O/H and N/H in an unambiguous way, because of the complications in calculating the hydrogen lines and the important cooling rate of the strong metallic lines. It seems that abundances in AGN are solar within a factor of $\sim$3, with indications of some overabundance of C, N, O and Fe, as in nuclei of spiral galaxies.

The density in the BLR is probably in the range of 10^8-10^{11} cm^{-3}. The lower limit is from the absence of broad forbidden lines and the upper from the observation of strong intercombination lines. For example, the critical density for CIII] 1909 is about 3 x 10^9 cm^{-3} and the line is among the strongest in the spectrum of most AGN. Other lines, (NIII] 1750, NIV]1486 ...) with critical densities of few x 10^{10}cm^{-3} are also observed with roughly their expected low density intensity. This suggests that much of the line emission is from a region where densities are below critical. Recently there have been

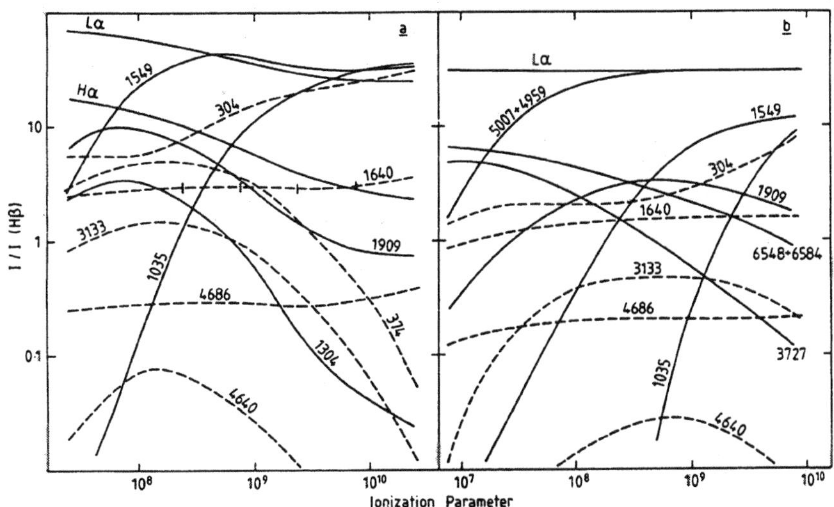

Fig. 3: Relative line intensities in "standard" photoionization models of AGN. Left: BLR model with a $F_\nu \propto \nu^{-0.5}$ ionizing continuum, and $N \sim 10^{9.5} cm^{-3}$. Right: NLR model, with $F_\nu \propto \nu^{-1.5}$ and $N \sim 10^4$ cm^{-3}.

some suggestions that part of the BLR is of much higher densities; up to 10^{12} cm^{-3}. This is related to the question of the central mass (Section 2.4) and the FeII spectrum (Section 4.1)

One of the most important results is the low value of the covering factor, as deduced from the equivalent width of several recombination lines. The best line to use is L_α, since under simple recombination conditions, and also in AGN clouds, the total number of L_α photons equals the number of ionizing photons. Under these conditions:

$$EW\ (L_\alpha) = \frac{1215}{\alpha} \left(\frac{3}{4}\right)^\alpha \left(1-\nu_{cut-off}^{-\alpha}\right) \frac{\Omega}{4\pi}\ \text{Å} \quad (\alpha \neq 0) \tag{4}$$

where $\nu_{cut-off}$ is a cut off frequency in Rydberg. The inferred value for many AGN ($\alpha \gtrsim 0.5$, $EW(L_\alpha) \sim 100$ Å) is $\Omega/4\pi \sim 0.1$.

Another way to study the covering factor is to observe the obscuration of the central source by the BLR clouds. Study of Lyman continuum absorption indicates that $\sim$10% of all objects show some obscuration (e.g. Baldwin and Smith 1983). Part of this may be intervening material (most of these observations are of high Z objects) and not in the BLR. The best data so far are on X-ray obscuration (e.g. Mushotzky 1982). They show that some of the lowest luminosity AGN are seen through a gas cloud with column density of $\sim 10^{23}$ atoms/cm^2. There are no such cases among the high luminosity AGN. Reichert et al.(1986) studied these cases and concluded that partial obscuration, by clouds with dimensions less than that of the optical continuum, are the most likely explanation. The covering fraction can reach, in some cases, 0.9 which is in conflict with the results of the photoionization analysis.

2.2.4. Comparison with observations. Older photoionization analysis of the BLR has been quite successful in explaining many line ratios and intensities observed in the spectrum of AGN (Davidson and Netzer 1979). Recent developments, especially the improvement of ultraviolet capabilities, show quite clearly a big discrepancy between models and observations. This was pointed out first by Baldwin (1977) and came to be known as the "L_α/H_β problem". While standard, pre-1980 models predict $L_\alpha/H_\beta \sim 50$, the observed ratio is $\sim$10. Careful study shows that the problem is in fact more general, i.e. there are too strong predicted ultraviolet lines (not only L_α).

The "L_α/H_β problem" caused renewed interest in the broad line spectrum and big improvement in the theoretical modeling. Models like those by Kwan and Krolik (1981) demonstrated that the observed L_α/H_β ratio can be produced in BLR clouds. However, careful analysis of weak helium and hydrogen lines clearly shows that models where L_α/H_β is close to its observed value fail to produce other line ratios. The problem is not fully solved yet and there are several different explanations, none very satisfactory. Other open questions are the Balmer decrement, the P_α/H_α ratio, the HeII 4686/H_β ratio and the strength of the high excitation lines of OVI and NV.

2.2.5. Dust and reddening. Dust is common in all known nebulae. Is it present also in AGN clouds? This is still an open question. The best reddening indicators, in low density nebulae, are the hydrogen lines. They cannot be used in the BLR clouds since we do not have a good enough theory for these lines. The best indicators for AGN clouds are the OI 8446/OI 1304 and HeII 4686/HeII 1640 line ratios (Netzer and Davidson 1979). The theoretical ratios are well known and they span a large wavelength range. Unfortunately, the lines are weak and difficult to measure. (See MacAlpine 1985). There are indications that AGN clouds may be reddened by up to $E_{B-V} \sim 0.25$ mag.

Continuum reddening is even more difficult to establish since the best signature in the galaxy, the 2200Å extinction feature, may not be present in different environments. The line and continuum reddening may not be identical, since this depends on the (unknown) dust location and the geometry. The smoothness of quasar's continua, and the similarity of slopes, suggest that continuum reddening cannot be very large. The location of the dust is another open question. There are suggestions that it is just outside the BLR (see Section 3) but this is not the only possibility. It is probably not mixed with the high density ionized gas, as this will reduce - dramatically - the strength of several resonance lines.

2.3. Cloud Motion and Dynamics

2.3.1. Important time scales. Several processes can change the state of the BLR gas, each with its associated time scale.

a. Ionization (or recombination) time. This is the time required to establish a new ionization balance, following a change in the ionizing flux. It is given by:

$$t_{rec} \simeq 10^{-4} \left(\frac{10^9}{N_e}\right) \text{ years} . \quad (5)$$

b. Evaporation time. Thermal evaporation of clouds in hot intercloud medium (see 2.3.2) may be important in the BLR. The exact rate is not well known since it is very sensitive to the strength and direction of the magnetic field. An estimate is

$$t_{evap} \simeq 0.1 - 10 \text{ years} . \quad (6)$$

c. Dynamical time. The crossing time of the BLR gives the lifetime of an unbounded system

$$t_{dyn} \simeq 250 \, L_{46} \quad \text{years} . \quad (7)$$

d. Sound crossing time. Typical clouds in the BLR have dimensions on the order of 10^{12-13}cm. Such clouds disintegrate, in the absence of confinement, on a time scale of

$$t_{sound} \sim 0.01 - 0.1 \text{ years} . \quad (8)$$

2.3.2. The two phase model. The emission line clouds are too small to be confined by self-gravity and the sound crossing time is much shorter than the dynamical time. Therefore, either clouds are being produced continuously, or they are confined by hot intercloud medium. The hypothetical confining medium has received much attention since it has been proposed by Mathews (1974). Extensive and comprehensive work on its properties can be found in Krolik McKee and Tarter (1981) and Kallman and Mushotzky (1985).

The general idea of the two phase model is of two distinct components, in pressure equilibrium but very different temperature, both heated by the same central source. The only important processes in the hot phase are Compton heating and cooling. They can produce, in AGN with large X-ray fluxes, temperatures of the order of 10^8K. Pressure equilibrium requires $N_e \cdot T$ to be about the same in both components, thus the density in the hot phase is four orders of magnitudes smaller than in the cold, line emitting phase. The basic physics relies on the finding that there are two possible solutions for the gas temperature over the range of interest in U (i.e. that range that gives a good match between the predicted and observed spectrum); one with $T \sim 10^4$K and the other with $T \sim 10^8$K. The hot phase gas produces the required confinement and explains the long life of the line emitting clouds. There are several difficulties in this model. First, the hot phase temperature may not be high enough and clouds may move through it supersonically. The temperature depends strongly on Compton cooling by the $\lambda < 912$ Å continuum, which is not observed. Second, drag forces will affect cloud motion and cause them to stop over a small distance. This is in conflict with the observed velocities. Also, other instabilities may develop and there may be a problem with τ (elec. scattering). The whole subject is currently under revision and further study.

2.3.3. Cloud motion and line profiles. The broad line width may be interpreted as due to motion of gravitationally bound clouds. An alternative that was considered by Mathews (1974), Blumenthal and Mathews (1975) and others (see review by Mathews and Capriotti 1985) is that the clouds are accelerated by radiation pressure.

Consider an optically thin cloud next to a source of ultraviolet photons. The gas is in ionization equilibrium and there are βN_e^2 ionizations and recombinations of hydrogen every second, each associated with a momentum transfer of $h\langle\nu\rangle/c$ to the cloud. The momentum transfer per mass density is

$$\frac{h\langle\nu\rangle}{c} \cdot \frac{1}{M(H)\ N_e} \tag{9}$$

and the equation of motion:

$$\frac{dv}{dt} = -\frac{GM}{R^2} + \frac{h\langle\nu\rangle}{c} \cdot \frac{1}{M(H)\ N_e} \cdot \beta N_e^{\,2} \ . \tag{10}$$

Thus, when radiation acceleration dominates, dv/dt is proportional to N_e. Simple calculation shows that this is indeed the case for the canonical values of M and R.

Acceleration by radiation pressure is much less efficient in optically thick gas. X-ray radiation can penetrate the cloud and the acceleration due to this depends on the column density. The question whether radiation acceleration or gravity wins depends on the column density and other parameters that are not very well determined. One of the attractions of the radiation acceleration mechanism is that it gives line profiles that are in good agreement with those observed.

Unfortunately, line profiles are not a very good diagnostic tool and several other models are as successful in producing the observed profiles. Another attraction is that it produces velocities in excess of the escape velocity, thus indicating a smaller central mass (see 2.4 and Section 4).

A simple test of the radiation acceleration model is the observed profile of L_α. This line is produced mainly in the H^+ part of the cloud and most line photons can only escape from the inner, illuminated part and not the neutral backside. The predicted line profile, for outward moving clouds, is very asymmetric with almost no emission in the blue wing (Ferland Netzer and Shields, 1979). This is not observed in AGN.

2.4. The Mass of the Central Object

A combination of the observed line profiles and the photoionization analysis enables a simple estimate of the central mass. There are two assumptions: a: A bound system of clouds, and b: U and N_e are known. The cloud-central source distance, R, is obtained from equation 3, and the mass from assumption a. For orbiting clouds:

$$M = \frac{v^2 R}{G} = \frac{v^2}{G}\left(\frac{Q}{4\pi U N_e}\right)^{1/2} \tag{11}$$

where v is a characteristic velocity obtained from the profiles.

A convenient way of expressing this result is by using the Eddington luminosity, $L_E = 1.3 \times 10^{38}(M/M_\odot)$ erg/sec. If the average ionizing frequency, in Rydberg, is $<\nu>$ and $N_{10} = N_e/10^{10}\text{cm}^{-3}$ then:

$$v \simeq 1700 \left[\frac{L_E}{L(UV-X)}\right]^{1/2} L_{46}^{1/4}(UV-X)\left[\frac{N_{10} U <\nu>}{c}\right]^{1/2} \text{ km/sec.} \tag{12}$$

Standard values for U ($\sim 3 \times 10^8$cm/sec) and N_e ($\sim 10^{9.5}\text{cm}^{-3}$) give $M \sim 10^8 M_\odot$ for faint Seyfert 1s and more than $10^{10} M_\odot$ for bright quasars.

The indirect estimate for the mass, given above, can be checked by referring to the recent observations of line variability that give a direct measurement of R (see M. Gaskell contribution). Such studies (e.g. Peterson et al. 1985) show that R is smaller than deduced by the photoionization analysis by at least a factor of 3. This means a smaller mass. The theoretical estimate of R depends on the product $N_e U$ and smaller R means larger $N_e U$. Several recent papers consider the possibility of much larger densities, up to 10^{12}cm^{-3}. There are no good photoionization models for such densities yet, and it is not known whether this is indeed the solution. Whatever $N_e U$ is, it must be larger than estimated from standard model, thus M in equation 11 is only an upper limit. Another reason why M may be an upper limit is that acceleration by radiation pressure may be important. Typical velocities in such cases are several times the Keplerian velocity, thus smaller mass is inferred.

2.5. The Mass Luminosity Relationship

Having estimated the mass of the central object we can use the observed luminosity to check for a possible M-L relationship. If all broad line AGN show similar line width, regardless of the luminosity, then $M \propto L^{1/2}$. (N_e and U are very similar in all objects). If line width depends on the luminosity, the relation is different. Several recent papers address this question (Wandel and Yahil 1985, Joly et al. 1985). Joly et al. find a marginal correlation of line width with luminosity, such that the FWZI (full width at zero intensity) of H_β is proportional to $L^{0.2\pm0.04}$. Assume for simplicity $V_{Keplerian} \propto L^{1/4}$, we get (see eq.12) $L \propto M$ or, using the experimental factors, $L/L_E \sim 0.01$.

The experimental correlation upon which this conclusion is based is very uncertain. For example, no correlation is found between the FWHM (full width at half maximum) of H_β with L or between the CIV 1549 line width and the luminosity (Joly et al. 1985). There are several selection effects that may bias the correlation and complete samples must be studied. Despite all this the best estimate of L/L_E is interesting since it implies a very large central mass and inefficient accretion. Using these numbers we find that the mass of a 10^{48}erg/sec quasar is $\sim 10^{12}$ $M_\odot$! The discussion in Section 4 shows why L/L_E may be larger than obtained by this simple method.

3. QUASARS, MINIQUASARS AND MICROQUASARS

So far we have considered mainly the brighter AGN. These are the ones fulfilling all four criteria in Section 1. Historically they were divided into Seyfert 1 galaxies ($10^{42} < L < 10^{45}$erg/sec) and quasars ($L > 10^{45}$erg/sec) but the great similarity in almost all properties suggests that they should really belong to one class. I prefer the notation "quasars" for the brighter and "mini-quasars" for the fainter objects in this group.

Many fainter objects that do not show strong broad lines certainly belong to the AGN group because of their nonstellar continuum and strong nuclear emission lines. Can we call them "microquasars"? The general success of photoionization theory, as applied to quasars and miniquasars, suggests that it may help to find the answer to this question. In the absence of broad lines one should use the next best, i.e. narrow lines. Information about the continuum could be useful too, but the optical continuum is faint compared with the stellar background and the ultraviolet one too faint for present UV detectors. The X-ray continuum is available, too, but most of these objects fall just below present sensitivity limits.

In this section I discuss three groups of AGN and show the similarity between them and the brighter objects of this class.

3.1. Narrow Emission Line Galaxies (NELG)

These are galaxies showing strong high excitation nuclear emission lines. Although they were known for years they have attracted much attention recently when it was discovered that their X-ray spectrum is very similar to that of the miniquasars. The similarity is both in continuum slope and in X-ray variability (e.g. Maccacaro et al. 1982, Mushotzky 1982). This raises the obvious question about the absence of BLR in these galaxies.

Careful spectroscopic studies of several such objects (e.g. Osterbrock 1981, Veron et al. 1980) clearly show a faint broad component in H_{α}, similar in width to the broad lines in quasars. A classification scheme for Seyfert galaxies, suggested by Osterbrock, is based on the relative strength of the narrow and broad components in the Balmer hydrogen lines. According to this scheme class 1.0 objects are those where no strong narrow component is seen, class 1.5 show a distinct but weak (up to $\sim$20% the total flux) narrow core and class 1.8 or 1.9 are those where the broad component contributes only 10% or less of the flux. Some of the strong X-ray NELG can definitely be classified as Seyfert 1.9.

A most interesting suggestion about the nature of NELG is that they are obscured, or reddened Seyfert 1 galaxies (Veron et al. 1981, Lawrence and Elvis 1982). A variable amount of dust, located somewhere between the BLR and the NLR, obscures the BLR and the continuum but not the NLR. Several interesting correlations confirm this idea. For example, the nonstellar continuum is redder and the broad Balmer lines weaker in edge-on spiral active galaxies, compared with the face-on ones. The observed differences are thus apparent and the large amount of extinction the only distinction between NELG and Seyfert 1 galaxies.

3.2. Seyfert 2 Galaxies

Type 2 Seyfert galaxies are those AGN showing strong narrow (400-1000 km/sec) lines with no trace of broad emission lines. The optical spectrum is similar to that of the NELG (which caused a lot of confusion in the classification) but the Seyfert 2s are much weaker X-ray emitters and do not show the weak broad H_{α} and the clear signs of extinction.

Until about two years ago it was believed that Seyfert 2 galaxies are the least active of the AGN group. Since then there were two exciting developments. Careful monitoring of Seyfert 1 galaxies showed, in two well-studied cases, a large reduction in the broad line flux and almost complete disappearance of this component (Alloin et al. 1984, Penston and Perez 1984). At minimum light Seyfert 1s look very similar to Seyfert 2s showing strong narrow lines. This raises the possibility that Seyfert 2 galaxies are those broad line Seyferts that are temporarily not active. There are some problems in this model, especially with the radio properties, but it is definitely an interesting possibility.

More exciting, in my opinion, is the recent discovery by Antonucci and Miller (1985) that the polarized light of NGC 1068 (a classical

Seyfert 2) shows clearly broad Balmer and FeII lines. This raised the possibility that the inner part of this galaxy contains a standard BLR, which is hidden by some material so that only the scattered (and therefore polorized) radiation reaches us. Such a possibility is illustrated in Fig. 4 below.

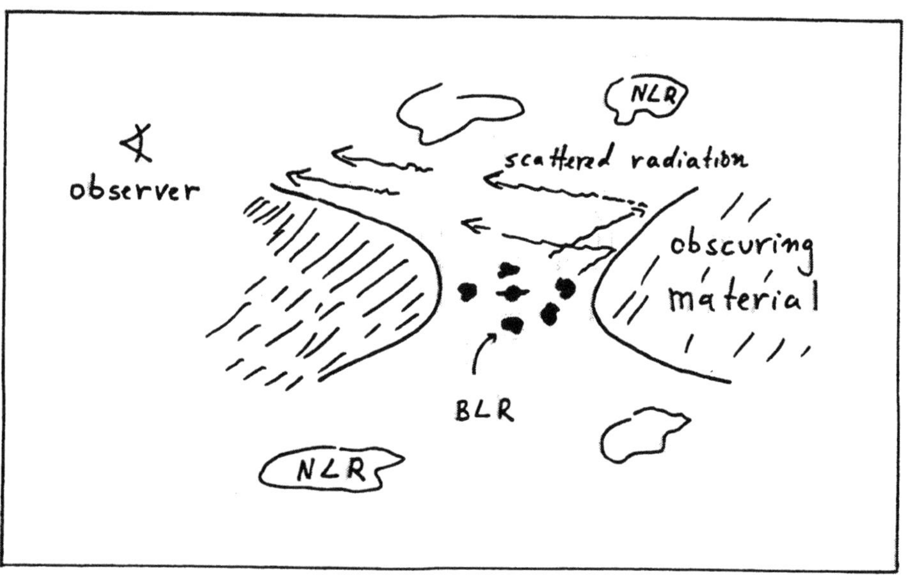

Fig. 4: A possible model for a Seyfert 2 galaxy. The BLR is obscured and only scattered radiation can reach the observer.

There are already several follow up studies emphasizing other aspects of this model. Snijders, Netzer and Boksenberg (1986) have made a careful study of the ultraviolet spectrum of NGC 1068 and found weak broad FeII lines, similar to the typical features in broad line AGN (see Sec. 4). Wilson (1986) and collaborators were able to show that the radio structure of Seyfert 2 galaxies is compatible with this model and that there are indications that the alignment of the NLR gas is in the same direction as that of the radio source.

These ideas, if confirmed, will prove that Seyfert 2 galaxies are "hidden Seyfert 1s", and are similar in nature to the brightest AGN.

3.3. Low Ionization Nuclear Emission Regions (LINERs)

Some galaxies show low ionization nuclear emission lines. In his detailed study of these objects Heckman (1980) defined a LINER on the basis of the relative strength of the oxygen forbidden lines and the Balmer lines. For example, in a LINER spectrum [OII]3727 is stronger than [OIII]5007, [OIII]5007/H_β < 3 and [OI]6300 / [OIII]5007 > 0.3. This is very different from the narrow line spectrum of Seyfert galaxies where [OIII]5007 / H_β > 8 and the high excitation lines are much stronger.

Up to 1/3 of all nearby galaxies show a LINER spectrum and their classification as AGN would make this class of objects much more common than thought.

The nature of LINERs has been an open question for several years. One suggestion, by Baldwin, Phillips and Terlevich (1981) is that they are shock excited nuclear regions more similar to supernova remnant than to photoionized nebulae. This idea is based on the observed strength of several emission lines, especially the strong [OIII]4363 that is a good high temperature indicator.

This idea was challenged by Ferland and Netzer (1983) who calculated photoionization models for low density gas and small ionization parameter. Such models can be used to show that narrow line ratios such as [OIII]5007 / [OII]3727 and [OIII]5007 / H_β form a single continuous sequence, from quasars and Seyfert 1s to Seyfert 2 and LINERs with only one variable: the ionization parameter. Photoionization by a powerlaw continuum, with low ionization parameter, can explain most of the observed line ratios in LINERs although there are still some unexplained features.

The last few years of research provided ample information on LINERs that cannot be discussed here. Better spectroscopy, radio and infrared measurements, and ultraviolet observations all point to the fact that these objects are indeed close in nature to the brighter AGN. I should only add here the discovery of weak broad H_α in many LINERs, and the strong X-ray continuum in some of them.

The optical UV continuum luminosity of LINERs is about two orders of magnitude lower than that of faint Seyfert 1s. Shouldn't we call them Microquasars?

3.4. Classification Schemes

The spectrum of broad line AGN has been discussed in Section 2. The continuity of properties can be extended to Seyfert 2s, NRLGs and LINERs using the narrow line properties, as discussed in this section. This suggests a possible, two parameter, classification scheme. Let us denote the luminosity class by Q (QI the brightest QIII the least luminous) and the spectral class by numbers, from 1.0 to 2.0, according to the ratio of the broad and the narrow components. The bright quasars are class QI 1.0, bright Seyferts QII 1.0 to QII 1.8 and LINERs QIII 2.0 or QIII 1.9. Varying along the sequence are the ionization parameter, the width of the narrow lines (and possibly also the broad lines) and other properties.

Like stellar classification, such schemes may tell us, eventually, something fundamental about the nature and evolution of AGN.

4. AGN CONTINUUM AND MASSIVE ACCRETION DISKS

4.1. AGN Continuum

Fig. 2 shows several distinct continuum features. An excess of radiation between 2000 and 4000 Å is very common and has been given the name: "the small bump". A large and flat component, at 1000 Å<λ<7000 Å, has been called the "big bump". These two bumps are the key to the understanding of quasars' inner structure.

The small bump has caused a lot of confusion and was attributed by some to a single black body with T∿27,000K. Recently, Wills, Netzer and Wills (1985) were able to show that most of the energy in this bump is due to thousands of FeII lines and the Balmer continuum. Fig. 5 shows an example of modeling the small bump in a typical quasar. The calculations are difficult and quite uncertain because of the complexity of the Fe^{+} atom. They indicate that most of the energy in the bump, above the powerlaw continuum, is from the BLR and not related to the innermost source. FeII lines are very intense in all broad line AGN and provide much of the cooling in the "warm" zone. Currently there is an "FeII problem" since no theoretical model predicts the large observed flux in these lines.

The big bump extends over a much larger wavelength range and can be described, in many objects, by $F_{\nu} \propto \nu^{-0.5}$. There is a noticeable steepening below L_{α} with $\alpha>1$. An interesting and important question is whether the observed λ<1000 Å continuum is intrinsic or only apparent. Several recent studies (Bechtold et al. 1984, Netzer 1985) discuss the possibility of a flat ultraviolet continuum which is steepened by intervening material. One possibility is that galaxies on the line of sight to the source cause the absorption. Another possibility is dust in the source itself.

There is a simple way to estimate the intrinsic slope of the ionizing continuum (Netzer 1985a). The number of L_{α} photons is about equal to the number of ionizing photons and the energy in all emission lines and diffuse continua is equal to the total energy absorbed by BLR clouds, which is almost the same as the total energy in the ionizing continuum. Combining the two, by using the observed flux from the BLR, we get the mean energy of an ionizing photon and thus the slope of the λ<912 continuum. The analysis shows that $\alpha<0.5$ if no line reddening is taking place and $\alpha \sim 1-2$ if E_{B-V} (lines) = 0.2 mag. This is much flatter than observed and supports the idea of intervening material. This kind of analysis was not possible until two years ago when it was realized that much of the flux in the small bump is due to line emission. Currently there are no photoionization models capable of explaining the total energy in emission lines if no reddening is assumed.

4.2. Massive Accretion Disks

The most popular (yet not the only) idea about the origin of the "big bump" is that it is due to thermal emission from the surface of a central accretion disk (e.g. Malkan 1983). Such disks are assumed to

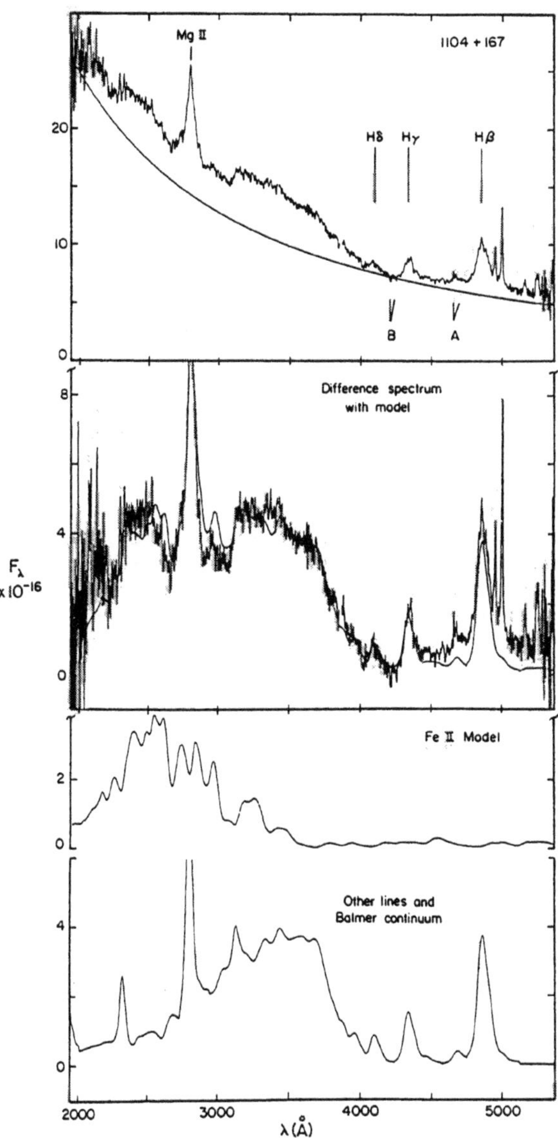

Fig. 5: The "small bump" in the spectrum of the quasar 1104+167.

Top: A powerlaw fit to the underlying continuum, showing the excess energy in the bump.

Bottom: Theoretical models for FeII and other features.

Middle: Model fitting of the powerlaw subtracted spectrum.

produce most of the observed blue continuum with flux up to the far ultraviolet and perhaps also into the X-ray.

The theory of massive disks is at such early stages that no good modeling is available. There is little, if any, understanding of the disk geometry, the structure of the atmosphere and corona, the instabilities, etc. All this is discussed by S. Phinney in this volume.

The classical work of Shakura and Sunyaev (1973) applies to standard thin disks. It can be used to find the radial dependence of the temperature, surface density and thickness of the disk for a given viscosity and accretion rate. The main unknown is the viscosity which depends on the turbulence and the magnetic fields. There are also relativistic effects that must be included at small radii and the influence of the centrally produced X-ray photons on the outer parts of the disks.

I shall not discuss the disk spectrum since this was covered by S. Phinney and since I consider the available calculations preliminary and inaccurate. In my opinion theoretical disk spectra cannot yet be used to fit the observed continuum and do not tell us much about its origin. Currently there is a great effort to improve the situation and better models are around the corner.

Are the observations good enough to test for the presence of disks? In terms of signal-to-noise and wavelength coverage, the answer is yes, at least for some well-studied AGN. There are, however, fundamental difficulties that prevent us from concluding one way or another. Reddening is a big unknown since a small amount of dust strongly affects the shape of the blue continuum. There is also an uncertainty in the exact shape of the non-disk infrared (λ>7000 Å) continuum and its contribution at short wavelengths. More important is the fact that we do not know the intrinsic shape of the λ<1000 Å continuum. In disk models this part determines the maximum disk temperature and the accretion rate. Since there are reasons to suspect that the observed continuum is different from the intrinsic one, no continuum fitting is very meaningful.

4.3. How to Detect Thin Accretion Disks

Despite the theoretical and observational uncertainties, there is a very promising way to detect AGN disks and learn about their properties. This is based on the special disk geometry and the different brightness seen by observers at different lines of sight. The apparent brightness of a thin disk depends on cos θ, where θ is the disk inclination angle. In a thick disk the dependence is different but in all cases the effect is large enough to be detected.

Perhaps the best way to detect geometrically thin disks is by examining the equivalent width of the strong ultraviolet lines of CIV1549 and L_α (Netzer 1985b). The total emergent flux in CIV1549 is isotropic and its observed flux does not depend on the viewing angle. The continuum flux, if emitted by a thin disk, is angle dependent. The line-to-continuum flux ratio (i.e. the equivalent width) is thus large, in edge-on disks and small in face-on ones. There is no way to detect this in a single object but a study of a large sample can show the effect.

A well known, yet never quite understood relationship in AGN, is the inverse dependence of the CIV1549 equivalent width on the continuum luminosity ("The Baldwin Effect" see Baldwin 1977b, Wampler et al. 1984). If quasars' ultraviolet continua are due to thermal emission from the surface of a thin disk, this will naturally explain the effect. The comparison of the observed and predicted equivalent width distribution tends to confirm this idea but more complete samples must be studied. There are alternative explanations for the "Baldwin Effect" but the thin disk idea certainly deserves thorough investigation. Thin disk, if present, may force us to re-evaluate some of the most fundamental ideas about the inner structure of AGN. The anisotropy of the radiation field causes angle dependent ionization such that clouds above the disk are exposed to a stronger ultraviolet flux and are more highly ionized. A possible configuration is that of an isotropic X-ray source in the centre of a thin ultraviolet-emitting disk. Fig. 6 shows the ionizing flux produced by such a source, at different angles. Although the exact shape of the emitted disk continuum is uncertain, the angular dependence is not, and several important dependences on θ are evident:

a: Angle dependent ionization parameter. High excitation lines at small angles and low excitation ones at large angles.

b: The observed ratio of the ultraviolet-to-X-ray flux is angle dependent.

c: Acceleration by radiation pressure is more efficient above the disk.

d: The covering fraction is angle dependent.

e: The hot-phase temperature may be angle dependent.

Fig. 6 shows one out of several possible configurations involving a disk. It demonstrates the need to include the angular dependence in future models of AGN. Such models are under study and detailed results will be available soon.

The special thin disk geometry may force us to change the estimates of the BLR size and the central mass. High density clouds near the plane of the disk can be close to the centre and still exposed to a weak radiation field. A low degree of ionization can thus be achieved at small radii. Larger ionization parameters (defined at some angles, say 60°) and smaller masses may thus be consistent with the observed spectrum (see Netzer 1986). A more general conclusion is that we must re-evaluate the models remembering that some of what we observe depend on our specific line of sight and not on the properties of the sources.

A final comment on thin disks. Several of the methods used to find quasars depend on their magnitude and colour. If thin disks are common, this introduces a selection effect that must be taken into account. For instance, magnitude limited surveys will discover more face-on disks near the survey limit. We may be missing a large fraction of the quasar population because of that.

4.4. Beaming

There is a new idea about AGN continua that may provide an alternative

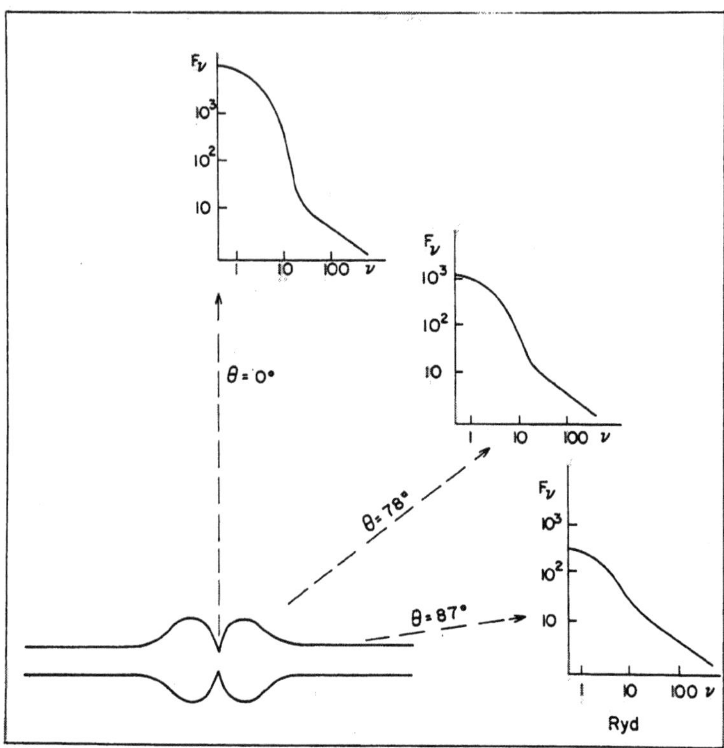

Fig. 6: Isotropic X-ray source in the centre of a thin disk emitting optical and ultraviolet radiation. The ionizing flux is shown at three angles relative to the surface of the disk.

to the thin disk model. This is the suggestion that the optical, ultraviolet and X-ray continua are beamed much like the radio source. I shall not discuss it in detail since several contributions in this volume already addressed the subject.

Beaming of the optical and X-ray continuum may help to explain some observed correlations of lines and continuum. It is discussed in two recent papers by Wills and Browne (1985) and Browne and Murphy (1986) and the reader is referred to these for more details. It is my feeling that either beaming or disks will prove to be wrong. At the moment I do not know which.

5. CONCLUSION

In this review I have tried to summarize our present understanding of

AGN. Photoionization modeling is the best tool for understanding the structure and dynamics of the BLR. Present models are complex and advanced. They are quite successful in explaining many of the observed spectral features but fail badly in others. There are several unsolved problems, such as the strength of the FeII lines and the hydrogen line spectrum. They may be related to the presence of a very high density component, dust or both.

Photoionization analysis provides the bridge between the most luminous AGN and the fainter ones. Objects at the faint end of this class show clear indications of activity. The BLR is definitely present in many of them but it is obscured by dust or by gas. All objects over a range of 10^{6-7} in luminosity form a continuous sequence with one variable: the ionization parameter.

Study of AGN continuum is not as advanced. In this review I have discussed accretion disks, and other continuum emitting processes are covered by other contributers in this volume. The information and discussion presented here was used to try and answer three questions that are at the forefront of AGN research. The mass of the central objects is the first question. It can be deduced from the photoionization analysis but new observations indicate that both the BLR and the central mass are smaller than estimated in this way. The M-L relationship in AGN must wait until the value of M is secured. There are some weak and marginal evidences for a simple $M \propto L$ dependence.

Have we discovered accretion disks in AGN? The theory of such disks is so uncertain that direct comparison of the observed and predicted disk spectra cannot be used to answer this question. I have pointed out an alternative and more promising way which makes use of the special disk geometry. The question whether we have already found thin disks in this way is similar to the question: "what is the true colour of the elephant"? and all participants in the Erice school know exactly what I mean.

References

Alloin D., Pelat, D., Phillips, M. and Whittle, M., 1984, Ap. J. 288, 205

Antonucci, R. R. J. and Miller, J. S., 1985, Ap. J. 297, 621

Baldwin, J. A., 1977a, MNRAS, 178, 67p

Baldwin, J. A., 1977b, Ap. J. 214, 679

Baldwin, J. A. and Smith, M. G., 1983, MNRAS, 204, 331

Barr, P. and Mushotzky, R. F., 1986, Nature, 320, 421

Baldwin, J. A., Phillips, M. M. and Terlevich, R., 1981, PASP, 93, 5

Bechtold, J., Green, R. F., Weymann, R. J., Schmidt, M., Estabrook, F.B., Sherman, R. D., Wahlquist, H. D. and Heckman, T., 1984, Ap. J. 281, 76

Blumenthal, G. R. and Mathews, W. G., 1975, Ap. J., 198, 517

Browne, I. W. and Murphy, D., 1986, MNRAS (in press)

Collin-Soufrin, S. and Dumont, S., 1986, Astron. Ap. 166, 13

Davidson, K. and Netzer, H., 1979, Rev. Mod. Phys. 51, 715

Ferland, G. J. and Netzer, H., 1979, Ap. J. 229, 274
Ferland, G. J. and Netzer, H., 1983, Ap. J. 264, 105
Ferland, G. J., Netzer, H. and Shields, G. A., 1979, Ap. J. 323, 382
Heckman, T. M., 1980, Astron. Ap. 87, 152
Joly, M., Collin-Souffrin, S., Masnov, J. L. and Nottale, L., 1985, Astron. Ap. 152, 282
Kallman, T. and Mushotzky, R. F., 1985, Ap. J. 292, 49
Keel, W. C., 1985, in "Astrophysics of Active Galaxies and Quasi-Stellar Objects (J. S. Miller, Edit.)
Krolik, J. H., McKee, C. F. and Tarter, C. B., 1981, Ap. J. 249, 422
Kwan, J. and Krolik, J. H., 1981, Ap. J. 250, 478
Lawrence, A. and Elvis, M., 1982, Ap. J. 256, 410
Maccacaro, T., Perola, G. C. and Elvis, M., 1982, Ap. J. 257, 47
MacAlpine, G. M., 1976, Ap. J. 204, 694
MacAlpine, G. M., 1985 (in "Astrophysics of Active Galaxies and Quasi-Stellar Objects", J. S. Miller, Edit.)
Malkan, M. A., 1983, Ap. J. 268, 582
Mathews, W. G., 1974, Ap. J. 189, 23
Mathews, W. G. and Capriotti, E., 1985, (in "Astrophysics of Active Galaxies and Quasi-Stellar Objects", J. S. Miller, Edit.)
Mushotzky, R. F., 1982, Ap. J. 256, 92
Netzer, H., 1985a, Ap. J. 289, 451
Netzer, H., 1985b, MNRAS, 216, 63
Netzer, H. and Davidson, K., 1979, MNRAS, 187, 871
Netzer, H., Elitzur, M. and Ferland, G. J., 1985, Ap. J. 299, 752
Netzer, H., 1986, MNRAS (in press)
Osterbrock, D. E., 1981, Ap. J. 249, 462
Penston, M. V. and Perez, E., 1984, MNRAS, 211, 33p
Peterson, B. M., Meyers, K. A., Capriotti, E. R., Foltz, C. B., Wilkes, B. J. and Miller, H. R., 1985, Ap. J. 292, 164
Reichert, G. A., Mushotzky, R. F. and Holt, S. S., 1986, Ap. J. 303, 87
Shakura, N. I. and Sunyaev, R. A., 1973, Astron. Ap. 24, 337
Shields, G. A., 1976, Ap. J. 204, 330
Snijders, M. A. J., Netzer, H. and Boksenberg, A., 1986, MNRAS, 222, 549
Veron, P., Lindblad, P. O., Zuiderwijk, E. J., Veron, M. P. and Adam, G. 1980, Astron. Ap. 87, 245
Wampler, E. J., Gąskell, C. M., Burke, W. L. and Baldwin, J. A., 1984, Ap. J. 276, 403
Wandel, A. and Yahil, A., 1985, Ap. J. Lett. 295, L1
Wills, B. J. and Browne, I. W. A., 1986, Ap. J. 302, 56
Wills, B. J., Netzer, H. and Wills, D., 1985, Ap. J. 288, 94
Wilson, A. S., 1986 (in "Observational Evidence of Activity in Galaxies", IAU Symp. No. 121)

IMPROVED ACCRETION DISK MODELS OF CONTINUUM EMISSION FROM ACTIVE GALACTIC NUCLEI

Wei-Hsin Sun and Matthew A. Malkan

Department of Astronomy
University of California, Los Angeles

ABSTRACT

We present calculations of the spectra of black hole accretion disks which may power Active Galactic Nuclei. We use a simple model which is more realistic than previous ones in two important ways. Firstly, we have accounted for the dominance of electron scattering opacity in the disk, by approximating the locally emerging flux with low-gravity non-LTE stellar atmospheres. This disk spectrum constructed with stellar atmospheres is harder than the disk spectrum from a sum of blackbodies. It also shows a moderately strong Lyman limit absorption edge, up to 0.25 dex in depth. Such an edge appears to have been observed in several intermediate redshift quasars. The Balmer absorption edge is somewhat weaker.

We have also included the strong general relativistic effects of gravitational focusing and disk inclination on the observed spectrum. The observed spectrum becomes significantly harder when the disk is viewed at higher inclinations. We find that the spectrum of an edge-on disk is approximately 5 times harder than that of the same disk viewed face-on. The inferred black hole masses and accretion rates change by comparable factors.

1. INTRODUCTION

Thermal radiation from an optically thick accretion flow onto a supermassive black hole is the most promising explanation for the strong optical and ultraviolet flux which dominates the energy output of quasars and active galactic nuclei (Malkan and Sargent 1982). The first comparison of disk spectra to broadband AGN energy distributions was made by Malkan (1983), who adopted the relativistic radial structure equations of Novikov and Thorne (1973) to calculate the flux produced by a disk at each radius. The integrated disk spectrum was calculated by summing blackbodies each with $T_{eff} = [F(r)/\sigma]^{1/4}$. It is assumed that the disk is optically thick, and that radial heat transport is negligible compared with heat loss through the surface. Since no knowledge of the internal vertical disk structure is required, disk spectra can be calculated without knowing the details of the viscosity. Spectra of face-on disks, corrected for the transfer effects of general relativity, were combined with a nonthermal power law, Balmer continuum and Fe II emission lines. Balmer continuum and Fe II line emission produce approximately 20% of the apparent 2400-3800Å continuum. The predicted spectra provided good fits to multi-wavelength UV/OPT/IR energy distributions (0.1 to 10 μm) of six quasars. However, in several cases the inferred accretion rate exceeded the Eddington limit for the fitted black hole mass by a factor of 1.2 to 2. Evidently the face-on blackbody disk models predict spectra which are too soft to provide a self-consistent fit to some of the observations.

Disks around nonrotating (Schwarzschild) holes with a dimensionless angular momentum-to-mass ratio $a/M=0$ or around rapidly rotating (Kerr) holes ($a/M=0.998$) both gave good fits to the data; in the latter case relativistic effects are especially important since the inner orbits of accreting materials are much closer to the hole. We adopt the extreme Kerr hole as more realistic, because accretion matter spins a black hole up to this limiting value in a relatively short time (Bardeen 1970).

W. Kundt (ed.), Astrophysical Jets and Their Engines, 125–128.

2. NON-LTE STELLAR ATMOSPHERES

In large accretion disks around massive black holes opacity is dominated by electron scattering, as in the atmospheres of hot supergiant stars. Flux is redistributed to shorter wavelengths, producing a spectrum with a color temperature higher than the effective temperature, especially in the far-UV. To make our disk spectra more realistic, we have replaced the local blackbody approximation with stellar atmosphere models, with surface gravities which are so low that they are nearly balanced by the outward force of radiation pressure (i.e., approaching the Eddington limit).

For temperatures from 15,000 to 55,000 K we use the non-LTE models of Mihalas (1972). However, we use Kurucz (1979) models with higher gravities (roughly 10 times larger than they should be at the Eddington limit) for effective temperatures from 5,500 to 12,500 K, since there were no better atmospheres available. Figure 1 compares the non-LTE atmosphere at an effective temperature of 55,000 K to a somewhat higher-gravity LTE model from Kurucz (1986) which includes line opacities. The two stellar atmospheres are significantly harder than the blackbody at high frequencies. The Kurucz model has a stronger Lyman jump due to the higher gravity, but is otherwise quite similar to the Mihalas model at this temperature. From 60,000 to 115,000 K we use the non-LTE atmosphere models from Husfeld et al. (1984). In these low-gravity models, the Lyman limit became quite weak and the HeII edge can even go into emission. We describe the spectra from hotter parts of the disk with an analytical approximation which simulates an electron-scattering atmosphere at high temperatures: $f_\nu = A\nu^2[\exp(h\nu/1.7kT)]$, with normalization $A=0.5(h/1.7kT)^3$. This approximation matches the existing models very well (Figure 2), and allows the estimation of spectra from disks with essentially the complete expected range of accretion rates and black hole masses. We included a total of 52 stellar atmospheres in our calculation. with effective temperatures ranging from 5,500 to 300,000K.

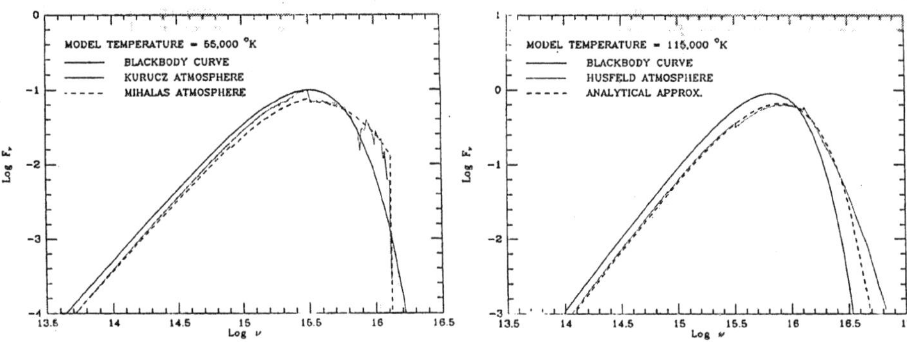

Figure 1. Stellar-atmosphere disk spectra are harder than local-blackbody disks.

Figure 2. The analytic curve gives a good approximation to the stellar atmosphere. Hydrogen and Helium II absorption edges become less prominent at high temperatures.

The disk based on stellar atmospheres is harder in the far-UV and soft X-rays than the sum of local blackbodies. At $\log \nu = 15.8$, their local luminosity is the same. At $\log \nu = 16.2$, the stellar-atmosphere disk produces twice as much luminosity as the blackbody disk. At $\log \nu = 16.4$ and 16.5, the differences are 30 and 200 times in local luminosity, respectively. Nevertheless, we find little difference in the best-fitting inferred black hole mass and accretion rate, between the predicted disk spectra with atmospheres and blackbodies, since the major differences are in the unobserved far-UV and soft X-rays. The biggest observable difference is the presence of a Lyman limit absorption edge which can be moderately strong, roughly 0.25 dex. As discussed below, such an edge appears to have been observed in several intermediate redshift quasars. The Balmer absorption edge is somewhat weaker, and more uncertain since it is formed primarily at temperatures for which we lack stellar atmospheres at sufficiently low gravities.

3. GRAVITATIONAL FOCUSING, DOPPLER SHIFTS, AND INCLINATION EFFECTS

Previous computations predicted the spectra of accretion disks viewed face-on. However, we should expect to observe accretion disks at various inclinations. The spectrum produced locally at the disk surface is transformed by gravitational focusing, Doppler shifts, and geometric inclination effects, which are strongest for observers at high inclination angles (Cunningham 1975).

For a fixed black hole mass of $5.5*10^{+8}$ M⊙ and accretion rate of 3 M⊙/yr (approximately 80% of Eddington rate) we varied the inclination angle from face-on (cos i=1) to edge-on (cos i=0). The resulting spectra, using the stellar-atmosphere

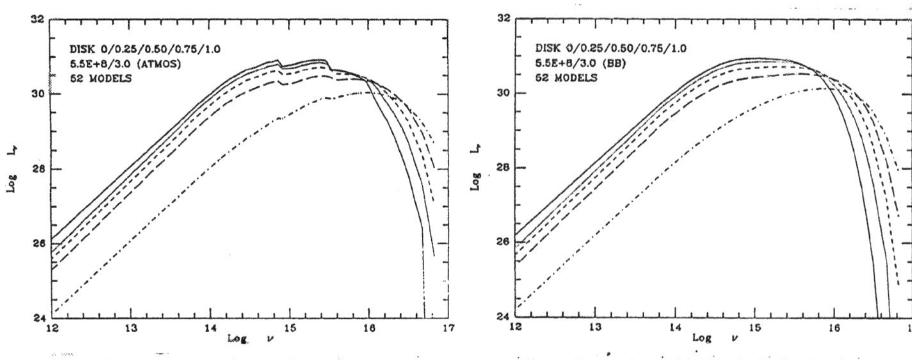

Figure 3a. An atmosphere disk viewed at different inclination angles. The edge-on disk is significantly harder than the face-on disk.

Figure 3b. A local-blackbody disk viewed at different inclination angles.

and local blackbody approximations, are displayed in Figure 3a and 3b, respectively. The apparent disk spectrum becomes harder when the disk is tilted from face-on to edge-on. At the lower frequencies ($\log \nu < 15$), the observed fluxes vary roughly as cos i. At a higher frequency ($\log \nu \sim 16$), the observed flux becomes independent of cos i. At even higher frequencies (up to soft X-rays), the more inclined disk appears more luminous. At a fixed accretion rate, the integrated energy is insensitive to inclination. However, the energy is redistributed to higher frequencies, and therefore the peak frequency shifts upward, when the disk is viewed at higher inclinations. From face-on to cos i=0.25, the entire disk spectrum is shifted up by roughly 80% in frequency. The spectrum of an edge-on disk is approximately 5 times harder than that of the same disk viewed face-on. This leads to comparable changes in the inferred black hole mass and accretion rate in fitting observed spectra with different inclinations. We also find that the Lyman limit absorption edge, which is prominent in a face-on case, is considerably weaker (even undetectable) in the inclined disk due to the smearing effect of Doppler shifts. The large strength of Lyman limit absorption edge might therefore be an indicator of disk inclination.

4. FITS TO THE OBSERVED SPECTRA

Figure 4 shows the fit to the observed continuum of 3C249.1 (1100+772, Z=0.311) with a face-on stellar-atmosphere disk with $\dot{M}$ = 0.6 M⊙/yr around a black hole with a mass of $4.8*10^{+8}$ M⊙. The observed spectrum appears to show a Lyman limit edge which is successfully modelled by the disk spectrum. Figure 5 shows the fit to 3C232 (0955+326, Z=0.533) with $M=1.2*10^{+9}$ M⊙ and $\dot{M}$=0.77 M⊙/yr using a blackbody disk. There is not such strong evidence of the Lyman edge in the observed spectrum. For both cases, the best-fit black hole mass and accretion rate are insensitive to the choice of blackbodies or atmospheres. However, these fits predict different far-UV and soft X-ray fluxes. Further, these fluxes are even more sensitive to the assumed disk inclination angle which also affects the strength of the Lyman edge. We find the best-fit accretion rates are 10% - 80% of the Eddington rate. We are currently in the process of fitting a larger sample of quasars.

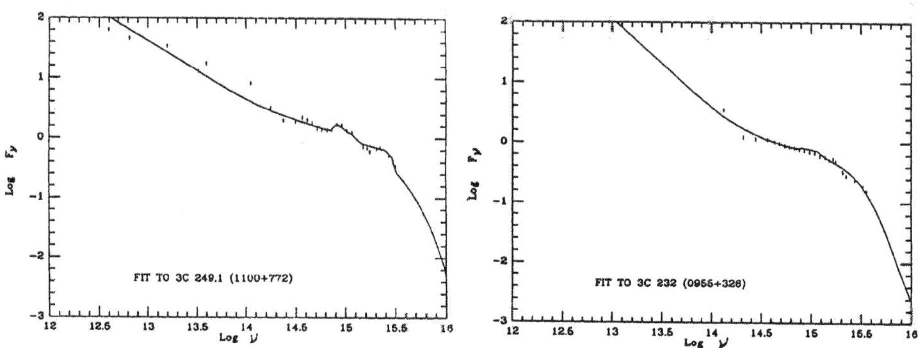

Figure 4. A face-on atmosphere-disk fit to 3C249.1.

Figure 5. A face-on blackbody-disk fit to 3C232.

5. CONCLUSIONS

(1) By (a) using non-LTE stellar atmospheres to describe the locally emerging disk flux and (b) accounting for the geometric and gravitational effects of disk inclination, we predict observed accretion disk spectra which are significantly harder than previous estimates. The peak frequencies may differ by a factor of 5 from face-on to edge-on disks.

(2) A moderately strong Lyman limit absorption edge may be present, especially in disks viewed face-on. This could be an indicator of the disk inclination angle.

(3) Reasonable fits to observed spectra with different assumed inclination angles can be obtained. However, the high-frequency tails of the spectra differ greatly. High-quality soft X-ray spectra are therefore very valuable in constraining the inclination angle.

6. REFERENCES

1. Malkan M.A. and Sargent W.L.W., Ap.J. 254, 22, 1982.
2. Malkan M.A. 1983, Ap.J. 268, 582.
3. Novikov, I.D., and Thorne, K.S. 1973, in "Black Holes", ed. C. Dewitt and B. Dewitt (New York: Gordon and Breach) p.343
4. Bardeen, J.M. 1970, Ap.J. 162, 71.
5. Mihalas, D. 1972, NCAR Technical Notes, "Non-LTE Model Atmospheres for B and O Stars", NCAR-TN/STR-76.
6. Kurucz, R.L. 1979, Ap.J. Suppl. 40, 1.
7. Kurucz, R.L. 1986, private communication.
8. Husfeld, D. et al. 1984, Astron. Astrophys., 134, 139.
9. Cunningham, C.T. 1975, Ap.J. 202,788.

VELOCITIES IN RADIO GALAXIES AND QUASARS

P.A.G. Scheuer
Mullard Radio Astronomy Observatory
Cavendish Laboratory
Cambridge CB3 OHE, U.K.

ABSTRACT: In this lecture I shall discuss the observational evidence relevant to the speeds of jets in radio galaxies and quasars. My conclusion is that the outer jets of weak (FR I) sources have non-relativistic speeds, while jets in strong (FR II) sources have speeds comparable with the speed of light, and perhaps very close to the speed of light.

1. STRONG AND WEAK RADIO SOURCES

There is a surprisingly sharp change between the morphologies of weak (radio power $P < 3.10^{25}$ W $Hz^{-1}sr^{-1}$) and strong radio sources. The difference was spelled out by Fanaroff & Riley (1974): the weak sources, Fanaroff & Riley's Class I, are brightest near the nucleus and become fainter and more diffuse at great distances; the strong sources (Fanaroff & Riley's Class II) are brightest near their extremities (where they generally have "hot-spots") becoming more diffuse nearer the centre. Since the work of Fanaroff and Riley, it has also become clear that the most recently accelerated population of relativistic electrons is found near the "hot-spots" in FR II sources, while the older populations, more affected by synchrotron loss, are nearer the galaxy or quasar; conversely, in FR I sources, the oldest electron populations are furthest from the galaxy. I shall often need to refer to these basic qualitative distinctions between FR I and FR II sources.

2. JETS ON LARGE AND SMALL SCALES.

The most direct evidence on the speeds of jets in extragalactic radio sources is the proper motion of structures in jets observed on the milli-arc-second scale by using Very Long Baseline Interferometry (VLBI). (See Fanti et al., 1984, and a summary of the current [1986.8] tally of superluminal sources by Porcas.) Using any conventional cosmological model to convert these proper motions into transverse velocities, we obtain very large speeds. The trouble is that we do not believe the answers, because they are many times c, the speed of light, even if we use a Hubble constant H of 100 km $s^{-1}Mpc^{-1}$. What most

W. Kundt (ed.), Astrophysical Jets and Their Engines, 129–136.

astronomers do believe is the "relativistic beaming hypothesis", i.e that the jets have velocities v close to c (and hence $\gamma \equiv (1-\beta^2)^{-\frac{1}{2}} \equiv (1-v^2/c^2)^{-\frac{1}{2}} \gg 1$), and that we observe preferentially the jets propagating at a small angle θ to the line of sight, because relativistic aberration and Doppler shift cause these to look much brighter than jets pointing in other directions. In agreement with this hypothesis, VLBI jets are observed to be one-sided; we should expect any relativistic counter-jet, pointing away from us, to appear extremely faint, and two jets in the plane of the sky would both appear very faint. According to the hypothesis, apparent transverse velocity is $v_{app} = v \sin\theta/(1-\beta \cos\theta)$, which can be as large as γv (reached when $\sin\theta = 1/\gamma$). With a few notable exceptions such as Prof. Kundt, astronomers are conservative creatures and would prefer to believe that, by the time the jets reach the less exotic environment of galactic scales and beyond (Kpc to Mpc), they consist chiefly of ordinary gas moving at ordinary speeds ($\beta \ll 1$), but this immediately leads to conflict with observation:

Problem 1 In powerful radio sources. i.e. Fanaroff & Riley's (1974) Class II, any large-scale jets that are observed are also one-sided. If the correct picture is

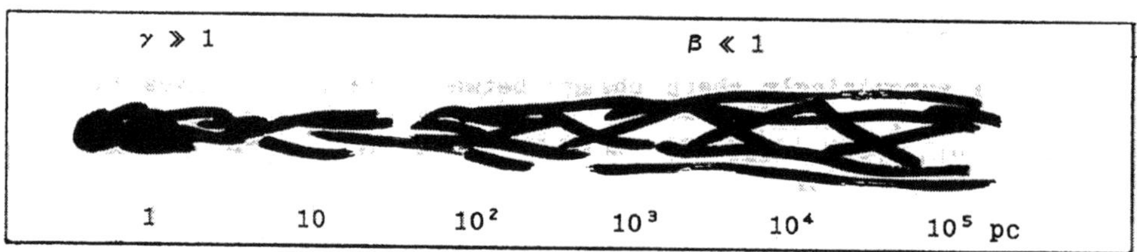

Figure 1. Decelerating jet on a logarithmic scale.

then the large-scale jets must be *intrinsically* one-sided. There are then two possibilities for the VLBI jets

(i) VLBI jets are also intrinsically one-sided. Then 50% of them point into the hemisphere away from us, and most of these will be too faint to detect. Therefore we should observe many sources with a large-scale jet but no VLBI jet. No such source is known.

(ii) VLBI jets are intrinsically two-sided; their observed one-sidedness is due to relativistic beaming. Then, in half the cases, the VLBI jet should point away from the large-scale jet. In fact, in every one of over 12 cases for which adequate data exist, the large-scale jet is a plausible continuation of the VLBI jet. (I have to choose words carefully because some of the jets appear to bend through large angles.)

Conclusion: *If VLBI jets are relativistic, then large-scale jets are relativistic too.*

Note that it is possible that the jets slow down somewhat from the large γ required to explain superluminal motions, but they must have a β of at least, say, 0.5 to account for the observed asymmetry of the large-scale jets.

3. LOWER BOUNDS TO JET SPEEDS: THE SPEEDS OF HOTSPOTS

Clearly the speed v_j of the jet (seen or unseen) must exceed the speed v_{adv} with which the hotspot at its end advances into the intergalactic medium. Thus it provides a modest lower bound for v_j.

(i) The observed radio emission from the hotspot requires at least the conventional minimum energy density u_{min} in relativistic electrons and magnetic fields, and hence exerts a pressure $\geq P_{min} = u_{min}/3$. The advance of the hotspot is determined by

$$\rho_{ext} v_{adv}^2 = \text{ram pressure} = P_{hotspot} \geq P_{min},$$

where the external density ρ_{min} can, in a few cases, be estimated fairly well from the thermal X-ray emission around the galaxy. For Cygnus A, Arnaud et al (1984) find ρ_{ext} corresponds to 5000 electrons m^{-3}, and conclude that $v_{adv} \geq 0.01c$. However, they use a value $u_{min} = 4.10^{-10}$ J m^{-3}; a more realistic value for the hotspots is 10^{-8} J m^{-3} (e.g.Hargrave & Ryle, 1976), leading to $v_{adv} \geq 0.05c$.

(ii) Pictures of the source at high and low radio frequencies provide estimates of the ages of the relativistic electron populations in various parts of the source. In Cygnus A, Winter et al. (1980) found that the 'synchrotron age' increased steadily from the hotpots towards the central galaxy: thus they were able to estimate the time taken for the source to grow to its present length and hence

$$\langle v_{adv} \rangle = \text{length/age} = 0.05c.$$

However, unlike the argument described in (i) above, this argument depends on the assumed value for the magnetic field; Winter et al. assumed equipartition between magnetic energy and relativistic electron energy. If one allows for a larger equipartition field when the source was younger, one should raise the estimate of mean growth rate to $\langle v_{adv} \rangle \simeq 0.1c$.

Notice the distinction between the two velocities:

v_{adv} = rate at which the hotspot pushes into the intergalactic medium;

$\langle v_{adv} \rangle$ = rate at which source length grows.

$v_{adv} > \langle v_{adv} \rangle$, because the jet may change direction (this is often called "precession", but I prefer not to imply a particular physical mechanism), and hence drill in turn into various places near the front of the source. I called this the "dentist's drill" model (Scheuer 1982), thereby probably doing a grave injustice to dentists. Notice also that neither of these velocities has any connection with the speed v_{flow} with which material flows through the hotspot.

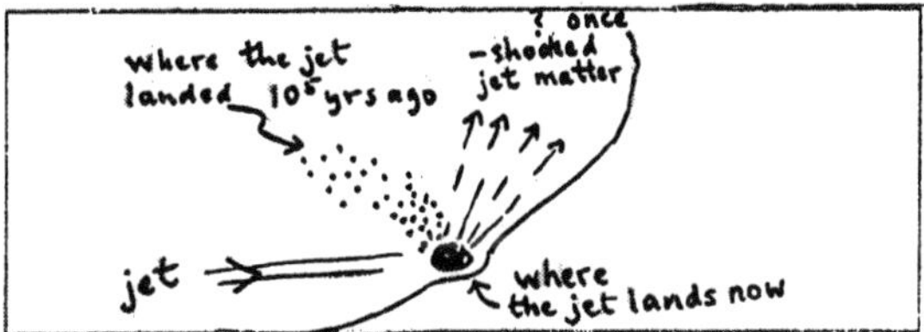

Figure 2. A possible interpretation of a hotspot.

4. UPPER BOUNDS TO JET SPEEDS

Several arguments have been used to set upper bounds to jet speeds; they are set out below, together with my doubts about their applicability.

(i) Bending. There are sources whose jets bend through very large angles. The observed brightness of a highly relativistic jet should change by orders of magnitude with angle to the line of sight; therefore this observation argues for non-relativistic speeds. The classic example is NGC 1265, a radio source in the Perseus cluster whose two-sided jets are swept back to leave a twin trail of radio emission. It and other similar sources are weak (FR I) sources. One FR II source whose jet has been traced continuously from pc to 100 kpc scales, bending through more than a right angle, is 3C120 (Walker, 1984). However, this is a superluminal source, whose jet probably starts close to the line of sight; the bending may therefore be grossly exaggerated by projection effects.

(ii) Two-sidedness. Even more basic is the fact that FR I sources are observed to have two jets of similar flux density on diametrically opposite sides of the nucleus.

As a result of (i) and particularly of (ii) above, it is plain that *the jets of FRI sources have non-relativistic speeds*, probably with $\beta < 0.3$.

There remains a question mark over the inner portions of FR I jets, which share many of the properties of FR II jets (magnetic field preferentially along the jet, small opening angle). If one wants to tell a unified story one would regard these portions as relativistic too; they would then pass through a strong shock, beyond which the jet becomes much slower, entrains some of the ambient gas, broadens out, and ultimately becomes a buoyant plume rather than a jet. Maps of "bent double" sources such as 3C465 (Eilek et al., 1984) suggest such a story particularly strongly (see Leahy, 1984). There is then the question whether these inner portions of FR I jets show the large flux density ratios between the two sides that one should expect for relativistic jets; I think that is an open question at present. A particular worry is that the bending of the jets in NGC1265 takes place mostly in the inner, very narrow, parts of those jets. One can argue (and some do) that some kind of friction leaves slowly-moving radio-emitting matter near the surfaces of relativistic jets, so that we can expect a little radio emission even from transverse or receding relativistic jets. This argument gains a little added plausibility in the case of FR I sources, since the ex-jet material moves on beyond the supposed strong shock, leaving the jet exposed to ambient gas, whereas in FR II sources the ex-jet material remains behind or flows backwards to shield the jet in a "cocoon" of very low density.

What of FR II jets? These are usually much less prominent and highly asymmetrical; indeed, in most cases they are observable on one side of the nucleus only. All this is just what one would expect of a randomly oriented set of intrinsically two-sided relativistic jets. However, physical arguments have been advanced in favour of non-relativistic speeds for FR II jets also, and they are given in (iii) and (iv) below.

(iii) Arguments from ram pressure. The decrease in percentage polarization of a jet at longer wavelengths is generally attributed to Faraday rotation. This yields an estimate for $\int n_e B_\parallel ds$ through the source (see my second lecture) and hence a lower bound for the electron density n_e if we assume that the magnetic field $B_\parallel$ along the line of sight is at most that corresponding to equipartition between magnetic and relativistic particle energy.

This procedure commonly yields a particle density n_e in the jet that is not much less than the particle density in clusters of galaxies as determined from their thermal X-ray emission. Pressure balance between the internal pressure of a hotspot and the ram pressures hitting it from the two sides requires

$$\rho_j(v_j-v_{adv})^2 = \rho_{ext}v_{adv}^2.$$

If we conclude from the depolarization measurements that ρ_j is not very different from ρ_{ext}, then v_j is not much greater than v_{adv}, and thus, using the estimates of v_{adv} from (i) and (ii) of section 3 we conclude that the jet is non-relativistic.

(iv) Lobe-filling arguments. All of the energy supplied through the jet in the lifetime t of the source must be accomodated in the "lobes", for very little can be radiated away. If the jet carries mass flux $\dot{m} = \rho_j \pi r_j^2 v_j$ (where ρ_j, r_j and v_j are the density, radius and speed of the jet), the energy delivered to the lobes is $\frac{1}{2}\dot{m}tv_j^2$, and therefore the gas pressure in the lobes is

$$\frac{2}{3}\left[\frac{\frac{1}{2}\dot{m}v_j^2 t}{V_{lobe}}\right] .$$

Balance between the pressure P_j in the jet and the surrounding lobe then requires

$$P_j = \frac{1}{3}\frac{\rho_j \pi r_j^2 v_j^3 t}{\pi r_{lobe}^2 \langle v_{adv}\rangle t}$$

(where we have substituted $\pi r_{lobe}^2 \langle v_{adv}\rangle t$ for the volume V_{lobe} of the lobe), and therefore

$$v_j^3 = \frac{3\,P_j}{\rho_j}\left[\frac{r_{lobe}}{r_j}\right]^2 \langle v_{adv}\rangle .$$

We can take estimates of P_j, ρ_j, (r_{lobe}/r_j) and $\langle v_{adv}\rangle$ respectively from minimum energy arguments based on the brightness of the jet, depolarization of the jet, radio maps, and the considerations of section 3. If we further assume that the lobes are in pressure balance with surrounding cluster gas, we can substitute the gas pressure (or an upper limit on it) from X-ray observations for P_j in the above formula. In this way, one obtains jet speeds $v_j \approx 0.1c$. These estimates are very insensitive to the assumptions, since the formula determines v_j^3. On the other hand, the assumptions may be grossly in error. I doubt very much whether there is pressure balance between the radio lobes and the surrounding gas (in powerful sources the lobes probably drive shocks outwards), but I should not expect the difference to be as large as a factor 1000. However, both (iii) and (iv) depend on a jet density derived from radio depolarization, and that may be totally wrong; as explained in my lecture on polarization, it is now almost certain that the depolarization is due to matter *outside* the jet.

5. THE POTASH-WARDLE JET.

Many quasars have very bright large-scale jets, which must have a large internal pressure, for the internal pressure must be at least P_{min}, one third of the minimum energy density derived from synchrotron radiation theory. This raises an awkward question, as pointed out by Potash & Wardle (1980) who had mapped the jet in 4C 32.69.

Problem 2. What prevents the jet from expanding sideways under its own pressure?

There are two possibilities:

(i) The jet is "free", so that only its own inertia restrains its expansion. Thus, if it has pressure P_{min} at some place, then one sound crossing time downstream it will have an expansion speed $\approx (P_{min}/\rho_j)^{½}$. Therefore the opening angle θ of the jet is at least $1/M_j$, where M_j is the Mach number of the jet. Now ram pressure balance at the far end of the jet requires that

$$\rho_{ext} v_{adv}^2 \approx \rho_j v_j^2 \approx P_j M_j^2 > P_{min}/\theta^2$$

where θ is the measured opening angle of the jet. For the jet mapped by Potash and Wardle this required either an embarrassingly large value of the intergalactic gas density ρ_{ext} or a relativistic value of v_{adv}.

(ii) The jet is confined by some external pressure. As Potash & Wardle pointed out, this cannot be the pressure of hot intergalactic gas, for that would again require embarassingly high densities, embarrassing because they are excluded by the lack of observable thermal X-rays at the source position. The standard resolution of this problem is that proposed by Potash & Wardle, that the jet carries a current, and the jet is thus confined sideways by hoops of magnetic field. (Dr Achterberg and Dr Benford will probably treat this mechanism more thoroughly.) There are at least two other quite natural explanations. The first is that the jet is roughly in pressure balance with the surrounding "cocoon" of very hot, shocked, ex-jet material, which is however *not* in pressure balance with the intergalactic gas, but rather drives a shock front into it. That is entirely satisfactory dynamically, but begs the question why the radio luminosity per unit volume should be so much greater in the jet than in a lobe at the same pressure: the particle acceleration processes would have to be more efficient (or produce a harder energy spectrum) in the jet, and most of the lobe pressure would have to be due to thermal or relatively low-energy particles. Of course this question applies not only to strong jets such as those in 4C 32.69 and 3C280.1, but, in a milder form, to every source in which we can see a large-scale jet. The second is that the jet is relativistic and pointing towards us; the intrinsic radio power is then much less than flux density x (distance)^2, and the computed P_{min} is a gross overestimate. Potash & Wardle considered this possibility, but concluded that making the jet relativistic led to at least as great a momentum flux. The appendix to their paper shows that this result comes from associating a proton with each radiating electron (to preserve electrical neutrality); the required number of protons moving with relativistic speed gave a large momentum flux. Their difficulty goes away if the jet consists of positron-electron plasma. This alternative interpretation has recently become more attractive because of the independent evidence for relativistic speeds in large-scale jets outlined in the next section.

6. THE CORRELATION BETWEEN DEPOLARIZATION AND LARGE-SCALE JETS.

Robert Laing has discovered a remarkable correlation. Since this lecture is about speeds I must mention it, even though Dr Laing will explain it himself in greater detail. What he finds is that, if a source contains a jet, the lobe on the source containing the jet depolarizes less than the lobe opposite the jet. It is possible to make up an *ad hoc* theory in which a jet removes plasma, by unknown means, from an extensive region around itself. However, the only tolerably coherent interpretation I know is Dr Laing's, which is that the jet is on the side of the source nearer to us, and that radiation from that side of the source therefore passes through less of the thermal plasma surrounding the source than radiation from the far side. In turn, the only reason why all observed jets should be on the side of the source facing us is that the jets are relativistic. Note that (as in Section 2 of this lecture) it is not necessary for the large-scale jets to have Lorentz factors of 5 or 10 to account for the asymmetry.

He also finds that very compact hot-spots occur on only one side of a source, if at all, and that where a jet has been found in the same source the compact hot-spot is on the same side of the source as the jet. (In a very few cases, such as Cygnus A, where there is an indication of a large-scale jet on both sides of the nucleus, the compact hotspot is on the same side as the more prominent jet.)

7. CONCLUSIONS

(i) The jets of FR II radio galaxies and quasars are relativistic up to and including the compact hot-spot. Superluminal motion provides the evidence for relativistic speeds on the 1-10 parsec scale, and the 1:1 correlation between depolarization and the side of the source on which a jet is observed provides independent evidence on 1-1000 kpc scales, while the very tight correlation explained in Section 2 ties the two scales together. The large-scale jets could have speeds as low as about 0.5c, but nothing strongly contradicts the simplest hypothesis that they continue with the same Lorentz factors that superluminal motions would indicate.

(ii) The same is probably true for some part of the jets of FR I sources close to the nucleus, but the direct evidence for this is weak, and will be hard to improve since the relevant inner parts of the jets are generally very faint. The outer, two-sided jets of roughly equal flux density correspond in some ways to the lobes of FR II sources, but quite probably entrain much more ambient gas. Their ages can be estimated by attributing the progressive steepening of the radio spectrum to synchrotron losses; the velocities estimated in this way are much less than c.

(iii) It is interesting to note that relativistic speeds for large-scale jets lead to excessive momentum flux for the jets of certain powerful sources if we require one proton mass per electron contributing to the synchrotron radiation. We shall have to take jets made of pair plasma seriously.

REFERENCES

Arnaud K.A., Fabian, A.C., Eales, S.A., Jones, C. and Forman, W.:1984. Mon. Not. R. astr. Soc., 211, 981.

Eilek, J.A., Burns, J.O., O'Dea, C.P. and Owen, F.N.:1984. Astrophys. J., 278, 37.

Fanaroff, B.L. and Riley, J.M.:1974. Mon. Not. R. astr. Soc., 167, 31P.

Fanti, R., Kellermann, K. and Setti, G. (eds.):1984. IAU symposium No. 110, 'VLBI and compact radio sources'.

Hargrave, P.J., and Ryle, M.:1976. Mon. Not. R. astr. Soc., 175, 481.

Leahy, J.P.:1984. Mon. Not. R. astr. Soc., 208, 323.

Porcas, R.: 1987. To be published in the proceedings of a workshop on superluminal radio sources held at Big Bear, California to celebrate Marshall Cohen's 60th birthday.

Potash, R.I. and Wardle, J.F.C.: 1980, Astrophys. J. 239, 42.

Scheuer, P.A.G.:1982. In IAU symposium No. 97,'Extragalactic radio sources', eds. Heeschen, D.S. and Wade, C.M.

Walker, R.C.:1984. In 'Physics of energy transport in extragalactic radio sources', NRAO workshop No.9, eds. Bridle, A.H. and Eilek, J.A.

Winter, A.J.B., Wilson, D.M.A., Warner, P.J., Waldram, E.M., Routledge, D., Nicol, A.T., Boysen, R.C., Bly, D.W.J. and Baldwin, J.E.:1980, Mon. Not. R. astr. Soc., 192, 931.

POLARIZATION AND MAGNETIC FIELD STRUCTURE

P.A.G. Scheuer
Mullard Radio Astronomy Observatory
Madingley Road
Cambridge CB3 OHE. U.K.

1. REVIEW OF THE BASIC PHYSICS

1.1. The polarization of synchrotron radiation

An ultrarelativistic electron (Lorentz factor $\gamma \equiv (1-v^2/c^2)^{-1/2} >> 1$) radiates seven times as much power with the E-vector perpendicular to the static magnetic field $\mathbf{B}_0$ as in power with the E-vector parallel to $\mathbf{B}_0$ (or the projection of $\mathbf{B}_0$ on the plane of the sky). However, the power in the two polarizations is distributed differently in frequency, so the ratio at a single frequency may not be quite 7:1. Averaged over a population of electrons distributed in energy according to the power law

$$dN = K\gamma^{-(2\alpha+1)}\, d\gamma \tag{1.1}$$

(which produces a radiated spectrum: $S_\nu \propto \nu^{-\alpha}$) the ratio of linearly polarized powers at a high frequency is $(3\alpha+4):1$. The "percentage polarization", measuring $100(S_{max}-S_{min})/(S_{max}+S_{min})$, is then $100(3\alpha+3)/(3\alpha+5)$; for α= 0.5, 1, 1.5 it is 69%, 75%, 79% respectively.

All these statements are independent of the angle between $\mathbf{B}_0$ and the line of sight, because a Lorentz transformation in the $\mathbf{B}_0$ direction will always take us to a frame in which the line of sight is perpendicular to $\mathbf{B}_0$. They are also almost independent of the pitch angle distribution of the electrons, because only electrons within about $\pm 1/\gamma$ radians of the line of sight contribute to the radiation towards us*. In fact, we normally expect the momentum distribution to be almost isotropic (in a

* Similarly, each electron radiates elliptically polarized radiation in our direction, but the right-handed elliptic from electrons passing just to one side of the line of sight is cancelled (to within ~ 1 part in γ) by the left-handed elliptic from electrons passing just to the other side of the line of sight. Circular polarization of a few parts per 1000 is detected from the cores of quasars, but it is not clear to what extent this represents intrinsic circular polarization at the $1/\gamma$ level and to what extent it is due to propagation effects.

W. Kundt (ed.), Astrophysical Jets and Their Engines, 137–149.

frame in which the magnetic field pattern is quasi-static) since markedly anisotropic distributions suffer numerous micro-instabilities.

1.2. Stokes parameters and the complex polarization parameter

While monochromatic radiation from a single source must have a specific elliptic polarization, radiation over a small but finite frequency band from a number of incoherent sources (zillions of relativistic electrons in our case) is only partially polarized. It is easy to see that the power measured by any polarimeter is a weighted sum of terms like $\langle E_a^*(t)E_b(t)\rangle$, (time averages over products of wave fields in two specified directions normal to the line of sight) and that these, in turn, can all be expressed in terms of such products referred to just two perpendicular directions, x & y:

$$\langle E_x^* E_x\rangle,\ \langle E_y^* E_y\rangle,\ \text{and}\ \langle E_x^* E_y\rangle.$$

Thus all the observable polarization properties are specified by the ***Stokes parameters***

$$\left.\begin{aligned} I &\equiv C.\tfrac{1}{2}\, \langle E_y^* E_y + E_x^* E_x\rangle \\ Q &\equiv C.\tfrac{1}{2}\, \langle E_y^* E_y - E_x^* E_x\rangle \\ U &\equiv C.\tfrac{1}{2}\, \langle E_y^* E_x + E_x^* E_y\rangle \\ &= \text{"Q for axes turned through } 45^\circ\text{"} \\ V &\equiv C.\tfrac{1}{2}\, \langle E_y^* E_x - E_x^* E_y\rangle \end{aligned}\right\} \qquad (1.2)$$

We shall want to use a constant C such that I represents W m^{-2} Hz^{-1} of total flux density. Q (U) describe the excess of linearly polarized power in P.A. 0° (45°) over that in P.A. 90° (135°). V, the real part of the product of E_y with iE_x, the perpendicular field $\lambda/4$ away, represents the circular polarization; we shall neglect it from now on as it is small for synchrotron sources.

For axes x', y' rotated through angle θ, it is easy to show ($E_x' = E_x \cos\theta - E_y \sin\theta$, etc) that

$$I' = I;\ Q' = Q\cos2\theta - U\sin2\theta,\ U' = U\cos2\theta + Q\sin2\theta\ .$$

Therefore it is sometimes convenient to use a complex polarization parameter

$$P \equiv Q + iU,$$

for which $P' = P\, e^{2i\theta}$. (1.3)

1.3 Faraday rotation

Propagation through a non-relativistic plasma (n_e electrons m^{-3}) containing a magnetic field B_o causes rotation of the plane of polarization through angle θ radians:

$$\theta = (\pi\lambda^2/c^3) \int \nu_p^2\ \nu_g\ B_\parallel\ dr = 0.81\ \lambda^2 \int n_e\ B_\parallel\ dr \qquad (1.4)$$

provided that the frequency c/λ is much larger than the plasma frequency ν_p and the gyro frequency ν_g. Here λ is the wavelength and $B_\parallel$ is the component of $\mathbf{B}_0$ along the element dr of the line of sight; in the numerical formula, λ, B, dr are in metres, Gauss, parsec respectively.

In a cluster of galaxies $n_e \simeq 10^2\ m^{-3}$; in a radio galaxy the magnetic field might be 10^{-5} Gauss, so we might get $\theta \sim 100\lambda^2 \sim 1$ radian at $\lambda = 10$ cm over a 10 kpc line of sight. If however, the magnetic field is much smaller outside the source and n_e is much smaller inside the source, and/or the magnetic field $B_\parallel$ has many reversals of direction - and all of these are likely - we expect much smaller rotations. Faraday rotation has two obvious consequences:

1) The observed polarization direction does not immediately tell us the preferred direction of the magnetic field in the source. However, by observing at several short wavelengths we can extrapolate to zero wavelength, where Faraday rotation vanishes, since $\theta \propto \lambda^2$. This works well on real observations.

2) The percentage polarization diminishes at longer wavelengths, because linearly polarized radiation at different depths (more precisely at different *Faraday depths* $0.81 \int_{\text{emitter}}^{\text{observer}} n_e B_\parallel\, dr$) into the source has been rotated by different amounts.

In practice we also have averaging over the finite beam size of the observations, resulting in further cancellation. For a given antenna array, the beam size is larger at longer wavelengths, so beware of subjective impressions of the trend of fractional polarization with wavelength; nevertheless, the trend is there when one examines sources observed at different wavelengths with the same angular resolution. Similarly, the mean polarization over whole sources (or large portions of them) is typically 0 to 10%; more recently, high resolution maps have shown polarizations more typically $\geq$25% in one beam area.

The course of depolarization with λ^2 should tell us about the distribution of n_e through the source. This does <u>not</u> work well on real observations.

1.4 What Brian Burn said, and what good and what harm has been done with it.

The most cited paper in this field is probably Burn (1966). Adding up contributions dP to the complex polarization parameter P at various Faraday depths τ_F along a line of sight we get

$$P = \int dP\, e^{-2i\tau_F\lambda^2} \tag{1.5}$$

so for a given model of the distribution of magnetic field B_0, n_e, and synchrotron emissivity we can predict $P(\lambda)$ for a given line of sight. This is essentially what Burn did, for a variety of geometries simple enough to allow analytic evaluation of (1.5):-

(i) a uniform slab with uniform $\mathbf{B}_0$

(ii) the same with a field $\mathbf{B}_0 + \mathbf{B}_r$, where $\mathbf{B}_0$ is uniform and $\mathbf{B}_r$ is random with some given $<|B_r|^2>$. (For spectral index $\alpha = 1$ this gives the same distribution of synchrotron emissivity as a stretched tangled field such as we shall consider later.)

(iii) the same as (i) and (ii), averaged over all the lines of sight in a spherical source.

In these models, the fractional polarization is roughly constant and the rotation $\propto\lambda^2$ until the rotation through the whole source is $\sim 90°$. At longer λ, the fractional polarization decreases (not necessarily monotonically) and the observable rotation angle wavers around the same value between 0° and 45° - very reasonably, since only the front part of the source, at rotations up to 45°, contributes to the polarized flux.

Adding up the synchrotron emissivities is straightforward but we have to be careful to understand *exactly* what Burn's model for Faraday depolarization is, since the scale length of B_r is crucial: scale length a over a distance L gives a rotation corresponding to an *effective* $B_{\parallel} \approx B_r\sqrt{a}/L$. Burn's calculation assumes $a\to 0$ and $B_r\to\infty$, keeping $B_r\sqrt{a}$ finite; an ensemble of rotations at a given depth into the source then has a Gaussian distribution. The problem is then tractable, but obviously unrealistic. I have tried to calculate what happens with a power-law hierarchy of scales of B, and while it is not difficult to predict what happens (e.g. for different beam sizes) in the high frequency limit, I have so far failed to get useful answers with Faraday rotation.

Burn also made a brave attempt to invert the Fourier transform relation (1.5) between $P(\lambda^2)$ and $dP/d\tau_F$, and to apply it to the Crab nebula, but as we do not know $P(\lambda^2)$ for $-\infty<\lambda^2<0$ he had to make symmetry assumptions which are unreal for any particular source.

Burn's models are very useful if treated properly, as a selection of illustrative models. They have sometimes been misused as 'predictions of the theory' rather than predictions for particular field geometries. For other kinds of field geometry, we can get quite different behaviour.

A prime example is the Crab nebula, where the low-density synchrotron-emitting nebula is criss-crossed by complex filaments with much higher n_e (optical line emission). A distribution of large Faraday rotations in the thicker filaments causes depolarization in some places at cm-wavelengths, leaving most of the source unaffected. At longer λ, thinner filamentary structures become important, and a larger part of the source is affected; this happens over a 100: 1 range in λ^2, throughout which almost all of the polarized radiation comes from parts of the nebula unaffected by Faraday rotation. Indeed, the observed rotation obeys $\theta\propto\lambda^2$ quite well, as would be expected for 'foreground' rotation by the interstellar medium.

2. PROGRESS AND REGRESS

2.1. Different kinds of ordered field.

If we take an isotropic tangle of field lines embedded in plasma (like a bowl of spaghetti) and stretch it (with two forks) or shear it (e.g. wind it around a fork) or squash it we get an anisotropic field geometry in which e.g. $\langle B_x^2 \rangle \gg \langle B_y^2 \rangle$ or $\langle B_z^2 \rangle$. but positive and negative B_x are equally common. Such a field will produce strongly polarized synchrotron radiation. but not much Faraday rotation - quite unlike Burn's mixture of uniform B_0 and random B_r . where B_0 will produce systematic Faraday rotation and is responsible for all the polarized emission. The fundamental difference between what I shall call *distorted random field* and truly *unidirectional field* is now much more widely appreciated than it used to be.

2.2 What is the field geometry in radio sources? I - Speculation

We are hampered by two things (at least!):

(i) Even the best observations do not distinguish clearly between *unidirectional* and *distorted random* field.

(ii) We have no theory of magnetic field generation. There is a large literature on dynamo theories, i.e. the generation of *unidirectional* field in systematically rotating bodies like Earth or Sun. However, these have no obvious relevance to bodies like radio galaxies. in which expansion by a factor $>10^6$ from the 'central engine' ensures that any initial angular velocity is reduced to less than one turn per source lifetime.

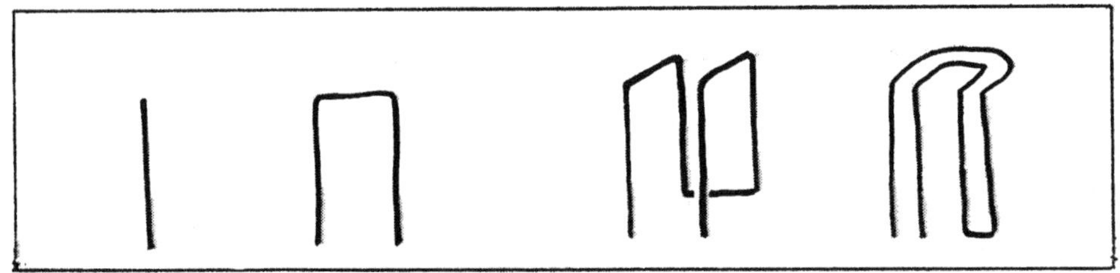

Figure 1.
Folding a magnetic field.

When I have asked distinguished experts about magnetic field amplification in turbulence *without* systematic rotation, they have generally shaken their heads thoughtfully and said little (cf. Moffatt 1978; sections 7.11 & 11.4). However, we can make some attempt to see what happens if we start with a bucket (size a_0) of almost perfectly conducting fluid (density ρ) permeated by a small seed field B_0, and we start to stir it with a wooden spoon, giving it speeds v_0 on the largest scale a_0. Eddies will fold the field over and simultaneously decrease its scale size: initially, we might expect the field to increase 4 times when the linear scale size is halved (folding about an axis parallel to B does not increase $|B|$ [see Fig.1] while folding about either of the two perpendicular axes doubles $|B|$). If B_0 is extremely small, it will soon be folded into such small scales that ohmic dissipation occurs, and we shall be left with not much more than the initial B_0. (Even large eddies eventually wind up B into small structures.) If B_0 is not quite so small, the folding to smaller scales stops at the scale size a where $B_a^2/2\mu_0 \approx \frac{1}{2}\rho v_a^2$, i.e. the magnetic forces suffice to stop the eddies.

Here is a specimen calculation - not to be taken too seriously.

We work one octave at a time.

Suppose at scale size a (i.e. range $a\surd 2$ to $a/\surd 2$)

field $\quad B_a/B_0 = (a_0/a)^\beta \quad$ (our initial guess is β=2)

eddy velocity $\quad v_a/v_0 = (a/a_0)^{1/\kappa} \quad$ (κ=3 for Kolmogoroff spectrum)

Therefore growth stops at $a = a_1$ such that

$$\frac{B_a^2}{2\mu_0} \approx \rho\, v_a^2 \tag{2.1}$$

or, in terms of the initial densities of magnetic and kinetic energy

$$u_{mag,o}\left[\frac{a_0}{a_1}\right]^{2\beta} \approx u_{K.E.,o}\cdot\left[\frac{a_1}{a_0}\right]^{2/\kappa}$$

$$\therefore\quad \left[\frac{B_a}{B_0}\right]^2 = \left[\frac{a_0}{a_1}\right]^{2\beta} = \left[\frac{u_{K.E.,o}}{u_{mag,o}}\right]^{\frac{2\beta}{2\beta + 2/\kappa}}$$

$$u_{mag} = (u_{mag,o})^{\frac{1}{\beta\kappa+1}}\,(u_{K.E.,o})^{\frac{\beta\kappa}{\beta\kappa+1}}\ . \tag{2.2}$$

For β = 2 and Kolmogoroff turbulence, this means

$$u_{mag} = (u_{mag,o})^{1/7}(u_{K.E.,o})^{6/7} \tag{2.3}$$

and if , for example, we started out with $u_{mag,o} = 10^{-14}\, u_{K.E.,o}$, then $u_{mag} = 10^{-2}\, u_{K.E.,o}$ and $a_1 = 10^{-3}a_0$. Of course this calculation is incorrect: apart from uncertainties in β and κ, some process (MHD "turbulence"?) must carry away the power input to scales small enough for dissipation; meanwhile, larger eddies (with bigger K.E. density) continue to wind up the field over a longer timescale. All the same, our attempt suggests that, within a few turnover times of the largest scale eddies, there will be a significant build-up of magnetic energy, plausibly to $10^{-3\pm1.5}$ of the kinetic energy. De Young's (1980) numerical work resulted in magnetic energy approaching equipartition after ~10 turnover times. If there is no further input of K.E., the magnetic field will decay, smallest scales first, with field line reconnection playing a larger part over somewhat longer timescales.

It is easy to imagine this kind of turbulent field generation in the jets of FR I radio galaxies, which look as if lot of entrainment were going on, and in the backflows in the lobes of FR II sources, and in the Rayleigh-Taylor unstable contact discontinuities in supernova remnants. It is much harder to see how it would work in the jets and hotspots of FR II sources.

What do observations tell us, for sources in which we claim to know something about the gas motion? In the sun, the dynamo process seems to dominate. but that is hardly surprising, since the time-scales are sufficiently leisurely to let field be expelled from small-scale structures. In shell-type supernova remnants turbulent field generation seems the best (or merely the only) candidate theory, and seems to work within the efficiency range our crude calculation suggested.

Two things are clear:- however the field is made, the fluid flow (into which the field is frozen) will distort the field, and the theme of much of the rest of this lecture will be that the polarization structure, in the high frequency limit, depends mostly on this distortion, and little (if at all) on the origins of the field. On the other hand, the rotation and depolarization depend strongly on the scale of field reversals, and hence on the origin of the field.

2.3 What is the field geometry of radio sources? II - observations. (See Bridle & Perley, 1984)

2.3.1 FR I sources.

Figure 2. Schematic diagram of preferred field directions in FR I sources

The preferred magnetic field direction is generally transverse to the jets, with 10-40% polarization (high frequency limit). However, in many cases the part of the jet nearest the nucleus has the magnetic field preferentially along the jet; the portion of jet with preferentially longitudinal field tends to grow with radio power, and takes over completely in FR II sources. In addition, there are indications that an FR I jet has some longitudinal field around its surface. All these features are illustrated by the magnificent jet in NGC 6251, on which Perley, Bridle & Willis (1984) made a comprehensive set of radio observations.

Most of this is easy to understand. Even if the jet propagates adiabatically at constant speed, conservation of longitudinal magnetic flux means that (longitudinal B) $\propto r^{-2}$ (r = radius of jet), while conservation of transverse magnetic flux means that (transverse B) $\propto r^{-1}$. Thus transverse magnetic field dominates more and more with increasing distance from the nucleus and hence increasing r. In fact, it is likely that FR I jets behave more like buoyant plumes, entraining external gas, slowing down, shearing at the edges (hence longitudinal field around the surface of the jet), and stirring the internal field, thus preventing total dominance of transverse magnetic field and at the same time helping to keep up its magnitude.

What is not yet quite clear to me is whether the observations of polarization *require* internal turbulence. Though the polarization is often strong, it is generally at the 25(±15)% level rather than the 70% that we should expect for a field stretched predominantly in one direction. If we look at a mat of tangled field stretched equally in the two directions transverse to the jet, we get the full fractional polarization $(3\alpha+3)/(3\alpha+5)$ only if we see the mat edge-on; at other viewing angles we see weaker polarization, calculated by Laing (1980). At 45° inclination, it is less than half of the maximum. Now we do not know at what angle we are viewing any given jet, so we must either resort to statistics (which generally produces many papers but little progress) or we can look for strong variations in the magnitude or direction of polarization within a short length of one jet. NGC 6251 seems to show such variations (Fig.15 of Perley et al., 1984), and this is prima facie evidence for internal stirring. But there are other interpretations: (i) there is clearly some longitudinal field near the surface of the jet; if the proportions of surface brightness due to surface and interior vary from place to place (as they almost certainly do) the fractional polarization will vary drastically; (ii) it is not certain that the 1662 MHz observations of Perley el al.'s Fig. 15 are free of depolarization effects, though they state that they find no evidence for depolarization between 1662 and 4885 MHz.

2.3.2 FR II sources.

The jets in FR II sources generally contribute only a small fraction of the total flux density (except in some quasars) and there are fewer detailed polarization measurements of them. They appear to have predominantly longitudinal fields. Their very narrow opening angles, and the fact that they are relatively dim and seem to preserve much of their momentum until they hit the external gas at a 'hotspot', strongly suggest that they have little interaction with their surroundings (except, perhaps, for a pressure,thermal or magnetic, that restrains them laterally). Therefore, although one can mutter words like 'shear', I do not think we can claim to understand the orientation of these fields - nor, for that matter, the preferentially longitudinal fields in the parts of FR I jets closest to the nucleus.

Such information as we have about the polarization of the hotspots themselves is consistent with a field preferentially transverse to the jet direction. Why am I so cautious about the data on the brightest and most obvious features of radio galaxies? It is because the polarization is often not particularly strong, and because there are still not many cases in which the hotspot structure is clearly resolved, so that we can be confident that we are seeing the polarized emission of something distinct from and largely uncontaminated by the polarized emission from the front of the 'lobe'. Perhaps the most detailed hotspot maps so far are VLA maps of Cygnus A at 0.1" resolution, which have 10-20 beamwidths across the two brightest 'hotspots' (depending on how one defines 'hotspot'). Here we see, in both Np and Sf hotspots, something of the order of 25% polarization, with electric vectors roughly along the axis of the source. In addition, we see indications of a tangential magnetic field around the leading edges of each hotspot, although it must be added that in the case of the Sf hotspot the structure with the tangential field looks like a detached filament.

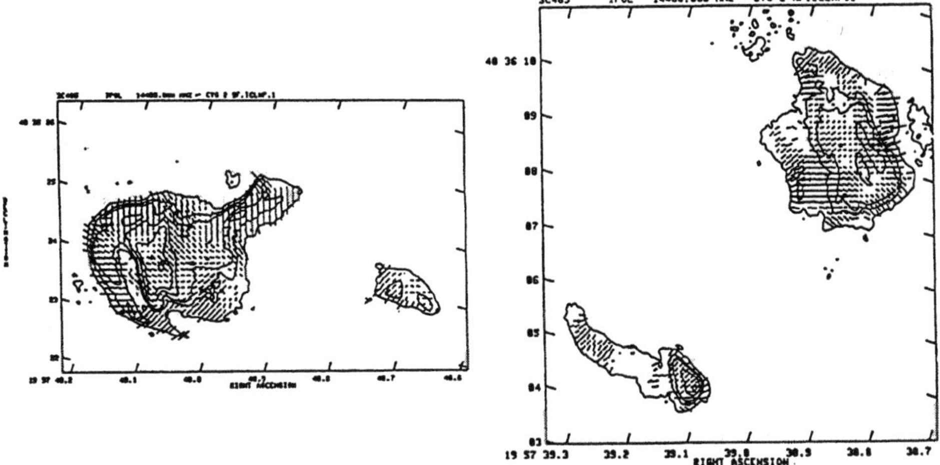

Figure 3. The polarization of hot spots in Cygnus A.

Once one goes beyond the hotspots, the structures and the polarization patterns are very varied and often complicated. Just one feature emerges very persistently in a large number of FR II sources: around the periphery of the radio lobes, the electric vectors tend to point outwards, like the spines of a hedgehog, implying that the magnetic field is either predominantly tangential, or (as Laing (1980) suggested) random field compressed onto the surface. We shall look into the interpretation of these patterns in more detail in section 2.5.

2.4 Where does depolarization occur?

Most of the systematic rotation of the polarization of a source occurs in the interstellar medium of our Galaxy. There is abundant evidence for that, in the striking variation of the magnitude of rotation measure with galactic latitude and the variation of its sign with galactic longitude.

It has been argued that the depolarization at low frequencies could also be attributed to small-scale structure in the Galactic rotation measure; the apparent depolarization would then be due to averaging, within the telescope beam, over portions of the source whose radiation was rotated through different angles. I have been very sceptical of this view, because there is no striking variation of depolarization with galactic latitude. Even more compelling is the fact that nearby sources with angular diameters of arc minutes do not obviously depolarize more than distant sources of the same type with angular sizes of arc seconds. However, recent measurements have shown that at low galactic latitudes, small-scale structure in galactic clouds may indeed be the most important cause of depolarization (Simonetti & Cordes, 1986).

This leaves two alternatives: either the depolarization is due to Faraday rotation within the source (as, for instance, in Burn's models), or else it is due to differential rotation within a telescope beam, but by magnetized plasma around the source (hot cluster gas?), not plasma in our Galaxy. The simplest assumption is that the depolarization is internal to the source, for we know that there must be a significant magnetic field there. For fairly uniform magnetic fields, whose magnitude corresponds to equipartition between magnetic energy and energy in relativistic electrons, that assumption leads to estimates of electron density comparable with the electron densities in clusters of galaxies as derived from X-ray observations. Higher densities are needed if magnetic field is mostly in very small-scale structures, with many reversals through the source. Recently there has been a tendency to think rather of depolarization by material around the source. In part this is due to observations like those of Perley et al. on NGC6251, where they find small-scale variations of rotation measure across the jet (their Fig. 21) which could well account for the depolarization found in earlier investigations made with lower angular resolution. That does not *prove* that the depolarization is due to external gas; small-scale magnetic field structure inside the jet would produce similar effects, though accompanied by some depolarization along each line of sight. However, the notion that the depolarization is due to external plasma is strongly reinforced by purely dynamical considerations which indicate that, at least in FR II sources, the density within the radio source is very much smaller than that of the surrounding gas. It is further reinforced by Laing's anti-correlation between depolarization and a jet on the same side of the source, discussed in his lecture and in my first lecture. That leaves us - i.e. you - with the problem of accounting for a substantial magnetic field in the gas around a radio galaxy, though perhaps rather smaller than the 'equipartition' field inside the source because the path length can be one or two orders of magnitude longer.

2.5 Numerical simulations of magnetic field and polarization structure, and some thoughts provoked by them.

The numerical simulations I am going to describe are based on the idea mentioned earlier, that the distortion of whatever field is convected down the jet, or made (by processes as yet unguessed at) in or just after the (presumed) shock at its end, may account for the polarization structure seen in the lobes of FR II sources, without requiring us to know much about the origin or nature of the input field. To make such a simulation, we first need a model of the fluid flow in a FR II source; several groups have computed such models, with codes of varying complexity and grids of various sizes, but they agree in their broad features. (Later I shall discuss the very few differences that are relevant here.) We used axisymmetric simulations computed by Anthony Williams , in which we implanted marker particles to follow the distortion of the matter. Thus one can see how a little cube containing frozen-in magnetic field distorts into a parallelopiped containing the same flux sheared appropriately, and how the initial relativistic particle spectrum changes because of the volume change (we assumed adiabatic changes). One can also work out the synchrotron emission from the distorted box, and, by averaging over the synchrotron emissions from a random distribution of *initial* magnetic field directions, one can predict the synchrotron emission from the distorted box if the initial field had no preferred direction. Having no better guide from theory, we did indeed compute what the polarization structure would be if the field started as an isotropic tangle of field lines. This turns out to be fairly easy only if the spectral index $\alpha=1.0$; we believe that this restriction is not at all critical, and introduces uncertainties small compared to most of the other unknown inputs. The question remains where to introduce the magnetic field (and the relativistic electrons). We have computed models in which either or both are introduced into the jet as it enters the computing grid, or just after the main shock. There is some physical justification for introducing the particles at the shock; the only reason for trying both alternatives for the magnetic field is the desire to find out how much difference the compression at the shock makes to the computed polarization structure. We assumed throughout that only the material from the jet (and not the shocked IGM) contributes to the radio emission.

The calculations were made by Alan Matthews. The diagram shows one of the results obtained when both field and particles were injected with the jet. Several features emerge very clearly:

(i) Even without particle acceleration at the shock, the hotspot dominates the map - simply because of the compression of particles and fields at the main shock.

(ii) Qualitatively, the simulation reproduces the commonly observed features of polarization strikingly well: the electric vectors stick out all around the periphery, and the magnetic field inside the hotspot is transverse to the jet axis.

(iii) Quantitatively, the simulation does too well: the computed polarizations are close to the uniform-field percentages over much of the map. This suggests that either there is some turbulence in the lobes, or that the anisotropy of the field is reduced by field line reconnection on very small scales (or both).

(iv) One of the features which is not immediately obvious from these diagrams, except by a strong limb brightening of the lobes, is the very large azimuthal magnetic field around the surface of the lobes. The cause is obvious: a very small loop around the jet axis becomes a large loop on the surface of the lobe as the jet is turned inside-out, and the azimuthal field is thus amplified by arbitrarily large factors.

(v) There is even an indication of the strange double structure observed in both hotspots of Cygnus A. The explanation appears to be related to the amplification of azimuthal field just mentioned: much of the emission from the hotspot comes from a ring just beyond the main shock, and as the field in this ring is largely azimuthal, it appears less bright where the predominant field is close to the line of sight.

To what extent should one believe these simulations? The fluid dynamical code used, though it contains flux corrections, still admits

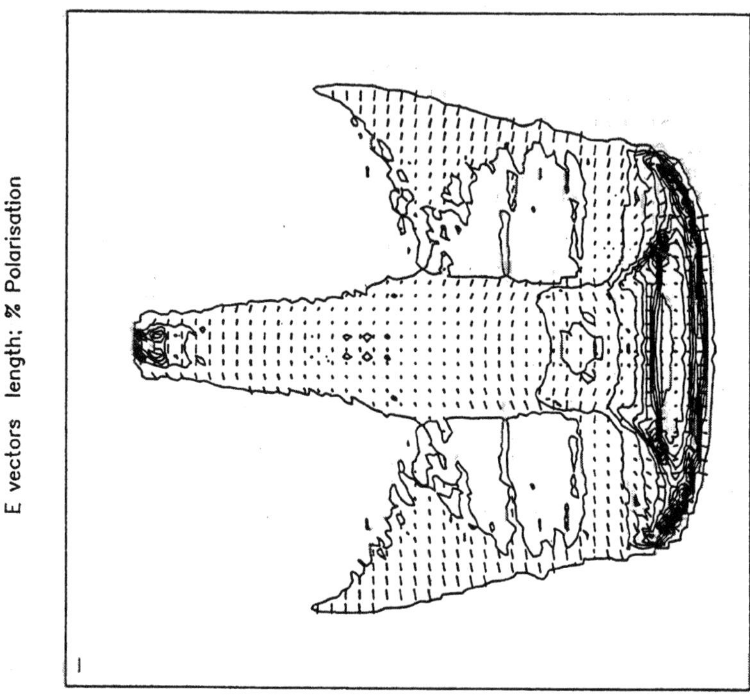

Figure 4. Polarization map predicted for a model source, viewed at 80° to the source axis. The model is a Mach 10 jet with the same initial pressure as and inital density 0.01 of the (uniform) ambient intergalactic gas. The small '|' near the bottom left-hand corner represents 100% polarization.

some momentum diffusion, and we must therefore beware of trusting the magnitude of the poloidal shear. (Tests to investigate this are in progress.) Similarly, the momentum diffusion inhibits Kelvin-Helmholz type instabilities at the contact discontinuity.

However, there is no way of blaming the computing for the azimuthal stretching of field. Our first reaction was to say that this is what one deserves for using precisely axisymmetric models, but on reflection I believe that it is a topological property of streamlines passing close to a stagnation point. It can be avoided only if the departure from axisymmetry is so great that the whole jet does a U-turn. We should also note that the stretching factor $\propto$(initial radius)$^{-1}$ does not produce large fields for a jet containing a quasi-uniform current, for the circulation of **B** around a small loop is then proportional to the initial area of the loop. However, if the jet is filled with small-scale fields, we have the possibility of generating a thin sheath in which the field strength is limited only by dissipative effects like reconnection, and by the field becoming dynamically important. I am not prepared to say that this is a route to the magnetic self-confinement of jets! More plausibly. the localized production of relatively large magnetic fields leads naturally to filamentary structures such as those that appear in the famous VLA map of Cygnus A (Perley et al.. 1984).

REFERENCES

Burn. B.J.:1966. Mon. Not. R. astr. Soc.. **133**. 67.

Bridle. A.H. and Perley. R.A.:1984. Ann. Rev. Astron. Astrophys.. **22**. 319.

De Young. D.S.:1980. Astrophys. J., **244**. 81.

Laing. R.A.:1980. Mon. Not. R. astr. Soc.. **193**. 439.

Moffatt. K.H.:1978. 'Magnetic field generation in electrically conducting fluids'. C.U.P.

Perley. R.A.. Bridle. A.H. & Willis.A.G.:1984. Astrophys. J. Suppl., **54**. 291.

Perley, R.A.. Dreher. J.W. and Cowan. J.J.:1984. Astrophys. J.. **285**. L35.

Simonetti. J.H. and Cordes. J.M.:1986. Astrophys. J., **303**. 659.

Do you really think that all jets have relativistic speeds?

NUMERICAL SIMULATION OF JETS

S.A.E.G.Falle
Department of Applied Mathematical Studies
University of Leeds
Leeds LS2 9JT

ABSTRACT. The reliability of published numerical simulations of fluid jets is assessed. It is shown that the resolution used in these calculations is much too low for the results to be of any use.

1. INTRODUCTION

In recent years a number of numerical simulations of fluid jets have appeared in the astrophysical literature (Norman, Smarr & Winkler 1982; Wilson & Scheuer 1983; Williams & Gull 1984; Smith, Norman, Winkler & Smarr 1985; Falle & Wilson 1985; Wilson & Falle 1985; Wilson 1987). The purpose of this work is to discover whether the morphology and other observable features of extragalactic jets can be understood in terms of a fluid model. Unfortunately this is not an easy thing to decide. We only see the synchrotron emitting electrons and have no direct way of observing the actual jet material or even deciding whether or not the jets behave like a fluid. As these issues are discussed elsewhere (see the paper by Kundt in this volume), I will not consider them here, but instead concentrate on the extent to which the simulations are reliable. This is a question which has been largely ignored in the Astrophysical literature although it is known from terrestrial experience that such simulations can be very misleading at the low numerical resolution used in most of the calculations (Glaz, Collela, Glass & Deschambault 1985). In order to illustrate some of the difficulties associated with such calculations I will first consider a simple one dimensional problem for which an analytic solution is known and then go on to discuss some of the special problems that arise in jet simulations.

W. Kundt (ed.), Astrophysical Jets and Their Engines, 151–161.

2. ONE DIMENSIONAL LINEAR ADVECTION

Consider the equation

$$\frac{\partial u}{\partial t} + \frac{\partial u}{\partial x} = 0 \quad , \tag{1}$$

which represents passive advection of a conserved quantity u . The characteristics are

$$x = t + \text{const.}$$

i.e. straight lines with slope 1 . If the initial conditions are

$$u(x,0) = u_o(x) \ ,$$

then the solution is

$$u(x,t) = u_o(x-t) \ .$$

Note that $u(x_1,t)$ depends only on u_o for $x < x_1$, i.e. information only propagates in the direction of increasing x .

2.1 Difference Approximations

To solve (1) numerically, we first set up a grid with spacings h,τ in the x and t directions respectively and define

$$u_{jk} = u(jh,k\tau) \ .$$

Since this is an evolution problem, a difference scheme must calculate u_{jk+1} for all j given u_{jk} for all j . A Taylor expansion gives

$$u_{jk+1} = u_{jk} + \tau\left(\frac{\partial u}{\partial t}\right)_{jk} + O(\tau^2)$$

$$= u_{jk} - \tau\left(\frac{\partial u}{\partial x}\right)_{jk} + O(\tau^2) \ , \tag{2}$$

since from (1)

$$\frac{\partial u}{\partial t} = -\frac{\partial u}{\partial x} \ .$$

Note that this is first order accurate in time. (2) provides an explicit algorithm for calculating u_{jk+1} provided we replace the x derivative by a suitable difference approximation.

Let us consider three such approximations:

a) Downwind $$\left(\frac{\partial u}{\partial x}\right)_{jk} = \frac{1}{h}(u_{j+1k} - u_{jk}) + O(h) \ , \tag{3a}$$

b) Upwind $$\left(\frac{\partial u}{\partial x}\right)_{jk} = \frac{1}{h}(u_{jk} - u_{j-1k}) + O(h) \ , \tag{3b}$$

c) Centred $$\left(\frac{\partial u}{\partial x}\right)_{jk} = \frac{1}{h}(u_{j+1k} - u_{j-1k}) + O(h^2) \ . \tag{3c}$$

All these are consistent, i.e. the discretisation error tends to zero as $h \to 0$. At first glance c) would appear to be superior to the others since it is a second order approximation. However, a) and c) are unconditionally unstable, which means that oscillations with wavelength $\sim 2h$ grow exponentially on a timescale τ. The solution given by such schemes does not tend to the exact solution as $h, \tau \to 0$.

2.2 Truncation Errors

Some insight into this behaviour can be gained by looking at the truncation error. The three schemes are actually higher order approximations to the equations

a) Downwind $$\frac{\partial u}{\partial t} + \frac{\partial u}{\partial x} = -\frac{1}{2}(h + \tau)\frac{\partial^2 u}{\partial x^2} \ , \tag{4a}$$

b) Upwind $$\frac{\partial u}{\partial t} + \frac{\partial u}{\partial x} = \frac{1}{2}(h - \tau)\frac{\partial^2 u}{\partial x^2} \ , \tag{4b}$$

c) Centred $$\frac{\partial u}{\partial t} + \frac{\partial u}{\partial x} = -\frac{\tau}{2}\frac{\partial^2 u}{\partial x^2} + \frac{1}{6}(\tau^2 - h^2)\frac{\partial^3 u}{\partial x^3} \ . \tag{4c}$$

Equations (4a) and (4c) contain a negative dissipation term for all positive values of τ and h. This is the reason for the instability of these schemes. Equation (4b) has positive dissipation for $h > \tau$, and negative dissipation if $h < \tau$. This is consistent with the fact that the scheme is stable if

$$\text{Courant number} = \frac{\tau}{h} < 1 \ . \tag{5}$$

This is called a Courant condition and some such restriction on τ will apply to all explicit schemes.

The upwind scheme can be shown to be convergent, that is $u_{jk} \to u(x,t)$ as $h \to 0$ provided the Courant condition is satisfied. However, for practical values of h, the dissipation term leads to diffusion of the solution as shown in figure 1.

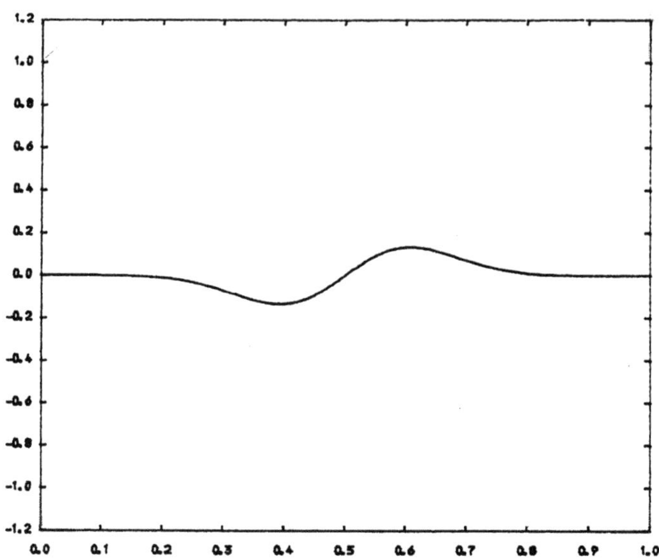

Figure 1. Single period of sine wave with initial height 1.0 and width 0.2 advected through 20 times its width with the upwind scheme at a Courant number of 0.5. (Courtesy of VNU press, from Falle, 1985)

2.3 Lax-Wendroff

The upwind scheme is only first order accurate, and so is too diffusive when used with practical values of h. This error can be reduced by using a higher order scheme.

The Lax-Wendroff scheme can be constructed by first calculating values at $j + \frac{1}{2}$, $k + \frac{1}{2}$ from

$$u_{j+\frac{1}{2}\,k+\frac{1}{2}} = u_{jk} + \frac{h}{2}\left(\frac{\partial u}{\partial x}\right)_{jk} + \frac{\tau}{2}\left(\frac{\partial u}{\partial t}\right)_{jk}$$

$$= u_{jk} + \frac{(h-\tau)}{2}\left(\frac{\partial u}{\partial x}\right)_{jk} \tag{6}$$

One can now approximate $(\partial u/\partial x)_{jk}$ by

$$\left(\frac{\partial u}{\partial x}\right)_{jk} = \frac{1}{h}(u_{j+1k} - u_{jk}) ,$$

and calculate u_{jk+1} from

$$u_{jk+1} = u_{jk} - \frac{1}{h}(u_{j+\frac{1}{2}\,k+\frac{1}{2}} - u_{j-\frac{1}{2}\,k+\frac{1}{2}}) .$$

With this scheme, u_{jk} can be shown to be a higher order approximation to

$$\frac{\partial u}{\partial t} + \frac{\partial u}{\partial x} = \frac{1}{6}(\tau^2 - h^2)\frac{\partial^3 u}{\partial x^3} - \frac{\tau}{8}(h^2 - \tau^2)\frac{\partial^4 u}{\partial x^4} . \qquad (7)$$

The first term on the right hand side is dispersive, while the second is dissipative if $h > \tau$. Hence the stability condition is again the Courant condition (5). The scheme is second order accurate in space and time, but the dispersive term dominates. For finite values of h this leads to ripples in regions where the gradients are large. This effect can be seen in figure 2. However, if u is smooth then the Lax-Wendroff is superior to the upwind.

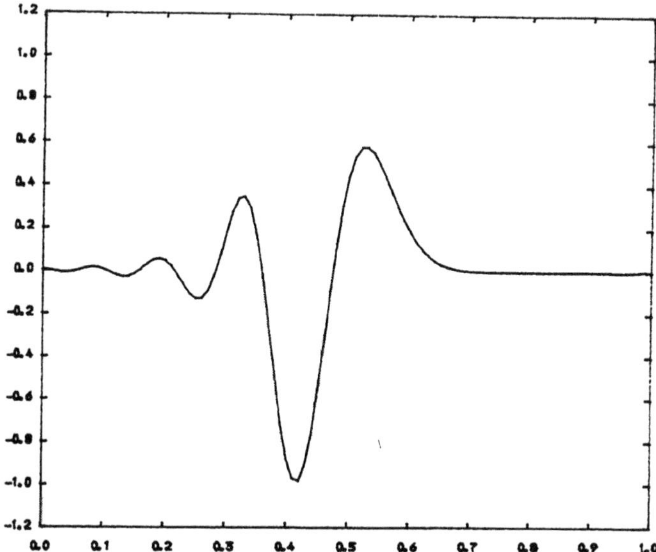

Figure 2. Single period of a sine wave with initial height 1.0 and width 0.2 advected through 10 times its width with the Lax-Wendroff at a Courant number of 0.5. (Courtesy of VNU press, from Falle, 1985)

2.3 Boundary Conditions

One of the difficulties with high order schemes is that they often need more boundary conditions than the exact equation. For example, suppose we want to solve (1) with a Lax-Wendroff in $0 \leqslant x \leqslant Jh$ (J is the number of mesh points). Then we must specify u_{J+1k} for all

k since the calculation of u_{Jk} requires the value of u_{J+1k}. But in the exact equation the value of u for $x \geq Jh$ has no influence on u for $x \leq Jh$.

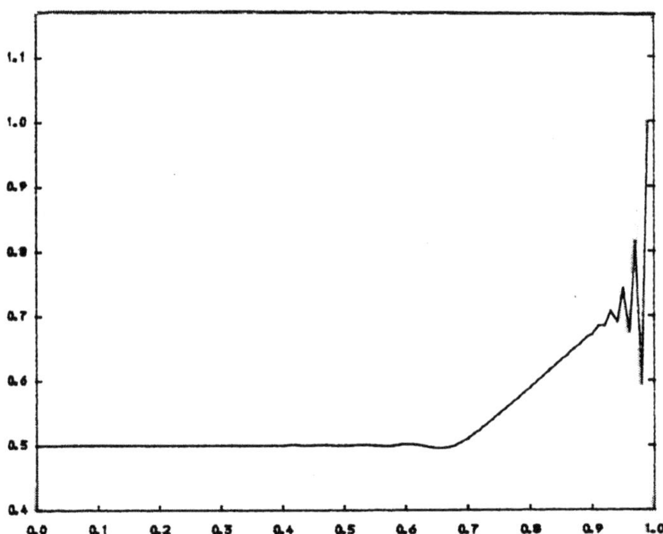

Figure 3. A ramp of width 0.6 connecting states with $u = 0.5$ and $u = 1.0$ advected off the grid at $x = 1.0$ with a Lax-Wendroff at a Courant number of 0.25. (Courtesy of VNU press, from Falle, 1985)

The consequences of this can be seen in figure 3, where we have imposed the boundary condition $u = 1.0$ at $x = 1.0$. An oscillatory error is introduced which decays away from the boundary like

$$e^{(j-J)\tau/h}$$

Provided $\tau/h \sim 1$, this is not a serious problem, but there are many situations where the Courant number at the boundary has to be small because of restrictions in other parts of the flow.

In contrast, the upwind scheme does not need u_{J+1k} and so only a boundary condition at $x = 0$ is required, just as for the exact equation. This is one of the reasons why upwind schemes are to be preferred.

2.4 Monotonicity

A scheme is said to be monotonic if

$$u_{jk} \text{ monotonic } \Rightarrow u_{jk+1} \text{ monotonic} .$$

This is an important property, firstly because it is true for the exact equation, and secondly because in non-linear problems it is necessary to ensure that unphysical solutions (such as rarefaction shocks) do not appear.

For schemes which are the same at all points, Godunov's theorem applies. This states that for linear advection only first order accurate schemes can be monotonic. This is illustrated by the fact that the upwind scheme, which is first order, is monotonic, while the second order Lax-Wendroff is not (see figure 2).

For a second (or higher) order scheme to be monotonic, there must be some switch which makes the scheme first order at points where it is not monotonic. This can either be done directly by introducing logical switches into the scheme, or by introducing some function which automatically reduces the order of the scheme at the relevant points.

2.4.1. Direct switching. In this approach some criterion is used to switch to a first order scheme at the appropriate points. Flux corrected transport (FCT) devised by Boris and Book (1976) is a scheme of this type. The algorithm consists of the following steps.

i) Compute a first guess u_{jk+1} using a monotonic first order scheme.

ii) u_{jk+1} can be used to calculate a second order correction.

iii) Correct u_{jk+1} to get u_{jk+1} which is second order accurate.

iv) Limit the second order correction if u_{jk+1} is not monotonic.

Although this idea is fundamentally sound, the limiter used in stage iv) is very ad hoc. This means that while FCT is excellent at propagating discontinuities, it tends to produce steps in smooth regions.

2.4.2 Averaging. Van Leer (1977,1979) uses a non-linear averaging function to reduce the order in regions where a second order scheme would be non-monotonic. This idea can most easily be illustrated by applying it to the Lax-Wendroff. From (6) we have

$$u_{j+\frac{1}{2}\, k+\frac{1}{2}} = u_{jk} - \frac{(h - \tau)}{2} \left(\frac{\partial u}{\partial x}\right)_{jk} .$$

Now set

$$\left(\frac{\partial u}{\partial x}\right)_{jk} = \frac{1}{h} \operatorname{av} (u_{j+1k} - u_{jk},\, u_{jk} - u_{j-1k}) ,$$

where $\operatorname{av}(a,b)$ is the averaging function. It has the properties

i) $\mathrm{av}(a,a) = a$,

ii) $\mathrm{av}(\alpha a,\alpha b) = \alpha\,\mathrm{av}(a,b)$,

iii) as $\dfrac{|a-b|}{|a+b|} \to 0$ then $\mathrm{av}(a,b) \to \frac{1}{2}(a+b)$

$\to \infty$ then $\mathrm{av}(a,b) \to a$ if $|a| < |b|$
$\to b$ if $|a| > |b|$.

This switches the scheme to first order when $\partial^2 u/\partial x^2$ is large.

2.4.3 A multigrid idea. Although both FCT and averaging do produce usable schemes, the switching criteria are ad hoc and have undesirable side effects. An alternative method of switching which uses estimates of the size of the truncation error is described by Falle (1985).

3. GASDYNAMICS

The one dimensional (Eulerian) equations are

$$\frac{\partial \underline{u}}{\partial t} + \frac{\partial \underline{f}}{\partial x} = 0 \qquad (8)$$

where

$$\underline{u} = \begin{pmatrix} \rho \\ \rho v \\ \dfrac{p}{(\gamma-1)} + \dfrac{1}{2}\rho v^2 \end{pmatrix}, \qquad \underline{f}(\underline{u}) = \begin{pmatrix} \rho v \\ p + \rho v^2 \\ v\left(\dfrac{\gamma p}{(\gamma-1)} + \dfrac{1}{2}\rho v^2\right) \end{pmatrix} \qquad (9)$$

For flows with p and v constant we get linear advection.
(8) can be written

$$\frac{\partial \underline{u}}{\partial t} + A\frac{\partial \underline{u}}{\partial x} = 0 ,$$

where the matrix A is the Jacobian of $\underline{f}$ with respect to $\underline{u}$

$$A_{ij} = \frac{\partial f_i}{\partial u_j} .$$

The eigenvalues of A are $v-a$, v, $v+a$ where a is the adiabatic sound speed. These eigenvalues are the characteristic speeds, so the characteristics are

$$\frac{dx}{dt} = v \pm a , \qquad \frac{dx}{dt} = v .$$

A genuinely upwind scheme diagonalises A locally and uses upwind differencing on each of the resulting equations. It is important to distinguish between a genuinely upwind scheme and donor cell advection since the latter merely uses upwind differencing for advection. It is only possible to impose the correct boundary conditions if the scheme is upwind.

4. THE TROUBLE WITH JETS

4.1 Vortex Rings

Strictly axisymmetric simulations of jets which are less dense than their surroundings suggest that there is a vortex ring at the jet head (Smith et.al. 1985). It seems that this vortex ring is periodically shed so that the flow near the head of the jet is not steady in the frame of the bow shock.

Now the induced speed v_i of an incompressible vortex ring is given by

$$v_i = \frac{C}{4\pi R}\left(\log \frac{8R}{a} - \frac{1}{4}\right) .$$

(Fraenkel 1970). Here C is the circulation, R is the radius of the vortex ring and a is the radius of the vortex core within which the fluid rotates as a solid body. If the boundary of the jet is a slip line of zero thickness, then a will be zero in the absence of viscous effects and the induced velocity will be infinite. So for incompressible flow the speed of the vortex and hence the flow near the jet head will be determined by the width of the shear layer at the boundary of the jet and the magnitude of the viscosity. Neither of these are correctly modelled in the simulations since both the width of the shear layer and the effective viscosity depend upon the numerical resolution.

For incompressible flow one would therefore have to use the correct physical viscosity to model the flow correctly. Fortunately for compressible flow the situation is not so bad. Moore (1985) has shown that compressibility reduces the speed of the vortex because it leads to a low density region in the vortex core. To get the correct vortex speed one therefore does not have to use the physical viscosity, which would in any case be impossible for these very high Reynolds' numbers. However, the numerical resolution must be high enough to ensure that there is a sonic line in the vortex core, because only then will the behaviour of the vortex be independent of the mesh size.

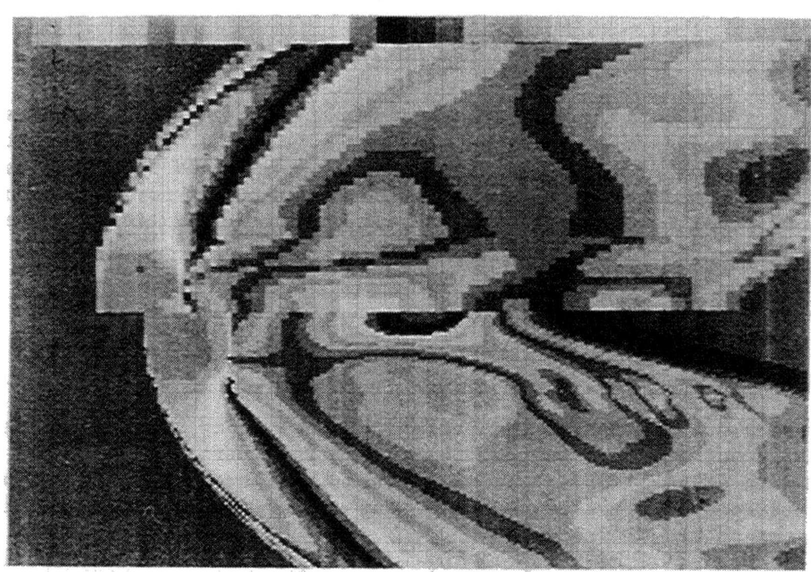

Figure 4 Time dependent simulation of a jet with initial Mach number 6 (relative to the jet fluid) and a density 0.1 times that of the external medium. 10 cells across jet (top), 20 cells across jet (bottom). Calculated using a first order Godunov scheme.

The effect of numerical resolution can be seen in figure 4 which shows the pressure contours at the head of a light jet for various values of the mesh spacing. As the mesh size is reduced, the vortex becomes more intense and as a result both the shape of the bow shock and the nature of the backflow changes. The highest resolution shown in figure 4 is twice that usually used in simulations, but it is nevertheless inadequate since there is no sonic line in the vortex core. In fact a reliable calculation would require at least double this resolution, that is 40 cells in a jet radius at input.

4.2 Turbulent Boundary Layers

At the high Reynolds' numbers that occur in Astrophysical jets the boundary between the jet and its surroundings is almost certainly turbulent. The rate at which this boundary layer grows depends upon the jet Mach number and the density contrast. If the Mach number and the external density are large enough, then the boundary layer grows slowly and it is quite possible for its thickness to grow more slowly than the jet radius. In that case the jet is not turbulent, but the flow at the working surface can still depend on the size of the boundary layer if it is large enough to affect the size of the vortex core. To correctly model the flow at the end of the jet one must therefore take into account the behaviour of the jet between the source and the working surface. To do this with the kind of numerical resolution required does not seem possible with present algorithms.

5. CONCLUSIONS

I have tried to indicate that the numerical modelling of jets is by no means straightforward. In fact most of the published simulations are worse than useless since the results are both quantitatively and qualitatively incorrect. The only sensible way to resolve these difficulties is to use modern numerical techniques and to check the results by comparison with terrestrial experiments whenever possible. Fluid models may or may not be appropriate for Astrophysical jets, but we cannot expect to make any progress if we use incorrect fluid dynamics to interpret the observations.

6. REFERENCES

Boris, J.P. & Book, D.L., 1976, J.Comp.Phys., 20, 397.

Falle, S.A.E.G., 1985, Cosmical Gas Dynamics, Ed. F.D.Kahn, VNU Press, p.149.

Falle, S.A.E.G. & Wilson, M.J., 1985, Mon.Not.R.astr.Soc., 216, 79.

Fraenkel, L.E., 1970, Proc.Roy.Soc.Lond., A316, 29.

Glaz, H.M., Collela, P., Glass, I.I. & Deschambault, R.L., 1985, Proc.Roy.Soc.Lond., A398, 17.

Moore, D.W., 1985, Proc.Roy.Soc.Lond., A397, 87.

Norman, M.L., Smarr, L. & Winkler, K-H.A., 1982, Astron. & Astrophys., 113, 285.

Smith, M.D., Norman, M.L., Winkler, J-H.A. & Smarr, L., 1985, Mon.Not.R.astr.Soc., 214, 67.

Williams, A.G. & Gull, S.F., 1984, Nature, 310, 33.

Wilson, M.J. & Falle, S.A.E.G., 1985, Mon.Not.R.astr.Soc., 216, 971.

Wilson, M.J. & Scheuer, P.A.G., 1983, Mon.Not.R.astr.Soc., 205, 449.

Wilson, M.J., 1987, Mon.Not.R.astr.Soc., 224, 155.

THE GAS DYNAMICS OF JETS

S.A.E.G.Falle
Department of Applied Mathematical Studies
The University of Leeds
Leeds LS2 9JT
U.K.

ABSTRACT. This article explores some of the consequences of assuming that both stellar and extragalactic jets behave like a fluid. General similarity arguments suggest that some of the observed morphology of extragalactic jets is due to the form of the ambient density distribution. The series of knots seen in both stellar and extragalactic jets appear to be related to the shock cell structure found in laboratory jets. It is shown that there can be quite a large pressure ratio between the jet and the outside medium even if there is no magnetic confinement.

1. INTRODUCTION

In this lecture I will look at what happens when a well collimated supersonic fluid jet interacts with its surroundings. Although I showed in my first lecture that numerical simulations of time dependent jets can be unreliable, it is still possible to predict some things about such flows and so get an idea of whether fluid models are consistent with the observed properties of stellar and extragalactic jets.

I will start with a number of general arguments which indicate how some of the observed morphology could arise and then go on to show that steady jet models can help us to get a quantitative understanding of the jets. To keep things as simple as possible I will ignore the dynamical effects of the magnetic fields although they must of course be present in the extragalactic jets. I will also suppose that the jets are collimated on very small scales so that their intermediate and large scale structure does not directly reflect the collimation process.

2. SIMILARITY ARGUMENTS

Suppose that at $t = 0$ a jet switches on in a medium with a spherically symmetric density $\rho_e(R)$ and pressure $p_e(R)$. If the jet is collimated on small scales then the only relevant jet parameters are its power P,

W. Kundt (ed.), Astrophysical Jets and Their Engines, 163–170.

mass flux Q, Mach Number M_j and initial solid angle Ω. As long as the pressures generated by the jet are large compared to p_e, we may ignore p_e and so the flow depends only on P, Q, Ω, M_j and ρ_e.

If ρ_e has a power law form

$$\rho_e = \frac{a}{R^\alpha} ,$$

then there is a characteristic length

$$l_c = \left(\frac{aP^{\frac{1}{2}}}{Q^{3/2}}\right)^{1/(\alpha-2)} .$$

Once the scale of the flow becomes large comparred to l_c, we expect it to be self similar with similarity variables r/t^β, z/t^β where r,z are cylindrical polar coordinates and

$$\beta = \frac{3}{\alpha + 5} .$$

The jet blows a cavity bounded by a contact discontinuity C outside which there is a shock S in the ambient gas. The shapes of the surfaces C and S are given by

$$C\left(\frac{r}{t^\beta}, \frac{z}{t^\beta}\right) = 0 , \quad S\left(\frac{r}{t^\beta}, \frac{z}{t^\beta}\right) = 0$$

where the functions $C(x,y)$ and $S(x,y)$ depend on Ω and M_j which are the only dimensionless parameters left in the problem. Provided the Mach number is high, the dependence on the Mach number should be weak, so things really depend only on Ω. Note that the flow at and exterior to C is self-similar in this way, but the flow interior to C need not be. To see how this works in the spherically symmetric case $(\Omega = 4\pi)$ see Dyson, Falle and Perry (1980).

The speed of the shock in the ambient gas is $v_s \propto t^{\beta-1}$, so the pressure behind this shock $p_s \propto t^{2\beta-2-\alpha}$. At C all lengths scale like t^β, so the jet radius scales like t^β. If the jet is adiabatic and the Mach number is high then the jet ram pressure at **C**: $p_r \propto 1/t^{2\gamma\beta}$. Now these two pressures must scale in the same way. If

$$2\gamma\beta > 2 + \alpha - 2\beta \qquad \text{Case I} ,$$

then p_r decreases too fast. So the jet must have shocked before it reached the surface C. This means either that the hot spot is recessed, or not there at all. This condition will be satisfied if

$$\alpha < 6/5 \qquad \text{for} \quad \gamma = 5/3 ,$$

$$\alpha < 4/5 \qquad \text{for} \quad \gamma = 4/3 .$$

On the other hand, if

$$2\gamma\beta < 2 + \alpha - 2\beta \qquad \text{Case II} ,$$

then the jet ram pressure decreases slower than the pressure behind the shock in the ambient gas. There is then no way to construct a similarity solution, but it is clear that the hot spots will be at the ends of the jet rather than recessed. Something like Scheuer's model C may well be appropriate in this case (Scheuer 1974). It is very tempting to identify Case I with FR I sources and Case II with FR II's.

3. OVERPRESSURE

There seem to be quite a few jets (e.g., M87) in which the minimum pressures in the radio knots along the jet are larger than the external pressure deduced from X-ray observations (Schreier et al. 1982). It is generally assumed that this means that the jet is either unconfined, or confined by magnetic forces. However, it is easy to show that the pressure inside a steady confined jet can be substantially higher than the external pressure if there are shocks in the jet.

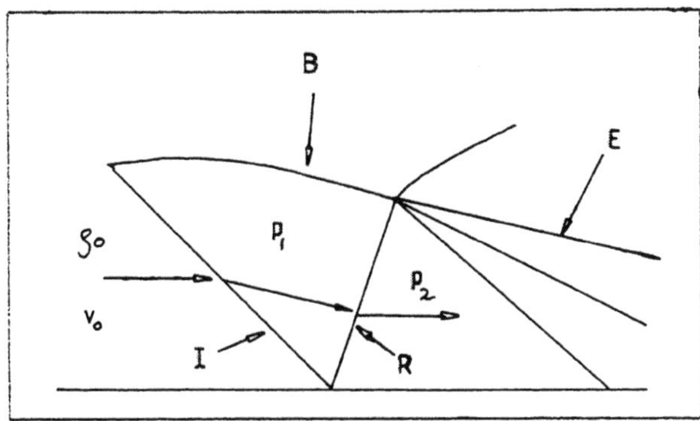

Figure 1. Flow through an incident reflected shock pair in the jet. I incident shock, R reflected shock, E expansion wave, B jet boundary.

If

$$\frac{p_1}{\rho_o v_o^2} \ll 1 ,$$

then the pressure ratio

$$\frac{\text{Jet Pressure}}{\text{Ext.Pressure}} = \frac{p_2}{p_e} = f(\gamma) = 12.5 \quad \text{for } \gamma = 5/3 .$$

This overpressure extends over a length $\sim 2M_j r_j$ where M_j and r_j are the Mach number and radius of the jet. The shocks in the jet can be due to reconfinement as discussed in Falle & Wilson (1985).

A large ratio between the jet and ambient pressures is still possible even if there are no shocks in the jet. To see this suppose that the external pressure is zero. As long as the jet exists at all, there must be some region of the jet which has non-zero pressure and so the ratio of jet pressure to external pressure is infinite. The reason is, of course, that such a jet cannot be uniform. If a jet emerges into a region in which the external pressure is zero, rarefaction waves will propagate into the jet and although the jet pressure eventually drops to zero, the distance over which this occurs scales like $2M_j r_j$ where M_j and r_j are the jet's initial Mach number and radius. This effect can be seen in figure 2 which shows the pressure in a steady jet propagating through an external pressure given by

$$p_e = \frac{1}{(1 + z^4)^{4/3}} .$$

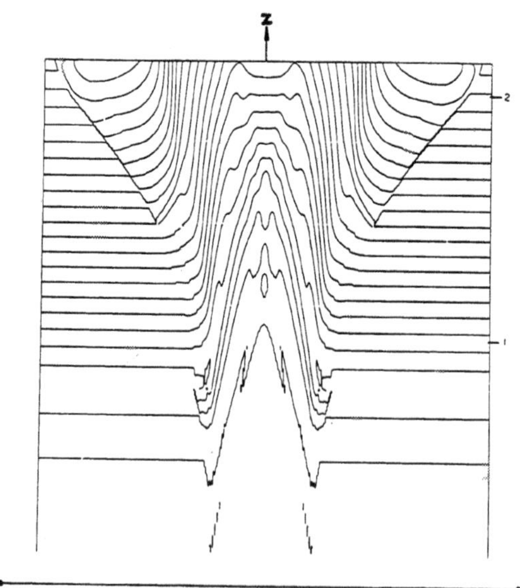

Figure 2. Pressure contours for an axisymmetric jet with initial Mach number 5.0 (Courtesy of Blackwell Sci. Publs., from Wilson and Falle, 1985).

4. BENT JETS

Suppose a jet emerges from the galaxy at right angles to the direction of motion of the galaxy through the intergalactic gas. The jet will be bent by the ram pressure of the external medium and its radius of curvature R will be given approximately by

$$\frac{\rho_j v_j^2}{R} = \frac{\rho_e v_e^2}{r_j} \; .$$

This comes from balancing the centrifugal force on the jet against the pressure gradient across the jet. The left hand side can be written in terms of the jet pressure

$$\frac{\gamma M_j^2 p_j}{R} = \frac{\rho_e v_e^2}{r_j} \; .$$

Since the pressure on the upstream side of the jet is $\sim \rho_e v^2$, there will be a shock in the jet unless $p_j \sim \rho_e v^2$. We therefore have

$$M_j^2 r_j = R \; .$$

However, it is easy to show that the bending of the jet will itself lead to a shock unless

$$M_j^2 r_j \ll R \; .$$

Since these two conditions are incompatible, we will get shocks in the jet unless the Mach number of the galaxy motion through the intergalactic gas is very low. If the galaxy is not moving at right angles to the jet, then v_e should be replaced by the velocity normal to the jet. There will therefore be shocks in the jet unless this normal velocity is very subsonic. Note that this is not what the numerical calculations of Williams and Gull (1984) show, but this is because their numerical resolution was too low to pick up the shocks.

5. SHOCK CELLS IN STEADY JETS

It is well known from terrestrial experience that shocks appear in steady jets which emerge from a nozzle into a region where the pressure is different from the jet pressure. In Astrophysical jets there is presumably no nozzle, but there is a considerable variation in pressure along the jets and if this variation is sufficiently rapid, then the effect is the same as that of a nozzle. The jet can remain in pressure equilibrium with its surroundings provided the scale height of the external pressure l_e is large compared with

$$l_j = 2M_j r_j \; .$$

This is the distance travelled by the jet fluid in the time it takes a sound wave to go from the jet boundary to the axis and back again. In Sanders (1983), Falle & Wilson (1985) and Wilson & Falle (1985) it is shown that shocks are formed in the jet provided $l_e < l_j$ and the external pressure is such that the jet opening angle would decrease if it was in pressure equilibrium with its surroundings. Falle & Wilson showed that a reasonable model of the knots in M87 could be constructed in this way, and a similar model seems to apply to NGC6251 (Wilson 1986).

Steady or approximately steady jets can occur in a number of ways. If the jet is much less dense than its surroundings, then it will blow a cavity and the walls of this cavity will move with a speed of order $v_j(\rho_j/\rho_e)^{\frac{1}{2}}$. Hence the shape of the cavity changes slowly compared to the flow time down the jet and the jet should be approximately steady, at least near the source. Alternatively, if the jet is denser than its surroundings, there will be no cavity and, once the jet has propagated a reasonable distance, things should relax to a steady state near the source.

All this does assume that the jet itself is not changing and there are no other external influences that might make it unsteady. What one wants is long straight jets since this at least means that the jet has been pointing in the same direction for a long time.

Apart from detailed modelling of sources like M87 and NGC6251, it is also possible to use steady jet theory to estimate the ram pressure in the jets from the spacing of the shock cells near the source. From laboratory experiments it is known that, for large Mach numbers, the spacing between the shock cells l is given by (Lewis & Carbon, 1964)

$$l = 1.38\,\gamma^{\frac{1}{2}} M_j r_j \left(\frac{p_j}{p_e}\right)^{\frac{1}{2}} .$$

For weak disturbances $p_j \sim p_e$ and is about the same as the pressure in the radio knots. The observed ratio of r to l gives the projected Mach number and so we can estimate the ram pressure from

$$p_r = \rho_j v_j^2 = \gamma M_j^2 p_j .$$

Since the true Mach number is always larger than the projected Mach number this gives us a lower limit on the jet ram pressure near the source. This can then be compared with the hotspot pressure which ought to be about the same provided the jet has not expanded too much before it reaches the hotspot. If this is done for Cygnus A and 3C111, one gets the following results:

Cygnus A
Hotspot pressure = 3×10^{-9} c.g.s., Ram pressure = 2×10^{-9} c.g.s.

3C111
Hotspot pressure = 4×10^{-10} c.g.s., Ram pressure = 2×10^{-10} c.g.s.

Although the agreement is not marvellous, it is within the observational uncertainties, which suggests that there is something in such models.

6. RELATIVISTIC STEADY JETS

Chiu (1973) has pointed out that the equations for steady flow are the same for relativisitic and non-relativistic flow. The only difference is that the equation of state is different in general and the density and pressure appearing in the relativistic equations are the proper density and pressure. If the gas is very hot, then the equation of state is also the same and a relativistic gas behaves like a non-relativistic gas with $\gamma = 4/3$. However, there is one important difference. Since there is an upper limit to the sound speed in a relativistic gas, the Mach number gives the gas speed directly.

$$M = \sqrt{2}\,\frac{\Gamma v_j}{c} .$$

Here Γ is the Lorentz factor.

This means that if we can determine the Mach number from the spacing between the shock cells in such a jet, then we have the jet speed. If this is done for the M87 jet using the spacing between knots D and E one gets $v_j = 0.989c$ if the jet is in the plane of the sky.

There are a number of reasons why the jets should be relativistic. Firstly, it seems consistent with the superluminal motions seen on parsec scales. Secondly, it offers a natural explanation for one-sidedness in jets. Thirdly, it means that the large minimum pressures estimated for some jets may simply be a result of Doppler boosting. Fourthly, if the jets actually consist of electrons and positrons, then the energy requirements become more modest since one does not have to worry about the large amount of invisible energy in the protons.

7. STELLAR JETS

Stellar jets seem to show many of the same morphological features as the extragalactic jets such as regular series of knots near the source, lobes and hotspots. So presumably if extragalactic jets can be understood in terms of a fluid model, so can stellar jets. If the observations actually measure the jet velocity (see Blome & Kundt for an alternative view) then the jets have velocities of order 200-400 km s^{-1}. The optical emission comes from regions with temperatures of about 10^4 K, so this means that the jets have Mach numbers in the range 20-40 which seems rather higher than those inferred for the extragalactic jets which are usually less than 10. Apart from this, estimates of the physical conditions in the jets suggest that radiative cooling is important so that, unlike the extragalactic jets, these jets do not behave adiabatically.

Falle, Innes & Wilson (1986) have applied steady jet models to stellar jets. In this case cooling can itself lead to shocks. What is required is that the cooling length should be small compared with the jet scale $2M_j r_j$. Then if the jet is initially hot, it will cool rapidly as it emerges from the source and its pressure will drop below the ambient pressure so that a shock is driven into it. In fact, the situation is much the same as that of a jet emerging into a region which is at higher pressure than that of the jet. It can be shown that the projected OI emission from such a jet agrees well with the observations of, for example, the jet that seems to be exciting HH34 (Falle, Innes & Wilson 1987).

8. REFERENCES

Blome, H.-J. & Kundt, W. 1987, IAU Symposium 122, ed. I.Appenzeller

Chiu, H.H., 1973, Phys. Fluids, 16, 825

Dyson, J.E., Falle, S.A.E.G. & Perry, J.J., 1980, Mon.Not.R.astr.Soc., 191, 785.

Falle, S.A.E.G., Innes, D.E. & Wilson, M.J. , 1987, Mon.Not.R.astr.Soc., in press.

Falle, S.A.E.G. & Wilson, M.J., 1985, Mon.Not.R.astr.Soc., 216, 79.

Lewis, L.H. & Carbon, D.J. 1964, AIAA Journal 2, 776

Sanders, R.H., 1983, Astrophys.J., 266, 73.

Scheuer, P.A.G., 1974, Mon.Not.R.astr.Soc., 166, 513

Schreier, E.J., Gorenstein, P. & Feigelson, E.D., 1982, Astrophys.J., 261, 42.

Williams, A.G. & Gull, S.F., 1984, Nature, 310, 33.

Wilson, M.J. & Falle, S.A.E.G., 1985, Mon.Not.R.astr.Soc., 216, 971.

Wilson, M.J., 1987, Mon.Not.R.astr.Soc., 224, 155.

COSMOLOGICAL EVOLUTION OF ACTIVE GALAXIES & QUASARS

J A Peacock
Royal Observatory
Blackford Hill
Edinburgh
EH9 3HJ

1. LUMINOSITY FUNCTIONS IN FRIEDMANN UNIVERSES

For convenience, we begin with a collection of the most useful formulae for relating luminosity functions in various wavebands to observations. For derivations see e.g. Weinberg (1972), Gunn (1978). These relations depend on the current values of the parameters H & Ω (we dispense with the usual zero subscript); where important, dependence of results on these values will be indicated, defining $h \equiv H/100\ \mathrm{kms^{-1}\ Mpc^{-1}}$. Often, the effect of varying h & Ω is an unimportant scaling and for simplicity numerical values will then be quoted assuming $h = \frac{1}{2}$, $\Omega = 1$.

For power-law spectra, flux density $S_\nu \propto \nu^{-\alpha}$, the monochromatic luminosity (usually L for optical astronomers, P for radio) is

$$P_\nu = S_\nu D^2\ (1+z)^{3+\alpha} \tag{1}$$

where D is angular-diameter distance. The usual expression for this is ill-defined as $\Omega \to 0$

$$D = \frac{c}{H}\ \frac{2}{\Omega^2\ (1+z)^2}\ [\Omega z\ -\ (2-\Omega)\ (\sqrt{1+\Omega z}\ -1)] \tag{2}$$

and may be transformed to

$$D = \frac{c}{H}\ \frac{z}{(1+z)^2}\ \left[\frac{1 + \sqrt{1+\Omega z} + z}{1 + \sqrt{1+\Omega z} + \Omega z/2}\right] . \tag{3}$$

For practical purposes (S in Jy and P in $\mathrm{WHz^{-1}\ sr^{-1}}$) the relevant value of $(c/H)^2$ is $10^{25.933}\ h^{-2}$. Note the units of P : equation (1) is really a surface brightness transformation, and there is no assumption of isotropic emission, so no factor 4π.

The other main formula for model building is that for comoving volume elements (where comoving distance is usually expressed in units of

W. Kundt (ed.), Astrophysical Jets and Their Engines, 171–183.

present-day Mpc). For an area of sky A sr, the differential volume element is $dV = AD^2 (1+z)^2 dr$, where dr is the element of comoving distance $dr/dz = (c/H)(1+z)^{-1} (1+\Omega z)^{-\frac{1}{2}}$. The integral volume out to redshift z is complicated except in two cases

$$V = \frac{8A}{3} \left(\frac{c}{H}\right)^3 \left[1 - (1+z)^{-\frac{1}{2}} \right]^3 \qquad \Omega = 1 \tag{4}$$

$$V = \frac{A}{8} \left(\frac{c}{H}\right)^3 \left[(1+z)^2 - (1+z)^{-2} - 4 \ln (1+z) \right] \qquad \Omega = 0 \ .$$

Note that it is usually sufficient to carry out analyses of luminosity functions by assuming a single value of Ω. If results for a different value of Ω are desired, they can be obtained directly: S & z in (1) are observables, so P scales according to D^2 and number density scales inversely with volume element, i.e. $\rho \propto D^{-2} (1+\Omega z)^{\frac{1}{2}}$. At a given redshift the translation to a different Ω simply involves moving the luminosity function horizontally and vertically by the appropriate amounts. Indeed, the translation to more general cosmologies can also be accomplished in this way, which is a justification for considering only Friedmann models: evolutionary studies cannot determine the geometry of the Universe.

One further useful quantity is the look-back time to a given redshift: $cdt = dr/(1+z)$. It is convenient to normalise this to the age of the Universe: $\tau \equiv t(z)/t(z = \infty)$. The following expression is exact for $\Omega = 0$ and $\Omega = 1$ and is a useful approximation (accurate to $\lesssim 1\%$ for $\Omega \lesssim 3$) in general.

$$\tau = 1 - (1+z)^{-(1+\Omega^{0.6}/2)} \ . \tag{5}$$

Finally, for dealing with optical samples, we need a conversion between optical magnitudes and flux densities. Such a conversion is given by Schmidt & Green (1983) but is not quite correct as it ignores the finite bandpass of optical filters. A thorough discussion is given by Matthews & Sandage (1963), but we need to scale their result to the most recent absolute calibration of Vega (Hayes & Latham 1975). For a $\nu^{-0.5}$ spectrum, the result is

$$B = -2.5 \log_{10} \left(\frac{S_\nu \ (4400\text{Å})}{3825 \text{ Jy}} \right) \tag{6}$$

and this depends slightly (~ 1% in flux density) on the spectral index. Using (6) and the definition of absolute magnitude as apparent magnitude at 10 pc, we have

$$M_B = -2.5 \log_{10} \left(\frac{P_\nu \ (4400\text{Å})}{10^{12.561} \text{ WHz}^{-1} \text{ sr}^{-1}} \right) \ . \tag{7}$$

2. RESULTS OF EVOLUTION STUDIES

2.1 The optical luminosity function.

The best determination of the evolving quasar luminosity function is that by Marshall (1986; 1987) which is very similar to the result of Shanks et al. (1986) and Boyle et al. (1987). The local luminosity function is given by a two power-law form:

$$\Phi\ (M_B) = \Phi_\circ\ [\ (L/L^*)^{2.6} + (L/L^*)^{0.25}]^{-1} \qquad (8)$$

where Φ is the (differential) comoving density of quasars per magnitude and L^* corresponds to $M_B = -21.8$. Evolution is parameterised by a shift in luminosity $L \propto (1+z)^k$ assuming an optical spectral index of 0.5. For $\Omega = 0$, $k = 3.5$ and $\Phi_o = 1331\ \mathrm{Gpc}^{-3}$; for $\Omega = 1$, $k = 3.2$ and $\Phi_o = 1512\ \mathrm{Gpc}^{-3}$. This form is largely based on observations of UVX quasars, and so cannot be applied at $z \gtrsim 2$. At low redshifts, Marshall incorporates data on Seyfert nuclei to deduce the low-luminosity portion of the function; the results should therefore be immune to incompleteness problems resulting from the non-stellar appearance of low-luminosity quasars. The luminosity function is plotted for various redshifts in Fig. 1. To indicate the regions where data exist, only (M_B-z) combinations with B < 21 are shown.

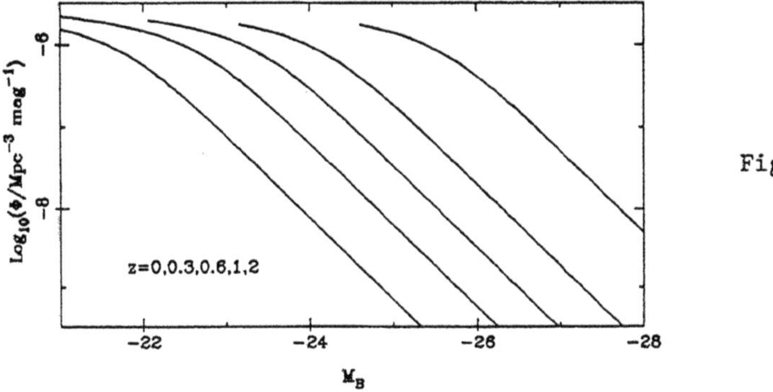

Figure 1

2.2 The Radio Luminosity function

The situation in the radio waveband is more complex than the optical, partly owing to the existence of radio galaxies having luminosities equal to those of quasars, partly because much larger datasets exist in the radio. These points have ruled out descriptions of the RLF as simple as (8).

Also, in the radio one must distinguish two populations with grossly different spectral indices: compact flat-spectrum sources with $\alpha < 0.5$ are treated separately from the steep-spectrum population. Peacock

(1985) has given limits on the RLF for these two classes; Fig 2 plots his (2.7 GHz) results in a form analogous to Fig. 1. The mean value of log ρ is shown where ρ is uncertain by a factor $\lesssim 2$. Note the units for the RLF: ρ is the comoving density of sources per unit $\log_{10}P$.

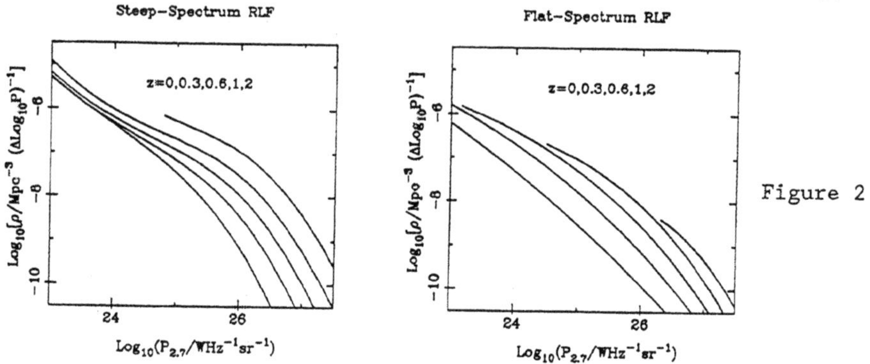

Figure 2

Figs 1 & 2 have some qualitative features in common: the luminosity functions are steeper at the bright end and the evolution appears differential in the sense that the density changes fastest at high luminosities. The similarity is closer when we consider the RLF for quasars alone: for $P_{2.7} \geq 10^{26}$ WHz^{-1} sr^{-1} essentially all flat-spectrum sources are quasars, compared with about 40% of steep-spectrum sources. At P ~ 10^{24}, however, only $\lesssim$ 1% of radio sources are quasars, so the quasar RLF must break to a very flat slope at low powers. We can illustrate this by deriving an RLF for quasars alone, which has not been attempted since Wills & Lynds (1978). We shall use the data described in Peacock & Dunlop (1986): 163 Parkes quasars with $S_{2.7}$ in the range 0.1 - 10Jy, but considering only $z < 2.2$ for direct comparison with the optical data. It turns out to be possible to fit a model RLF very similar in form to the optical one (8). The local RLF is given by

$$\rho(P) = \rho_o[(P/P*)^a + (P/P*)^b]^{-1} \tag{9}$$

where $(a,b,P*) = (0.0,\ 1.6,\ 10^{25.0}$ WHz^{-1} sr^{-1}) for steep-spectrum quasars and (0.6, 2.0, $10^{25.5}$ WHz^{-1} sr^{-1}) for flat-spectrum. Again, as for optical quasars, luminosity evolution $P \propto (1+z)^k$ is assumed: k turns out not to be significantly different for the two spectral types. For Ω=1, k=2.8 and ρ_o=10 Gpc^{-3} (steep-) or 2 Gpc^{-3} (flat-); for Ω=0, however, a model of the above form could not fit the data. This is not surprising: pure luminosity evolution cannot work for all values of Ω (cf. Section 1).

2.3 X-Ray Evolution

Studies of evolution of X-ray selected quasars and active galaxies are limited by small datasets. Fortunately, there appears to be a simple connection between the X-ray and optical wavebands: there is a close correlation between X-ray and optical luminosities, with spectral indices in the range ~ 1.4 ± 0.2 (where $\alpha_{ox} = 0.384 \log_{10} (L_o/L_x)$). The spectral index correlates weakly with optical luminosity, so that $L_x \propto L_o^{0.8}$. Furthermore, there is no evidence that α_{ox} depends on redshift: given an evolving optical LF, predictions for the results of X-ray surveys can be made readily. Unfortunately, such predictions appear to differ from observation by a factor ~2, in the sense that too many objects are predicted. The reason for this is unclear and likely to remain so until we can obtain new X-ray data. For a review of the evidence on X-ray properties of quasars see e.g. Avni and Tananbaum (1986).

The main complication to the above picture is that radio-loud quasars have higher X-ray luminosities (by a factor ~3) than radio-quiet (at a given optical luminosity). The interpretation here is that there must be an additional X-ray component related to the radio emission, with a synchrotron or synchrotron self-compton origin (see e.g. Kembhavi et al. 1986). No studies of whether this additional component evolves have been carried out.

3. INTERPRETATION OF EVOLUTION

The most elementary example of physical interpretation of cosmological evolution is one that has been with us for decades in the form of the struggle for supremacy between luminosity- and density-evolution models. It has been argued that the success of the former class of model in accounting for the optical quasar data is evidence that quasars are long-lived objects that simply decay gradually over a Hubble time. To see how seriously to take such a claim, consider the formalism describing conservation of objects on the P-z plane. In its simplest form (Cavaliere et al. 1983), this assumes that quasars of a given luminosity all evolve similarly:

$$\dot{\rho} + \frac{\partial}{\partial x} (\rho \dot{x}) = S\ (P,z) \tag{10}$$

Where $x \equiv \mathrm{Log}\ P$, time derivatives are donoted by dots and S is some unknown "birth function" (assumed to be zero in luminosity evolution models). Even in the simplified case of $S = 0$, the interpretation of (10) is nontrivial. For luminosity evolution $\rho\ (x,\ t) = \Phi\ (x - f(t))$ and the solution of (10) is $\dot{x} = \dot{f} + c/\rho$ where c is a constant of integration. Thus, we can certainly choose the solution $c = 0$ and have uniform dimming ($\dot{f}$ is negative) but it is equally possible that in fact all sources <u>brighten</u> with time (c large and positive). This brightening is differential in that $\dot{x}$ must increase with P, which would probably be unphysical for radio galaxies (e.g. Baldwin 1982), but is less obviously ruled out for radio-quiet quasars (see Blandford 1986). Such a basic problem as uncertainty of the sign of $\dot{x}$ should indicate that drawing conclusions from (10) without additional physical constraints will be near impossible. We may also note that the basic equation may in any case not always apply: if source lifetimes are short, then the LF could be in a quasi-steady state, with sources of a given P equally likely to brighten or fade. In this case, (10) must be modified to include Fokker-Planck-type terms containing second derivatives in P.

The danger in taking luminosity evolution models too seriously can be domonstrated more directly. We have seen above that radio quasar data can be fitted by a luminosity evolution model, apparently implying ages $\sim H^{-1}$, whereas for extended radio sources we have direct evidence from spectral ageing for much shorter lifetimes. The most spectacular example of this is probably Cygnus A, which is only about 4.5×10^6 $h^{-3/7}$ years old (Winter et al. 1980): this source has certainly not faded slowly over a Hubble time. We cannot obtain as definite answers for radio-quiet quasars, but the example of Cygnus means we should be careful about assuming them to be long lived.

Even though it is clear that powerful radio sources are being "born" today, the main idea of luminosity evolution can be restated in a way that is still relevant: is nuclear activity a phase that every galaxy passes through, or is it recurrent in only a few special galaxies? The next Section considers this question.

4. CONSTRAINTS ON ACTIVE GALAXY LIFETIMES

4.1 Black hole masses

Rees (1984) has reviewed models for powering active galaxies by black holes and some important results are summarised here. First, for isotropic emission and static emitting material, the Eddington Limit relates the minimum mass to total energy output $\dot{E}$;

$$M_{min}/10^8 M_{\odot} = \dot{E}/10^{39.1} \text{ W} \quad . \tag{11}$$

We now need $\dot{E}$ for quasars, which is usually obtained by noting νS_ν peaks in the optical, applying a rather uncertain "bolometric correction", g, and assuming isotropy. This works out to be

$$\dot{E} = 10^{38.5} \text{ g } 10^{-0.4\ (M_B+25)} \text{ W} \tag{12}$$

where $g \simeq 10$ is a reasonable value.

In pure luminosity evolution models, the total energy emitted, and hence mass accreted can be found. Taking an efficiency ϵ (see Lynden-Bell 1978), the implied black hole mass from the LF discussed in Section 2.1 (considering z<2 only) is

$$\frac{M}{M_{\odot}} = 10^{9.3} \epsilon^{-1} h^{-1} g\ 10^{-0.4(M_B+25)} \quad (\Omega=1) \tag{13}$$

and about a factor 2 higher for Ω=0. For canonical values (h,g,ϵ) = (0.5,10,0.1) this yields the enormous mass of $10^{12.4}$ $M_{\odot}$ for 3C273. This is inconsistent with the value ($\sim 10^{10.5} M_{\odot}$) estimated using variability & line-width data to weigh the hole (Wandel & Mushotzsky 1986). Note that equations (11) and (13) imply that local quasars have $L \sim 10^{-3.2}$ L_{EDD} (a factor ~10 below the results of the above authors): these objects could have been much more luminous in the past without exceeding the limit (11). Lastly, the best direct mass estimate is ~ $10^{8.8}$ $M_{\odot}$ for NGC4151 (Ulrich et al. 1984). With $M_B \simeq -19$, this result is not greatly in conflict with equation (13), which would imply $M \sim 10^{9.2} M_{\odot}$.

A more general argument was given by Soltan (1982). He showed that the total emitted energy density of quasars is independent of h & Ω. Taking the Ω = 1 LF from Section 2.1 and integrating over all M_B & z < 2, we get a mass density in black holes

$$\rho_{BH} = 10^{11.7} \, g\epsilon^{-1} \, M_{\odot} \, \mathrm{Gpc}^{-3} \quad . \tag{14}$$

To relate this to what we expect to see in galaxies, note that Smith et al. (1986) have shown that underlying galaxies in quasars are luminous, with a median $M_B \simeq -22$, and that this is ~ independent of nuclear luninosity. From Felten (1977) this translates to a mass per galaxy of ~ $10^{6.2}$ $g\epsilon^{-1}$ h^{-3} $M_{\odot}$. This corresponds to all luminous galaxies undergoing a quasar phase, and yields a canonical mass for "dead" quasars of $10^{9.1}$ $M_{\odot}$. From Section 2.1, the current density of quasars with $M_B < -18$ (the faintest limit studied by Smith et al.) is ~ 0.5 times that of galaxies with $M_B < -22$. Thus, about ¼ of galaxies with $M_B < -22$ produce quasars, and the figure of $10^{9.1}$ $M_{\odot}$ is ~ independent of whether quasars are long-lived or occur briefly in all galaxies.

4.2 Galaxy interactions and triggering of quasars.

If quasars have lifetimes much shorter than H^{-1}, a mechanism for generating new quasars is required, and one long-standing suggestion is that interaction & merging of galaxies is involved. Some reasonably direct evidence for this is provided by the observations of Stockton & Mackenty (1987). They used narrow-band filters to probe the distribution of gas via the OIII 5007Å line around a sample of quasars. In ~40% of their objects, gas was found to extend over several kpc (further than any starlight from the underlying galaxy), often displaying a characteristic morphology suggestive of the tidal tails seen in pairs of interacting galaxies. The asymmetric and chaotic structure of the emission-line regions led Stockton and Mackenty to argue that gas was in a non-steady state which would only persist for $\sim 10^8$ years.

The evidence of Stockton & Mackenty, although impressive, is only circumstantial (as they point out). A more definite argument concerns the statistics of multiple nuclei in radio galaxies. Multiple nuclei are commonly seen in cD galaxies, and this has been used to argue that such objects form via mergers (Hoessel 1980). Radio galaxies tend to be either of cD type (if FRI: Fanaroff and Riley 1974) or slightly less luminous, but still obeying an optical magnitude-core radius relation characteristic of mergers (Lilly & Prestage 1987). Care is needed in this argument, because many of the observed "multiple nuclei" are merely chance projections of a separate cluster galaxy (Tonry 1985). Nevertheless, some multiple systems are certainly real dynamical entities; the most striking example is probably 0255+05 (3C75) which has two nuclei, each with a twin-jet FRI radio source (Owen et al 1985).

The merging picture has been used by Roos (1985) in an attempt to explain the evolution of active galaxies. He assumes that all galaxies contain massive black holes which are fuelled by tidal disruption of stars resulting from a merger. In a hierarchical clustering picture, the merger rate was higher in the past and this produces a luminosity evolution picture roughly consistent with what is observed. It is hard to know how seriously to take such a model, involving as it does a number of assumptions which are hard to test. At the very least, though, it serves as a stimulus to further searches for evidence of current/recent interactions involving active galaxies.

5. THE REDSHIFT CUTOFF

5.1 Optical illusions

Now that evolution around $z \sim 1$ is determined with reasonable precision, the question of greatest current interest is what happens at earlier times, $z \gtrsim 2$? Probing evolution at high redshift is always going to be hard: for $z \gg \Omega^{-1}$, $S_\nu \propto (1+z)^{-(1+\alpha)}$, while $dV/dz \propto (1+z)^{-3/2}$. The flux density declines in a quasi-Euclidean manner, but the volume elements at high redshift quickly become too small to offer much encouragement to observers. In addition, high-redshift quasars may be obscured in a variety of ways. Shortward of Lyα, the continuum becomes blanketed by absorption due to intergalactic clouds of neutral hydrogen. At a given column density, the comoving density of these clouds evolves as $dN/dz \propto (1+z)^\gamma$, where $\gamma = 2.17 \pm 0.36$ (Murdoch et al. 1986). The optical depth in these clouds is then $\tau \propto (1+z)^{1+\gamma}$ and reaches unity (i.e. 1 mag. suppression of the continuum) at $z \simeq 3$; searches in the blue are clearly likely to be inefficient.

Other forms of obscuration are possible, but the strength of the effect is less certain. Ostriker & Heisler (1984) argue that dust may prevent us seeing high-z quasars. For opacity $\propto \lambda^{-1}$, the optical depth to dust is

$$\tau = \tau^* \int \frac{(1+z)^2 \, dz}{\sqrt{1+\Omega z}} \qquad (15)$$

where τ^* may possibly be as high as 0.1-1 in the blue.

Finally, with intergalactic material being largely ionised, obscuration by Thomson scattering or bremsstrahlung absorption needs to be worried about. The Thomson optical depth is

$$\tau = 0.04\ \Omega_{IGM} \int \frac{(1+z)\,dz}{\sqrt{1+\Omega z}} \tag{16}$$

whereas that due to bremsstrahlung is

$$\tau = 0.0002\ h^3\ \nu_6^{-2}\ \Omega_{IGM} \int \frac{(1+z)^2 dz}{\sqrt{1+\Omega z}}\ T_6^{-3/2} \tag{17}$$

where T_6 is temperature in 10^6K, ν_6 frequency in MHz and Ω_{IGM} the contribution of ionised hydrogen to Ω (see Longair and Rees 1973). Even with Ω_{IGM}~1, radio emission is not going to be affected at $z \lesssim 10$; dust and Ly-α absorption will similarly be unimportant for optical searches carried out in the red.

5.2 Observations

Reports of a redshift cutoff existed some time before they were justified; throughout the 1970s, understanding of how optical searches were biased against high-redshift objects slowly improved (e.g. Carswell and Smith 1978). The first hard evidence of a cutoff was probably the 4m grism survey of Osmer (1982). He surveyed 5 deg^2 for objects in the range $3.7 < z < 4.7$ and found none to a limit $R \simeq 19$. More stringent limits have been set recently by the 5m CCD grism searches carried out by Schmidt, Schneider and Gunn (1986;1987). These covered $3 \lesssim z \lesssim 4.5$ to (redshift-dependent) limits equivalent to $B \simeq 22$ over 0.91 deg^2 and $B \simeq 20.5$ over 7.84 deg^2. again finding no quasars. The precise significance of these non-dections is a variable quantity: Schmidt et al. use extrapolations of the evolving LFs of Schmidt and Green (1983) to predict that ~100 quasars should have been found, which seems impressive. However, the Schmidt and Green LF seems a poor fit to the data (Marshall 1985), and in any case the result may simply be telling us that the true evolution at z>2 is slower than the extrapolation. The most conservative calculation one can do is to take the LF of Section 2.1 and assume that this is unchanging at z>2. In this case, the predicted numbers for the 5m searches are 7(Ω=0)/13(Ω=1) for the large-area survey and 19(Ω=0)/16(Ω=1) for the deeper one. This provides evidence for a moderate supression of the LF at $z \simeq 4$ (a factor ~10) at the typical absolute magnitudes probed by these surveys ($-28 \lesssim M_B \lesssim -25$), in agreement with the results of multicolour work (Koo 1986). There are, however, suggestions that this cutoff may be luminosity dependent. Schmidt et al. and Koo (1986) find that the lack of high-redshift quasars in their surveys becomes significant beyond $z \simeq 2.5$ (this does depend as above on how the sums are done), whereas for the highest luminosities ($M_B \simeq -29$) evolution continues beyond $z \simeq 3$. Osmer and Smith (1980) found no cutoff up to z = 3.25 and Hazard and collaborators have recently found several bright high-redshift quasars: they estimate a surface density of >0.1 deg^{-2} with $3.5 < z < 3.8$ and R $\lesssim 18$ (Hazard et al. 1986), as compared with the minimal prediction (as above) of 0.006(Ω=0)/0.012 (Ω=1) deg^{-2} to B < 19. Note that all these high-redshift B limits are theoretical equivalents (ignoring obscuration) of searches in the red. The position will remain uncertain because grism surveys are somewhat insensitive over $2 \lesssim z \lesssim 3.5$ (the carbon lines are inconveniently weak). The best direct probe of the optical redshift cutoff will probably come from multicolour surveys.

Studying the cutoff in the radio is more straightforward, partly because the redshift-dependent selection effects are simpler and partly because radio-loud quasars are rarer. this latter point means that we only have good statistics over about one decade of radio power. This is illustrated in Figure 3, which shows all the quasar data used by Peacock and Dunlop (1986).

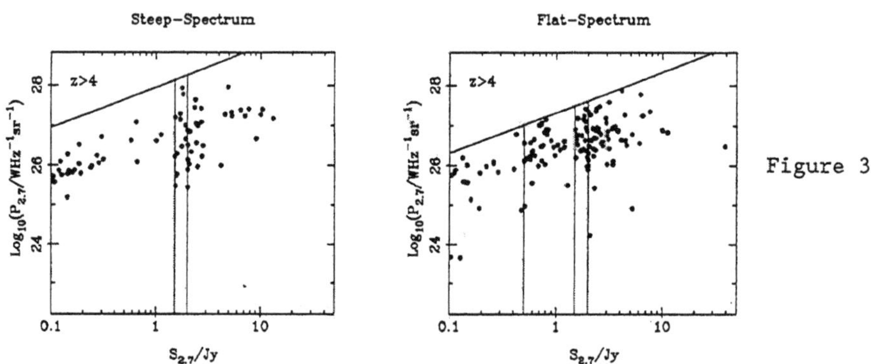

Figure 3

We can clearly only make a meaningful statement about quasars with $P_{2.7} \sim 10^{27}$ WHz^{-1} sr^{-1}. This is best achieved by using a modification of the standard V/V_{max} test, firstly by combining all samples into one coherent dataset to obtain the generalised statistic V_e/V_a (Avni and Bahcall 1980) and secondly by considering a redshift band: analyse the distribution in volume of objects at $z>z_o$ only. The sample shown in Figure 3 has 39 quasars with $z > 1.9$; for this redshift band $< V_e/V_a >$ is 0.387 (Ω=1) or 0.415 (Ω=0) as compared with the expected value for a constant comoving density: $< V_e/V_a > = 0.5 \pm (12n)^{-\frac{1}{2}}$. This provides evidence for a decrease in comoving density beyond even quite a moderate redshift. Dunlop et al. (1986) estimate from this result that the quasar RLF decays as $\rho \propto (1+z)^{-3}$ beyond z~2: z=4 radio-loud quasars are likely to be hard, but not impossible, to find.

5.3 Interpretation

It is clearly premature to attempt any detailed explanation for the goings-on at z>2 when there remain so many observational uncertainties. Nevertheless, there exist a variety of general points which are likely to be relevant and it is worth listing them here.

First, even if there is a redshift cutoff, it need not indicate a lack of objects at high redshifts. Consider equation (10): in the limit of very short source lifetimes the first term may be neglected and we obtain

$$\rho(P,z) = \dot{x}^{-1} \int S(P,z)dx \qquad (18)$$

This expresses the commonsense fact that, for a constant birth rate, the faster sources go through their lifecycles (i.e. high $\dot{x}$), the lower the density seen at a given time will be. We can arrange for ρ to be lower at high redshift either by triggering fewer objects or by killing them more quickly.

Mechanisms in the latter category are hard to describe simply as we must first specify how the active nuclei are fuelled before we can talk about how to reduce the fuelling. Blandford (1986) has described a model of this sort in which all galaxies contain a black hole accreting at the Eddington limit. Blandford postulates an ad hoc decline of the mean fuelling rate with redshift so that a progressively smaller proportion of black holes are currently active. In this model, the redshift cutoff is achieved by a further arbitrary function: the black hole birth rate, which is supposed to be small for $z \gtrsim 2$. Although simple, this model does reproduce qualitatively the evolution of the optical luminosity function. In particular, the redshift cutoff is luminosity dependent: total numbers at high z are suppressed, but the luminosity evolution of Section 2.1 continues, as the break luminosity in equation (8) corresponds to a black hole which is accreting at the full fuelling rate. This model is currently the most promising of those claiming a physical interpretation of cosmological evolution.

Extended radio emission may be inhibited at high redshift, simply because the Universe is denser. Inverse Compton losses on the microwave background are a factor $(1+z)^4$ higher (Scheuer 1974). At first sight, the denser IGM would aid radio emission by suppressing adiabatic losses, but a more important effect may be a radical alteration of source morphology. Barthel (1986) has shown that at $z \gtrsim 1.5$, steep-spectrum quasars become more distorted and smaller; he attributes this change to their denser environments. Certainly, a dense IGM appears to have a strong effect at low redshift in causing the FR division (see e.g. the discussion in Prestage and Peacock 1987). It is however worth bearing in mind that the environments of quasars at $z \sim 1$ are not yet fully understood, but they may lie in rich clusters at these redshifts (Yee and Green 1986). If so, the IGM density will be lower at high redshift until we reach the epoch of cluster formation, at which point other considerations arise.

The most interesting aspect of the redshift cutoff is that it may provide a constraint on theories of galaxy formation. This is an immensely complex area which is not worth trying to discuss in any detail here. The main point is that the redshifts of interest for the cutoff are low: z~2. If we want to produce the cutoff by a reduction in "birthrate" then this implies that a critical stage in the formation of structure in the Universe was reached quite recently. This may be either the formation of clusters (if we want to trigger black holes via interactions) or of galaxies themselves. The best way of following up such a suggestion is the direct one - examine in detail the structure and environment of active galaxies at $z \gtrsim 2$. It is interesting that at least one radio galaxy at these redshifts bears a passing resemblance to a primaeval galaxy (Spinrad et al. 1985).

REFERENCES

Avni, Y. and Bahcall, J.N., 1980. Astrophys. J., 235, 694.
Avni, Y. and Tananbaum, H., 1986. Astrophys. J., 305, 83.
Baldwin, J.E., 1982. IAU Symp. No. 97, "Extragalactic Radio Sources", eds. D.S. Heeschen and C.M. Wade (Dordrecht: D. Reidel), p21.
Barthel, P.D., 1986. IAU Symp. No. 119, "Quasars", eds. G. Swarup and V.K. Kapahi (Dordrecht: D. Reidel), p181.
Blandford, R.D. 1986. IAU Symp. No. 119, "Quasars", eds. G. Swarup and V.K. Kapahi (Dordrecht: D. Reidel), p359.
Boyle, B.J., Fong, R., Shanks, T. & Peterson, B.A. 1987. Mon. Not. R. astr. Soc. in press.
Carswell, R.F. and Smith, M.G., 1978. Mon. Not. R. astr. Soc., 185, 381.
Cavaliere, A., Giallongo, E. & Vagnetti, F.,1983. Astrophys.J., 269, 57.
Fanaroff, B.L. & Riley, J.M., 1974. Mon. Not. R. astr. Soc., 167, 31p.
Felten, J.E., 1977. Astr. J., 82, 861.
Gunn, J.E., 1978. Proc. 8th Saas-Fee Course, "Observational Cosmology", Geneva Observatory, p1.
Hayes, D.S. & Latham, D.W, 1975. Astrophys. J., 197, 593.
Hazard, C., McMahon, R.G. and Sargent, W.L.W., 1986. Nature, 322, 38.
Hoessel, J.G., 1980. Astrophys.J., 241, 93.
Kembhavi, A., Feigelson, E.D. and Singh, K.P., 1986. Mon. Not. R. astr. Soc., 220, 51.
Koo, D.C., 1986. "The structure and Evolution of Active Galactic Nuclei", eds. G. Giuricin, F. Mardirossian, M. Mezzetti and M. Ramella, (Dordrecht: D. Reidel), p317.
Lilly, S.J. & Prestage, R.M., 1987. Mon. Not. R. astr. Soc., in press.
Longair, M.S. and Rees, M.J., 1973. Cargèse lectures in Physics, Vol. 6, Gordon and Breach.
Lynden-Bell, D., 1978. Physica Scripta, 17, 185.
Matthews, T.A. & Sandage, A.R., 1963. Astrophys. J., 138, 30.
Marshall, H.L., 1985. Astrophys. J., 299, 109.
Marshall, H.L., 1986. The Structure & Evolution of Active Galactic Nuclei", eds. G. Giuricin, F. Mardirossian, M. Mezzetti & M. Ramella, (Dordrecht: D. Reidel), p627.

Marshall, H.L., 1987. Astrophys. J., in press.
Murdoch, H.S., Hunstead, R.W., Pettini, M. & Blades, J.C., 1986. Astrophys. J., in press.
Osmer, P.S., 1982. Astrophys. J., 253, 28.
Osmer, P.S., and Smith, M.G., 1980. Astrophys. J. Suppl., 42, 333.
Ostriker, J.P. & Heisler, J., 1984. Astrophys. J., 278, 1.
Owen, F.N., O'Dea, C.P., Inoue, M. & Eilek, J.A., 1985. Astrophys. J., 294, L85.
Peacock, J.A., 1985. Mon. Not. R. astr. Soc., 217, 601.
Peacock, J.A. & Dunlop, J.S., 1986. IAU Symp. no 119, "Quasars", eds G. Swarup & V.K. Kapahi, (Dordrecht: D. Reidel), p455.
Prestage, R.M. and Peacock, J.A., 1987. Mon. Not. R. astr. Soc., submitted.
Rees, M.J., 1984. Ann. Rev. Astr. Astrophys., 22, 471.
Roos, N., 1985. Astrophys. J., 294, 486.
Scheuer, P.A.G., 1974. IAU Symp. No. 74, "Radio Astronomy and Cosmology" ed. D.L. Jauncey (Dordrecht: D. Reidel), p343.
Schmidt, M. & Green, R.F., 1983. Astrophys. J., 269, 352.
Schmidt, M., Schneider, D.P. and Gunn, J.E., 1986. Astrophys. J., 306, 411.
Schmidt, M., Schneider, D.P. and Gunn, J.E., 1987. Astrophys. J., in press.
Shanks, T., Fong, R., Boyle, B.J. and Peterson, B.A., 1986. IAU Symp. No. 119, "Quasars", eds. G. Swarup and V.K. Kapahi (Dordrecht: D. Reidel), p37.
Smith, E.P. Heckman, T.M., Bothun, G.D., Romanishin, W. & Balick, B., 1986. Astrophys. J., 306, 64.
Soltan, A., 1982. Mon. Not. R. astr. Soc., 200, 115.
Spinrad, H., Filippenko, A.V., Wyckoff, S., Stocke, J.S., Wagner, R.M. and Lawrie, D.G., 1985. Astrophys. J., 299, L7.
Stockton, A. & Mackenty, J.W., 1987. Astrophys. J., in press.
Tonry, J.L., 1985. Astrophys. J., 291, 45.
Ulrich, M.H., Boksenberg, A., Bromage, G.E., Clavel, J., Elvius, A., Penston, M.V., Perola, G.C., Pettini, M., Snijders, M.A.J., Tanzi, E.G. & Tarenghi, M., 1984,. Mon. Not. R. astr. Soc., 209, 479.
Wandel, A. & Mushotzky, R.F., 1986. Astrophys. J., 306, L61.
Weinberg. S., 1972. "Gravitation & Cosmology", Wiley, New York.
Wills, D. & Lynds, R., 1978. Astrophys. J. Suppl., 36, 317.
Winter, A.J.B., Wilson, D.M.A., Warner, P.J., Waldram, E.M., Routledge, D., Nicol, A.T., Boysen, R.C., Bly, D.W.J. & Baldwin, J.E., 1980. Mon. Not. R. astr. Soc. 192, 931.
Yee, H.K.C. and Green, R.F., 1986. IAU Symp, No. 119, "Quasars", eds. G. Swarup and V.K. Kapahi (Dordrecht: D. Reidel), p481.

UNIFIED BEAMING MODELS AND COMPACT RADIO SOURCES

J A Peacock
Royal Observatory
Blackford Hill
Edinburgh
EH9 3HJ

1. BASICS OF RELATIVISTIC BEAMING

Our interest in beaming models comes from the presumed existence of bulk relativistic flows in the nuclei of active galaxies. It is worth noting that this was in fact a theoretical prediction (Rees 1967), although (as we shall see) there is still some debate about precisely what is moving. We can begin with something uncontroversial: the relativistic transformation of specific intensity (or surface brightness).

$$I_\nu\ (\delta\nu_o) = \delta^3\ I_\nu\ (\nu_o) \tag{1}$$

where δ is the doppler factor caused by relative motion of source and observer. Equation (1) may be derived in an elementary way by transforming photon number densities and energies plus the relativistic aberration of solid angle elements. A more direct approach is to note that I_ν/ν^3 is proportional to the photon phase-space density, which is a relativistic invariant and is conserved along light rays (by Liouville's theorem). If we now consider an optically thin (spherical) blob moving with velocity $\beta(c=1)$ at an angle $\cos^{-1}(\mu)$ to our line of sight then, <u>if</u> the emission is isotropic in the blob's rest frame, the received flux density clearly also transforms as in (1). The doppler factor is $[\gamma(1-\beta\mu)]^{-1}$ and so, for a power-law spectrum with flux density $S_\nu \propto \nu^{-\alpha}$, we have

$$S = S_o\ (1-\beta\mu)^{-(3+\alpha)} \tag{2}$$

where S_o is the value at $\mu=0$ (in the plane of the sky). For a quasi-continuous jet formed out of finite-lifetime blobs, the number of blobs observed at a given instant scales as δ^{-1} and hence the appropriate index in (2) becomes $(2+\alpha)$. The 'standard' model consists of a pair of such jets oppositely directed; however, life is simpler if we follow Lind and Blandford (1985) and neglect the receding component. We can then obtain a useful expression for the probability distribution of the beamed flux density. In general there will be an isotropic

W. Kundt (ed.), Astrophysical Jets and Their Engines, 185–196.

component in addition to the jet: we define R to be the ratio of <u>one</u> side of the jet to the isotropic component, at $\mu=0$. The amplification relative to $\mu=0$ is therefore $A= 1+R[(1-\beta\mu)^{-(2+\alpha)} - 1]$ and (since μ is uniformly distributed)

$$P(>A) = \frac{(1-\beta)}{\beta}\left[\left(\frac{(A_m+R-1)}{(A+R-1)}\right)^{1/(2+\alpha)} -1\right] \tag{3}$$

where the maximum amplification $A_m = 1+R[(1-\beta)^{-(2+\alpha)} - 1]$. This function is plotted in Figure 1 for various values of γ and R ($\alpha = 0$ is taken): note that $A_m \gtrsim 100$ is required before the 'characteristic' $A^{1/(2+\alpha)}$ form becomes established.

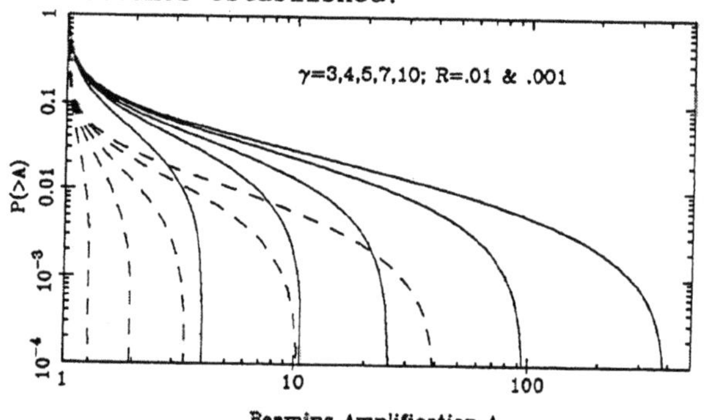

Figure 1

Equation (3) is a severe idealisation and some complications are discussed by Lind and Blandford (1985). The most natural of these is to recognise that if the jet is in any way non-steady, the observed emission will arise behind shocks. The effective beaming (post-shock) speed is then lower than the shock speed - which is the pattern speed observed with VLBI. If the jet consists of relativistically hot material ($P = \rho c^2/3$) then we have the exact relation between the Lorentz factors of the shock and the post-shock material:

$$\gamma_{post\text{-}shock} = \gamma_{shock}\, 2\, (9-\beta_{shock}^{-2})^{-1/2} \tag{4}$$

(neglecting the velocity of the unshocked jet material). If we consider only strong shocks, then we have the approximate relation

$$\gamma_{post\text{-}shock} \simeq \gamma_{shock}\left(\frac{2}{\Gamma} -1\right)^{\frac{1}{2}} + \frac{(\Gamma-1)}{\Gamma(2-\Gamma)} \tag{5}$$

(valid to $\lesssim$ 10 percent for γshock $\gtrsim 5$) where Γ is the quasi-specific-heat ratio; $\Gamma = 5/3$ for cold plasma, 4/3 in the ultrarelativistic limit. See Blandford and McKee (1976) for details of relativistic shocks. Equation (5) tells us that the effective Lorentz factor for beaming could easily be only ~ half that inferred from superluminal motion.

Further complications considered by Lind and Blandford include non-planar shocks and optical depth effects. The latter are especially important: for synchrotron emission, the optical depth scales as $\nu^{-(\alpha+5/2)}$ and the jet is likely to become optically thick when seen end-on. These effects thus tend to broaden the beaming cone considerably beyond the canonical $\theta \sim 1/\gamma$; we can probably still use a redistribution function of the form (3), but the relevant Lorentz factor is a parameter uncertain by a factor $\gtrsim 2$.

Less conventional explanations for superluminal motion include the 'light-signal' model and electron streaming models. In the former (e.g. Lynden-Bell 1977), the appearance of moving VLBI knots is generated by a pulse of radiation being scattered off a slowly-moving jet, in which case the emission is not strongly beamed. This model has fallen out of favour recently as it cannot explain why VLBI jets are one-sided (the cores are known to be stationary: Bartel et al. 1984). However, if current ideas about intrinsic one-sidedness persist (e.g. Bridle 1984), this model may revive, although there will still be difficulties explaining the lack of inverse-Compton X-rays in some sources (Marscher and Broderick 1981). Streaming models are perhaps even more speculative: here, bulk plasma flow is replaced by a highly anisotropic electron distribution with electrons following magnetic field lines at very low pitch angle. Such a configuration will indeed be generated by synchrotron losses, but scattering will tend to re-isotropise the distribution. Our knowledge of field configurations in active nuclei is too poor to be certain if such a model is possible, but it is certainly attractive in terms of its low energy requirements. Beaming in this model is again weak as the cone angle of visibility is set by the field geometry, not electron Lorentz factor (see e.g. Coleman 1986).

It should be clear, then, that beaming will be a difficult ghost to lay to rest, simply through our ignorance of the realistic P(>A) function. In what follows, equation (3) will be assumed, in order to see if the simplest picture can be made consistent.

2. UNIFIED MODELS

The essence of all beaming theories is to ask what a compact radio source would look like when any putative beamed component is turned away from our line of sight: what is the unbeamed parent population? There have been three attempts to answer this question, which we now consider.

Note that, unless otherwise stated, $H_0 = 50\ \mathrm{kms^{-1}\ Mpc^{-1}}$ and $\Omega_0 = 1$ are assumed throughout.

2.1 The Scheuer-Readhead Model

This unified model is almost certainly incorrect; nevertheless, it was such a classic paper that we can profit greatly from studying its downfall. Scheuer and Readhead (1979) suggested that compact radio quasars when unbeamed would appear as radio-quiet quasars. By assuming that all the radio emission was beamed - i.e. equation (3) with $R = 0$ - they were able to predict the radio flux-density distribution for an optically-selected sample: $P(>S) \propto S^{-1/(2+\alpha)}$, assuming a small dispersion in plane-of-sky flux densities. This prediction was not confirmed: the detection rate appeared to be a function of redshift (Smith and Wright 1980) and, more crucially, Condon et al. (1981) found that many quasars were not detected by the VLA at sub-mJy levels, implying a much slower variation of $P(>S)$ with S. In retrospect, these objections did not really refer to the main idea of the Scheuer-Readhead model. All the above authors were working in terms of a paradigm introduced by Schmidt (1970) whereby radio and optical luminosities of quasars correlated. The beaming model was therefore used to predict the distribution of radio-optical flux ratios. The results of Smith and Wright and Condon et al. can equally well be regarded simply as evidence against the Schmidt hypothesis and, indeed, Peacock, Miller and Longair (1986) have used similar arguments to show that radio and optical luminosities probably do not correlate.

The real objection to the Scheuer-Readhead model comes from more detailed observations of compact radio quasars. These turn out to have $\gtrsim 1$ percent of their total flux in the form of extended steep-spectrum emission (Perley et al. 1982; Schilizzi and de Bruyn 1983; Browne & Perley 1986). The morphology of this emission is such that it seems most unlikely to be beamed: the unbeamed counterparts of compact quasars would still be quite strong radio sources - much more luminous than the typical radio-quiet quasar.

2.2 The Orr-Browne Model

The steep-spectrum haloes discussed above led Orr and Browne (1982) to the idea that compact quasars were simply normal double radio sources seen end-on. In particular, they assumed that the unbeamed counterparts remained quasars, rather than appearing as radio galaxies (see Section 3 for discussion of the possibility that the optical quasar light may be anisotropic). Orr and Browne used this idea to predict the source counts for flat-spectrum quasars, given those for steep-spectrum quasars, with a fair degree of success. They did not check that their model also yielded the correct redshift distributions for flat-spectrum quasars, but this has been verified by Kapahi and Kulkarni (1986).

We can generalise these analyses to consider the full evolving luminosity function. Consider the integral luminosity function resulting from the application of beaming amplification to the plane-of-sky parent luminosity function N_0:

$$N(>L) = N_o(>L) + \int_{L/A_{max}}^{L/A_{min}} P(>A) \; dN_o \; . \qquad (6)$$

A_{max} and A_{min} are the beaming amplifications within which the source is classified as steep- or flat-spectrum. Orr and Browne give a complicated prescription for this division; we shall simply take $1 < A < 2$ to correspond to a steep-spectrum source, $2 < A < A_m$ for flat-spectrum. The results are highly insensitive to the exact division adopted. Since realistic values of A_m are $\lesssim 100$ (cf. Figure 1), we can take N_o to be locally a power law $N_o\,(>L) \propto L^{-\delta}$, in which case

$$\frac{N_{fs}}{N_o} = 1 + \int_2^{A_m} P(>A) \;\; \delta A^{(\delta-1)} \;\; dA \qquad (7)$$

and similarly for N_{ss}. Figure 2 plots the ratio N_{fs}/N_{ss} as a function of δ.

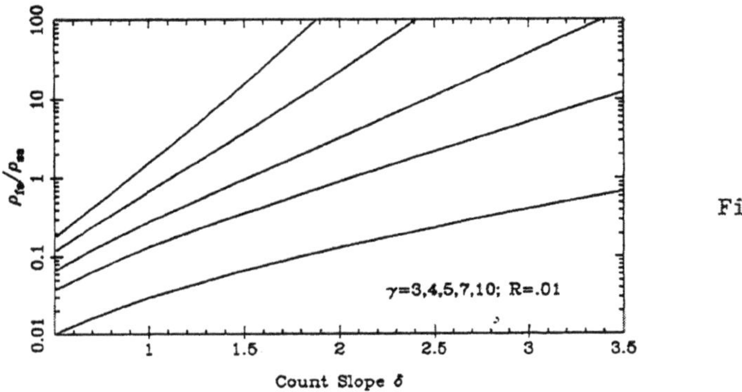

Figure 2

If $\delta \gtrsim 1$ and $A_m \gtrsim 5$ then the large - A_m result is accurate to a factor $\lesssim 1.5$:

$$\frac{N_{fs}}{N_{ss}} \simeq \frac{(1-\beta)}{\beta} \; \frac{A_m^{\delta}}{\delta(2+\alpha)} \simeq \frac{R^{\delta}(2\gamma^2)^{\delta(2+\alpha)-1}}{\delta(2+\alpha)} \; . \qquad (8)$$

Orr and Browne argue that the appropriate value of R for steep-spectrum quasars is approximately 0.012 $((1+z)\nu_{obs}/5\mathrm{GHz})$ - allowing for a factor of 2 difference in their definition of R. This is probably not a good approximation for $(1+z)\nu \gtrsim 10$ GHz as the central components will then start to go optically thin and R will eventually cease to change with increasing frequency. Note that radio galaxies of the same total power as steep-spectrum quasars have by comparison very small observed values of R (~ 0.001): these objects can nevertheless generate substantial amplifications given high Lorentz factors ($A_m \simeq 40$ for $R = 0.001$, $\gamma = 10$ and $\alpha = 0$). We should therefore not be surprised at the existence of core-dominated radio galaxies with superluminal motion (e.g. 3C120: Walker 1984).

To test the predictions of Figure 2 most stringently we require the evolving radio luminosity function, since the unified scheme has to apply at all redshifts. The most recent determination of this is by Peacock (1985), but it is not precisely what we need as it does not distinguish radio galaxies from quasars. An RLF for quasars alone is given by Peacock (1987: this volume). The data for both steep- and flat-spectrum classes can be described by a luminosity evolution model with comparable degrees of evolution for each class. This latter point is slightly in conflict with the unified model, which predicts increasing dominance of flat-spectrum sources at high redshift. However, as mentioned above, the $R \propto (1+z)$ scaling on which this is based is probably unrealistic. It will therefore suffice to compare ρ_{ss} and ρ_{fs} at one redshift (we take $z = 1$, where most of the data lie). For $\Omega = 1$, Figure 3 plots firstly the two RLFs and secondly ρ_{fs}/ρ_{ss} versus slope of the steep-spectrum RLF.

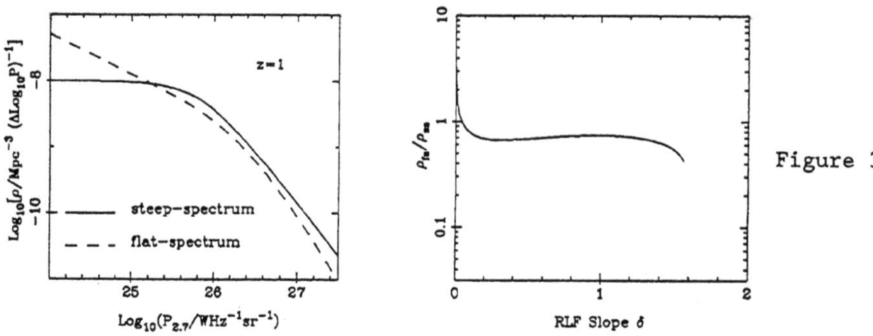

Figure 3

The interesting thing about this plot is that the two RLFs are virtually identical for $P > 10^{25}$ WHz^{-1} sr^{-1}. The flat-spectrum RLF does not decline for lower powers as would be predicted; it is not clear how much of a problem this is, as there are few objects in this region. Most of the data give us only one point: $\rho_{fs}/\rho_{ss} = 1$ at $\delta = 1.6$: this is as expected for a beaming model with $R = 0.01$, $\gamma = 5$, which is essentially the Orr-Browne conclusion. Any departures in detail from the model may not be too much of a problem: it is unrealistic to suppose that there is no spread of (R,γ) values or that these are totally uncorrelated with power. Orr and Browne suggested that the model γ value could be regarded as the mean of some distribution; we disagree. From (8), taking canonical figures of $\delta=2$, $\alpha=0$, we get $\rho_{fs}/\rho_{ss} \propto \gamma^6$: the model γ value will tend to be close to any upper cutoff in a distribution.

What further tests of the Orr-Browne model do we have? Some impressive arguments in favour were provided by Wills and Browne (1986), who found a strong correlation between broad emission-line velocity widths and radio core dominance, in the sense expected if the line-emitting clouds are confined to a disk: compact quasars have narrower lines. Wills and Browne also make the good point that any comparison of extended and compact sources which aims to test the unified scheme must be made at constant extended power. Thus, for example, the differences between the respective underlying nebulosities found by Boroson et al. (1985) may simply reflect a dependence on extended power.

Miller (1984) presented an argument against the Orr-Browne scheme, based on the fact that (for a given X-ray power) compact quasars have radio cores on average ~30 times stronger than extended quasars. If the X-ray emission is unbeamed, this rules out the beaming by a factor ~1000 required in the Orr-Browne model. This argument has been countered by Browne and Murphy (1987), who conclude that the X-ray emission in quasars does indeed contain a beamed component. They are able to come to this conclusion because radio maps of high dynamic range are now available for almost all the X-ray quasars studied by Miller (1984) and the correlation of X-ray power with core dominance at constant extended radio power can be studied directly: the core-dominated quasars are on average a factor ~10 more X-ray luminous.

A final potential test of the Orr-Browne model concerns the cluster environments of compact sources. This is probably the best test of all since it is about the only property which can be guaranteed to be isotropic (nuclear line and continuum emission can be obscured by dust lanes). We know the local galaxy density for the parent extended sources, and that this decreases with increasing extended luminosity (Longair and Seldner 1979; Prestage 1985; Prestage and Peacock 1987), so that there is a clear prediction for what should be found for compact sources. The problem at present is that studies have been limited to $z < 0.1$-0.2 by lack of deep galaxy counts and so only compact sources with $P_{2.7} \lesssim 10^{24}$ $\mathrm{WHz^{-1}\ sr^{-1}}$ have been studied. However, Prestage and Peacock find the cluster environments of these objects to be very sparse, possibly less dense than for FRII sources (classical doubles: see Fanaroff and Riley 1974) and certainly lower than FRI sources, which are the putative parents of such weak core-dominated sources (although the redshift limit means we are not yet able to compare many objects of the same extended luminosity). This result requires confirmation and extension to higher redshifts (and hence powers) using CCDs to generate faint galaxy samples. If rich environments continue not to be found, this will be a strong argument against both the Orr-Browne model and the unified picture for blazars discussed next.

2.3 Blazars

The least ambitious (and most plausible) unified scheme was suggested by Blandford and Rees (1978). The term 'blazar' covers objects with highly polarised ($\gtrsim$ 3 percent) rapidly variable (~ 1 day) optical nuclear emission (see Angel and Stockman 1980). These properties make a static isotropic model extremely hard to sustain: either there will be a compton catastrophe if the relativistic electrons and magnetic field are in equipartition or, if not, rapid synchrotron cooling will produce thermal electrons which wash out the polarisation. Blandford and Rees therefore concluded that the case for bulk relativistic flow was strong. In forming a unified scheme for blazars, we may need to recognise the original division of the Angel-Stockman term into two classes: BL Lac objects or OVV quasars according to the absence or presence of strong broad emission lines. Blandford and Rees suggested that the former category be regarded as beamed versions of ordinary radio galaxies.

This hypothesis became testable given measurements of the extended flux density of BL Lacs. Browne (1983) showed that the local density of BL Lacs with extended $P_{1.4\ \mathrm{GHz}} \gtrsim 10^{22.6}\ \mathrm{WHz^{-1}sr^{-1}}$ was $\gtrsim 2 \times 10^{-8}\ \mathrm{Mpc^{-3}}$, which Browne claims to be about 0.017 times the density of elliptical radio galaxies with such luminosities. If correct, such a proportion of BL Lacs would be consistent with beaming into a cone of semi-angle γ^{-1} with $\gamma \simeq 5$ (although Browne's figure appears to be incorrect: see below). A more severe test (as in Section 2.2) would be to see if the overall luminosity function for BL Lacs behaves as predicted, but we have little idea how these objects evolve with cosmological epoch. Existing lists of Blazars (e.g. Angel and Stockman 1980) contain a surprisingly high proportion of bright low-redshift objects, and this has led some authors (e.g. Woltjer and Setti 1982) to ascribe a low degree of evolution to the blazar population. Existing blazar samples are of course grossly incomplete, but such a bias to low redshifts may be expected in beaming models. This is because it is well known that radio galaxies change their emission-line properties an an extended radio power threshold corresponding to the Fanaroff-Riley (1974) transition between classical double sources (FRII: $P_{1.4} \gtrsim 10^{24.5}$ $\mathrm{WHz^{-1}sr^{-1}}$) and more diffuse objects: only FRII sources have strong nuclear emission lines (Hine and Longair 1979). At low redshifts, blazars will tend to have extended emission below the FR threshold - forming a classical line-less BL Lac. This argument is supported by the data of Antonucci and Ulvestad (1985), who give extended luminosities for all the Angel and Stockman objects. Figure 4 shows histograms of extended luminosity for objects with known redshift, distinguishing those objects which Angel and Stockman class as strong lined from those which are not.

There is a clear separation at about the FR borderline, in line with the above argument. The remaining objects of unknown redshift are presumably largely of the lineless class: if they have $z \lesssim 0.2$ almost all will fall into the FRI class.

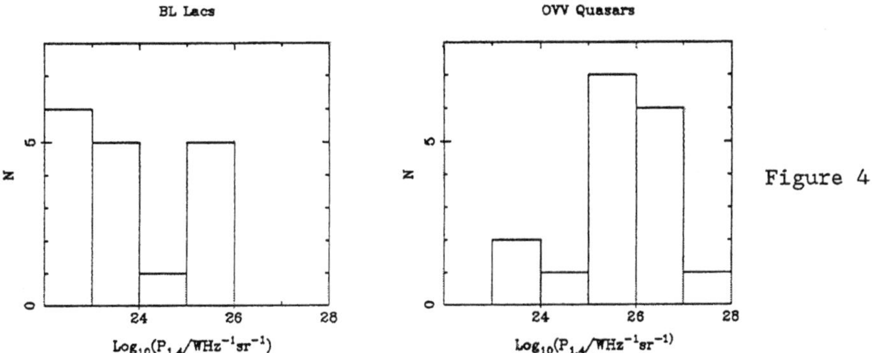

Figure 4

In addition to luminosity effects, the bias of BL Lacs to low redshift is assisted because FRI sources have relatively strong cores ($R \sim 0.01$) and beaming dominates more easily. Indeed, the 12 objects in Antonucci and Ulvestad's list that are core dominated by a factor > 50 all have extended $P_{1.4} < 10^{24.8}$ WHz^{-1}sr^{-1}. There is a problem here in that the most compact sources are core dominated by a factor ~ 1000, implying Lorentz factors $\gtrsim 10$ for $R = 0.01$, in conflict with the estimate from the space density. However, this calculation seems dubious: Browne adopts a space density of radio galaxies with $P_{1.4} > 10^{22.6}$ WHz^{-1}sr^{-1} of 1.3×10^{-6} Mpc^{-3}, whereas the luminosity function of Windhorst (1984) yields 1.1×10^{-5}, so that in fact only 1 object in ~ 500 is a BL Lac. It is difficult to reconcile such a low fraction of BL Lacs with such high maximum amplifications: the formal solution of $\gamma = 90$, $R = 4 \times 10^{-6}$ is nonsensical. Probably we need to abandon the assumption of a universal Lorentz factor. Suppose we had $P(>\gamma) = \gamma^{-x}$ and let us seek limits to x. From (3) we have for $A_m \gg 1$ $P(A{>}2) \simeq R^{1/(2+\alpha)}$; since $A_m \simeq 4\gamma^4 R$, the total probability of $A{>}2$ is $\sim R^{x/4 + 1/(2+\alpha)}$. If we allow for a (likely) substantial incompleteness in the BL Lac samples and take $P(A{>}2) \simeq 10^{-2}$, then $R = 0.01$ and $\alpha = 0$ imply $x \simeq 2$. For an illustrative model, this is quite a sensible answer, as it allows ~ 10 percent of BL Lacs to be very highly beamed ($\gamma \gtrsim 10$) without producing vast numbers of core-dominated objects.

More direct support for the idea of elliptical radio galaxies as the parent population for blazars comes from the observation of low-level blazar activity in the nuclei of FRI sources like Cen A (Bailey et al. 1986): polarised nuclear continuum may be detectable in many FRI sources (at least in the infrared, where extinction is less important). This will probably be the best way of testing the unified blazar model in the future.

3. ALTERNATIVES

We have painted ourselves into something of a corner over the last two Sections. The arguments in favour of both the Orr-Browne and

Blandford-Rees unified schemes appear quite impressive, but they may be inconsistent. This is because one can regard essentially all flat-spectrum quasars as blazars: in the radio ($\nu \gtrsim 10$ GHz), they display the characteristic variable polarised emission that defines this class. In the optical/infrared, variable polarised continuum is seen in $\gtrsim 50\%$ of objects (Meisenheimer, private communication); the exceptions are generally those of high optical luminosity where the synchrotron component is washed out by the 'blue bump' (the presumed accretion-disk feature seen at ~ 2000Å in the rest frame). We therefore have two competing alternatives for the parent population of flat-spectrum quasars: either extended steep-spectrum quasars or extended radio galaxies. Deciding between these on the basis of statistics of number densities will be hard: the extended emission about flat-spectrum quasars is usually of FRII luminosity (Browne and Perley 1986; Section 2.3) and the numbers of FRII quasars and radio galaxies are comparable (Section 2.2). There is then no conflict between the Orr-Browne model and the idea of BL Lacs being beamed FRI galaxies: we are instead concerned with the class of parent object for luminous compact quasars such as 3C273.

In trying to answer this question, we are faced with another puzzle - what is the relation between the FRII sources which are galaxies and those which are quasars? A recent review of this problem was given by Owen (1986): the extended structures of both classes are tantalisingly similar and both appear to possess central cores and one-sided jets, but the latter are a factor ~10-100 weaker in radio galaxies. The similarity in appearance and number strongly argues for a connection between the two classes, and there are really only two natural alternatives: relation by a difference in time or a difference in orientation. In the first case, the nuclear quasar activity would be of shorter duration than the main radio outburst, with a radio galaxy spending ~50% of its time in an 'excited' quasar state. To explain why we always see the strong one-sided jet with the nucleus, we would require the flaring to last much longer than the travel time along the jet. If we use orientation to explain the difference, then the approximate equality of radio galaxies and extended quasars would imply $\mu > \frac{1}{2}$ in (2) for quasars. For the standard model, this implies beaming by a factor > 4 with respect to $\mu=0$ (or a factor $\simeq 2$ lower if we include the counterjet), rising rapidly with μ. This could make quite an attractive picture, with galaxies for $\mu \lesssim 0.5$, extended quasars for $0.5 \lesssim \mu \lesssim 0.9$ and core-dominated quasars for $\mu \gtrsim 0.9$, were it not for one problem: how do we switch on the quasar light for $\mu \gtrsim 0.5$? We cannot do it by beaming: there is a beamed component in the light of compact quasars only (see above). Nevertheless, there is evidence for anisotropic light in radio galaxies (e.g. Cygnus A: Pierce and Stockton 1986) and the most natural way of making it visible over ~ 2π sr is by obscuration. Searches for obscured quasars in radio galaxies (e.g. Fabbiano et al. 1986) would seem highly worthwhile. It may seem contrived to have two mechanisms for anisotropy in the radio and optical, but we may have little alternative. Evidence is now arriving that superluminal motion may be found often in cores of quasars selected via their extended lobes (Zensus and Porcas 1986). On a conventional picture, this forces us to have beams pointing towards us, and something like obscuration is then the only obvious way to fix up the optical properties.

There are two ways of looking at the above discussion. We can be positive and note that beaming models come close to unifying a wide range of phenomena in an appealingly simple way; real life is messy enough that we shouldn't be surprised if one or two details don't appear to fit in at first sight. Alternatively, we can regard any sort of ad hoc tinkering with the model as a modern version of mediaeval epicycles - we may be working with a fundamentally unsound picture. Certainly, the number of tests of unified models is depressingly small and it is still possible that core-dominated quasars will have to be understood on their own terms as a distinct physical class. Indeed, we already know of one class of powerful radio source which we are happy not to fit into any unified picture: the compact steep-spectrum sources such as 3C48 (e.g. van Breugel et al. 1984). In the end, probably the best thing to do is to continue detailed studies of putative beamed objects. For example, the beautiful MERLIN maps of 3C273 (Davis 1986) with their >50:1 asymmetric structure may place stronger constraints on unified models than any study of the grand sweep of population statistics.

REFERENCES

Angel, J.R.P. & Stockman, H.S., 1980. Ann. Rev. Astron. Astrophys., 8, 321.

Antonucci, R.R.J. & Ulvestad, J.S., 1985. Astrophys. J., 294, 158.

Bailey, J., Sparks, W.B., Hough, J.H. & Axon, D.J., 1986. Nature, 322, 150.

Bartel, N., Ratner, M.I.,Shapiro, I.I., Herring, T.A. & Corey, B.E., 1984. IAU Symp. No. 110, "VLBI and compact radio sources" eds. R. Fanti, K. Kellermann & G. Setti, (Dordrecht: D. Reidel), p113.

Blandford, R.D. & McKee, C.F., 1976. Phys. Fluids, 19, 1130.

Blandford, R.D. & Rees, M.J., 1978. Pittsburgh Conference on BL Lac Objects, ed. A.M. Wolfe, p.328. Univ. Pittsburgh Press.

Boroson, T.A., Persson, S.E. and Oke, J.B., 1985. Astrophys. J., 293, 120.

Bridle, A.H. 1984. NRAO Workshop No. 9, "Physics of Energy Transport in Extragalactic Radio Sources", eds. A.H. Bridle & J.A. Eilek, p.135.

Browne, I.W.A., 1983. Mon. Not. R. astr. Soc., 204, 23P.

Browne, I.W.A., & Perley, R.A., 1986. Mon. Not. R. astr. Soc., 222, 149.

Browne, I.W.A. and Murphy, D., 1987. Mon. Not. R. astr. Soc., submitted.

Coleman, C.S., 1986. "The Structure and Evolution of Active Galactic Nuclei", eds. G.Giuricin, F. Mardirossian, M. Mezzetti & M. Ramella, (Dordrecht: D. Reidel), p521.

Condon, J.J., O'Dell, S.L., Puschell, J.J. & Stein, W.A., 1981. Astrophys. J., 246, 624.
Davis, R.J., 1986. IAU Symp. No. 119, "Quasars", eds. G. Swarup and V.K. Kapahi (Dordrecht: D. Reidel), p211.
Fabbiano, G., Willner, S.P., Carleton, N.P. and Elvis, M., 1986. Astrophys. J., 304, L37.
Fanaroff, B.L. & Riley, J.M., 1974. Mon. Not. R. astr. Soc., 167, 31P.
Hine, R.G. & Longair, M.S., 1979. Mon. Not. R. astr. Soc., 188, 111.
Kapahi, V.K. & Kulkarni, V.K., 1986. IAU Symp. No. 119, "Quasars", eds. G. Swarup & V.K. Kapahi, (Dordrecht: D. Reidel), p207.
Lind, K.R. & Blandford, R.D., 1985. Astrophys. J., 295, 358.
Longair, M.S. and Seldner, M., 1979. Mon. Not. R. astr. Soc., 189, 433.
Lynden-Bell, D., 1977. Nature, 270, 396.
Marscher, A.P. and Broderick, J.J., 1981. Astrophys. J., 247, L49.
Miller, L., 1984. IAU Symp. No. 110, "VLBI and compact radio sources", eds. R. Fanti, K Kellermann and G. Setti (Dordrecht: D. Reidel), p189.
Orr, M.J.L. & Browne, I.W.A., 1982. Mon. Not. R. astr. Soc., 200, 1067.
Owen, F.N., 1986. IAU Symp. No. 119, "Quasars", eds. G. Swarup and V.K. Kapahi (Dordrecht: D. Reidel), p173.
Peacock, J.A., 1985. Mon. Not. R. astr. Soc., 217, 601.
Peacock, J.A., Miller, L. & Longair, M.S., 1986. Mon. Not. R. astr. Soc., 218, 265.
Perley, R.A., Fomalont, E.B. & Johnston, K.J., 1982. Astrophys. J., 255, L93.
Pierce, M.J. and Stockton, A., 1986. Astrophys. J. 305, 204.
Prestage, R.M., 1985. PhD Thesis, Univ. of Edinburgh.
Prestage, R.M., and Peacock, J.A., 1987. Mon. Not. R. astr. Soc., submitted.
Rees, M.J., 1967. Mon. Not. R. astr. Soc., 135, 345.
Scheuer, P.A.G. & Readhead, A.C.S., 1979. Nature, 277, 182.
Schilizzi, R.T. & de Bruyn, A.G., 1983. Nature, 303, 26.
Schmidt, M., 1970. Astrophys. J., 162, 371.
Smith, M.G. & Wright, A.E., 1980. Mon. Not. R. astr. Soc., 191, 871.
van Breugel, W.J.M., Miley, G.K. and Heckman, T.M., 1984. Astr. J., 89, 5.
Walker, R.C., 1984. NRAO Workshop No. 9, "Physics of Energy Transport in Extragalactic Radio Sources", eds. A.H. Bridle and J.A. Eilek, p20.
Wills, B.J. and Browne, I.W.A., 1986. Astrophys. J., 302, 56.
Windhorst, R.A., 1984. PhD Thesis, Univ. of Leiden.
Woltjer, L. & Setti, G., 1982. Proc. Vatican Study Week, "Astrophysical Cosmology", Pont. Acad. Scient. Scripta Viria 48, p293.
Zensus, J.A. and Porcas, R.W., 1986. IAU Symp. No. 119, "Quasars", eds. G. Swarup and V.K. Kapahi (Dordrecht: D. Reidel), p167.

MAGNETIC EQUILIBRIA OF JETS

Gregory Benford
Department of Physics
University of California
Irvine, CA 92717, USA

ABSTRACT. It seems likely that streams of matter ejected from rotating magnetospheres must inevitably carry away currents, with return currents driven inductively at the head of a jet. Propagating equilibria probably involve a compressed current region surrounded by a plasma halo carrying a return current. Magnetic fields will generally be twisted, with the azimuthal field B_θ confining the hotter synchrotron-bright inner region, where the axial field B_z is strongest. Magnetic confinement can come to the fore as the external galactic pressure diminishes with jet propagation distance. Synchrotron losses themselves can lead to equilibria which have currents very nearly aligned with B, producing states of lessening net radial force. This system can be bent by external pressures which act through the insulating B_θ blanket around the jet core, keeping actual particle scattering low because pressures are mediated magnetically. Observations show that jets displaying B_z polarization are bright and emerge from only one side of galactic cores, while two-sided sources are relatively dimmer and show B oriented perpendicular to the jet axis. This strongly suggests that establishment of magnetic equilibria is critical far down in the core. We may be able to deduce the features of this core by determining the jet current, since this is an invariant of the flow. Cascade of magnetic energy down to smaller lengths may fuel particle scattering and hence synchrotron emission, as well. Magnetic fields may thus enter into every major feature of jets.

There is some evidence that jets from active galactic nuclei are not confined by a passive intergalactic gas density (Bridle and Perley, 1984). The usual argument is that the internal jet pressure must be at least that given by the equipartition argument (relativistic particle pressure roughly equals magnetic field pressure). When this minimum energy density cannot be confined by the external gas, a natural confining mechanism is ordered, self-generated magnetic fields. This demands that a net current flow in the jet. (Benford, 1978; Chan and Henriksen, 1980; Bicknell and Henriksen, 1980)

Unfortunately, a confining azimuthal B_θ which wraps the jet is difficult to observe. If the jet is not well resolved, the helical

W. Kundt (ed.), Astrophysical Jets and Their Engines, 197–204.

configuration can "blend out" (Bridle et al., 1986).

Also, seeing B_θ far out from the center is sensitive to the radial distribution of synchrotron electrons. Concentration of them at the compressed center would suppress the B_θ signature. Geometry also comes into play. There is only a restricted range of geometries for which synchrotron emission from a helical field produces a polarization perpendicular to the jet axis at its edges (Laing, 1981). To do so requires (a) that $B_\theta >> B_z$ (tight winding), and (b) that the jet lie nearly in the plane of the sky. Also, tangled field configurations can ape ordered ones in the convolutions of synchrotron emission. A saving note is that Faraday rotation may dispel illusion. Disordered fields cannot produce an organized rotation measure gradient across a jet. A confining field, though, will yield a reversal in sign of the integral of B_θ along an azimuthal coordinate. This should give a clear signature unless the jet is pointing nearly at us. It demands enough thermal matter to make the B_θ-dominant region Faraday thick at frequencies where resolution is possible with existing instruments, 0.3 to 3 GHz. In this range synthesis-array sizes and sensitivities, together with ionospheric effects, allow arcsec polarimetry resolution. (Bridle, Perley and Henriksen, 1986).

A net current I_j must come from the central engine or else be supplied by some dynamo along the jet path. This seems unlikely because the only source in the jet must be an inherent helicity in the flow. But as a jet expands, helicity diminshes and cannot make fresh I_j. Thus jets cannot generate much confining field once they have expanded to more than a few times their original radius.

Consider, then, how much current a central accretion disk must carry in order to produce a standard-strength magnetosphere. We envision (Fig. 1 from Camenzind, 1986) a disk field B_θ which reverses sign as we pass through the disk axially, since jets flow out both polar regions. Because the magnetic field B_θ emerging from an accretion disk at radius R* must reverse sign in the disk, a radial current $I_r = 10^{19}$ Amp $R^*_{16}(B_\theta/100G)$ must flow. This current seemingly flows into the corotation region of the inner magnetosphere of a collapsed central object and may connect with the high latitude region where a Goldreich-Julian density must exist, $n_{GJ} \cong 10^{-5}(B_z/100G)/R^*_{16}\ cm^{-3}$.

Electric fields will drive away a current $I_j = n_{GJ}ecA_j$ at a speed $\approx$ c. This outflowing current in a jet has a luminosity

$$L_j = 3\ 10^{31}(I_r/10^{19}A)\gamma_j\ erg/s$$

if I_j carries away I_r. This is not a high luminosity even if the jet energy factor $\gamma_j \sim 10^5$. If $A_j \approx 2x10^{31}\ cm^2$, we insure that $I_j \sim I_d$ near the object. (As an example at smaller scale, for a pulsar with a disk nearby, $B_\theta \sim 10^6 G \sim B_z$ will require an area $A_j \sim 10^{13}\ cm^{-2}$, through which flows $\sim 10^{14}$ Amp. This is enough to provide corotation current for the Crab pulsar.)

Confining an extragalactic jet of pressure P_j and radius R requires $I_j \approx 10^{19}$ Amp $(P_j/10^{-12})^{1/2} R_{kpc}$. So we see that the

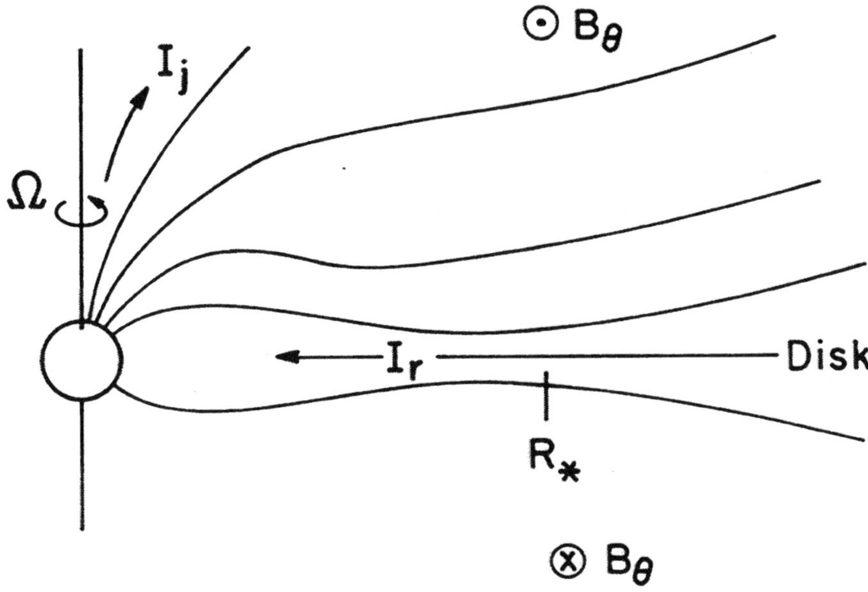

Figure 1. A corotating magnetosphere formed by magnetic fields either attached to a central rotor or to the disk orbiting a rotor. In the disk a radial current I_r must flow to support the change in sign of B_θ which occurs axially at the disk layer. Plasma centrifugally driven out at the polar zone carries a density at least that required by co-rotation. For some conditions $I_j \sim I_r$, the outflow jet current. The return current from this circuit may find its way back to the disk and comprise I_r.

confining current I_j agrees roughly with the magnetic field in the disk of $B_\theta \sim 100$ G. Thus conventional disk models allow I_j of about the right magnitude, and this I_j will create a closed circuit as it propagates out. How does it generate the return current, I_r?

Clearly a jet will prefer to drive I_r through a cocoon area A_c exceeding the jet area A_j. Since $|I_r| = n_r v_r e A_r \cong |I_j| = n_j v_j e A_j$, the energy in the return current E_r is

$$E_r = E_j \ (nA)_j/(nA)_r$$

so to minimize E_r/E_j a jet should seek a dense, large area cocoon, $(nA)_r >> (nA)_j$. A fat, dense cocoon emerges from the numerical simulations of Smith et al. (1985), if $n_j/n_r << 1$. Also, a primary reason to invoke currents is to provide magnetic confinement of jets that appear to have inadequate confinement by external gas pressure. A hoop field B_θ provides extra pressure on the central jet, of average level (Achterberg, these proceedings)

$$\langle P_j \rangle \approx 2P_{ext}(A_r/A_j) \quad .$$

So to minimize the energy invested in driving I_r, and to simultaneously provide magnetic confinement, we want fat, high density cocoons which carry the return current. But how can I_r be spread over $A_r \gg A_j$?

Figure 2 shows a jet current driving an induction field E_z. An important point is that the jet is always flared at the head because of (1) scattering by the dense plasma, and (2) inadequate B_θ to

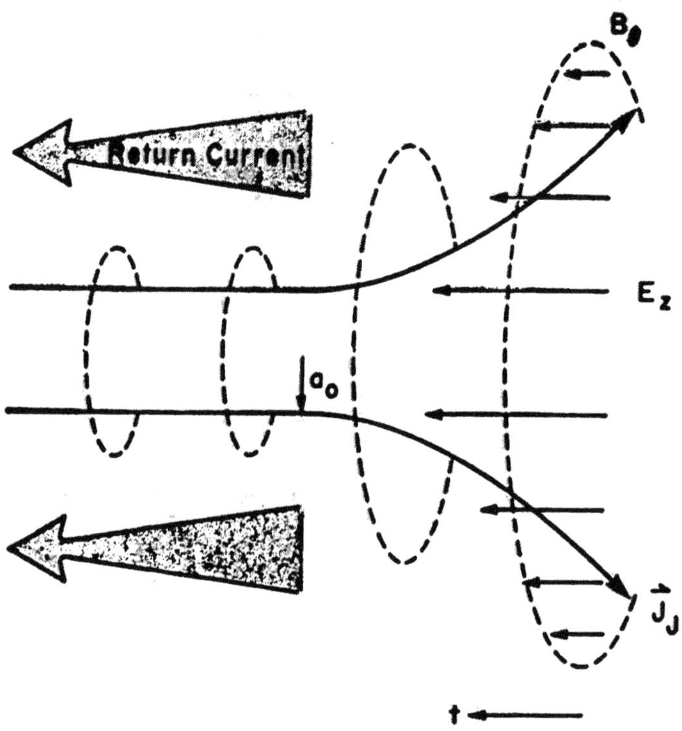

Figure 2

restrain the outward pressure. This is crucial to establishing the return current by the inductive field E_z, which arises from dI/dt.

But the perfect MHD picture would lead to the return current being driven only in a skin depth, $c/\nu_p \cong 10^8$ cm within the jet radius. This means the jet will be a sheet pinch, with B_θ distributed inside the jet, growing with radius, but then dropping to zero within the skin depth. There is nothing inherently wrong with this picture, but it is probably unrealistic in the face of imperfect conductivity and effects of beam expansion (see below). Further, we know that gross fluid vortices characterize the backflow of jet matter from the head into the cocoon (Smith et al., 1985). It seems natural to suppose such vortices will carry current into the cocoon, dispersing it. Then the relevant scale for spreading I_r is the same as the dynamic radius of the hot cocoon, which often has $A_r >> A_j$.

Is it plausible that such currents arise and yield stable regimes? We know nature does use currents to make beams propagate. Lightning produces fields of $10^3 - 10^4$G, making a major contribution to the pressure (atmosphere plus magnetic) of $\sim 10^6$ dyne/cm^2. Lightning travels about 10^4 times its radius before it strikes the ground or a cloud, which provides the conducting path for charge balance. The discharge is unstable for hose-like sidewise motions. One does not see lightning strokes wriggling in neat sinusoids because the growth time is comparable to the total stroke time, about 1 sec.

The Crab Nebula must return the currents emitted by the pulsar, over a scale > light year (where the "wisps" occur). The Crab current is 3×10^{14} Amp (derived from the Goldreich-Julian corotation currents); this is 10^{10} times the current in a lightning bolt. Confining such quasar jets as 4C32.69 requires a current $I \sim 1000$ times this.

In astrophysical cases, the jet charge is taken away in a time R_j/c, and the voltage drop at the end of the jet is $\sim 10^6 (\nu/\nu_p)L$ volts where ν is the collision frequency associated with the turbulence, ν_p the plasma frequency, and L is the depth of the resistive layer (not necessarily the "working surface") in units of 10 kpc. Thus we don´t expect more than an MeV/particle from this zone, and the voltage drop is less important than the collisonal losses.

MAGNETIC NOZZLES

We expect that a pinched jet will be most luminous where pressure is highest; i.e., the axis, where B_z dominates. Thus the confining B_θ is pushed conveniently off to the dim cocoon region, where it is hard to observe.

Still, if our aim is not to merely "study the weather" (Rees, 1982), then invoking currents necessarily tells us something about the birthing region itself. Currents make loops, and any J_z seen at 100 kpc distances must have come from the magnetosphere or nozzle. Clear evidence of B_θ thus tells us something we can use in exploring acceleration models in the core.

A valuable hint may lie in Bridle´s (1982) noting that $B_\parallel$-dominated jets are bright and one-side, while $B_\perp$-dominated jets are dim and two-sides. I suggest that:

(a) the bright sources have a larger seed $B_\parallel$ to shear-amplify (and perhaps larger shear) so $I(\nu) \propto B_\parallel^2$ is higher;

(b) the seed $B_\parallel$ is close to the value which would blow open the nozzle in even the strongest jets, $\sim 10^4$ G;

(c) the other side of the galaxy has seemingly minor difference in confining gas density, or a slightly higher $B_{\parallel}$, which takes that nozzle beyond the blowout limit. This jet is then unconfined at small radius and spreads its energy over a region, lowering its surface luminosity below what we can see. (Further out, focusing occurs to give distant hot spots).

The B_{θ}-dominated jets, on the other hand have smaller seed $B_{\parallel}$, do not disrupt their nozzles, and because their seed $B_{\parallel}$ is not close to the blowout field, can tolerate fluctuations in conditions on both sides of the core. (Fig. 3).

This demands that asymmetry deep within the engine persist for many dynamic time scales. By relating the failure to focus to one parameter, $B_{\parallel}/B_{\theta}$, which governs confinement at small radii, this picture connects several observed systematics. The obvious way to test this idea is through VLBI polarization, if we can see a $B_{\perp}$-dominated region closer in than the $B_{\parallel}$-strong regime.

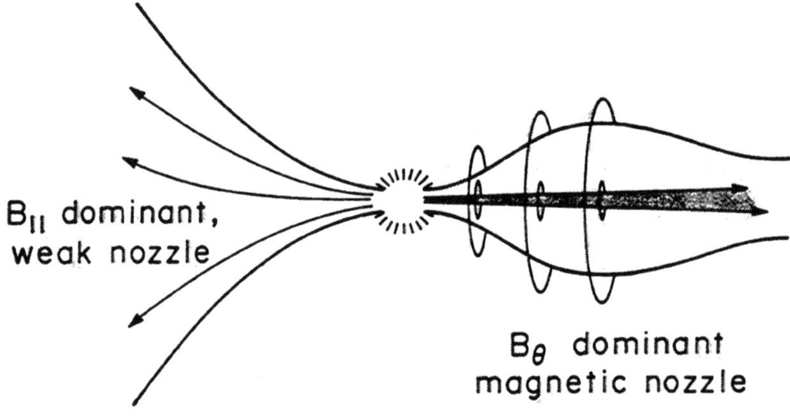

Figure 3.

On the $B_{\parallel}$-dominant side, though there will be some B_{θ}, the expansion rate $d\theta/dt$ will be large. Recollimation can occur downstream, producing lobes. On the confined B_{θ} side, $d\theta/dt$ is low. Because a helical field geometry with significant B_{θ} can tolerate some sidewise displacement without disruption, some misalignment between jets on the pc and kpc scale is natural and self-correcting.

There are grave difficulties in making the original Blandford and Rees nozzle picture work. The nozzles seem to be 100 pc or so long (the scale height of the galactic gas), and yet VLBI doesn't seem to show us these features (though, of course, the story isn't fully in yet). I suggest that a way to shorten the scale length for self-built nozzles is to invoke a shorter distance -- the self-pinch length. Perhaps including a dominant B_{θ} in calculations such as Siah and Wiita's (1986) could yield magnetic nozzles of the needed size. In such environments, synchroton losses are probably vital. Such nozzles could not last forever. Magnetic diffusion times in geometries with ratio β = plasma pressure/magnetic pressure, ion temperatures T in GeV and collison frequencies ν comparable to the cyclotron frequency are

$$\text{diffusion time} \approx 10^7 \text{ yr } R_{14}^2 \; B_4 \; \beta_{-2} \left(\frac{\nu_c}{\nu}\right)_1 T^{-1}$$

which might imply a comparable flip-flop time. The main point is that magnetic nozzles may be implied by studies of confinement at 100 kpcs away. This motivates continuing studies of jets far from the "action".

ELECTROMAGNETIC JET INTERACTIONS

Outside the cocoon, an observer sees no net current since $I_r = - I_j$. This means an intergalactic magnetic field B_g cannot exert any $\vec{I}_j \times \vec{B}_g$ force on a jet unless B_g penetrates the cocoon and samples a fraction of I_j not shielded by I_r. This demands that the same process of head spreading which distributes I_R must entrain B_g as well, as seems plausible.

When a jet emerges from a galactic halo and senses an oranized B_g which can turn it, the jet will curve in the region where B_G is ordered. The radius of curvature will be

$$\rho_c = 125 \; R_j \left(\frac{n_j}{10^{-4}}\right)\left(\frac{v_j}{1000 \; \frac{km}{s}}\right)^2 \frac{1}{(B_{-3} B^*_{-6} \sin\theta)}$$

where B* is the intergalactic field and B the jet self-confining field. There is some evidence that some highly bent sources may be distorted in this way (J. Eilek et al., 1984).

The principal dynamic aspect of an ordered magnetic field is to serve as a buffer between the slings and arrows of the external environment and the beam. The elastic quality of magnetic fields can allow perturbations to deform the outer regions of a jet-cocoon system, and deflect impinging flows, without permitting abrupt faces to act upon the actual jet. This can make the synchrotron-luminous core appear more serene than the environment would normally allow.

Further, synchrotron emission itself can affect the evolution of magnetic equilibria. Radiation exhausts particle energy perpendicular to the ambient field, forcing $\vec{J}$ more nearly parallel to $\vec{B}$ (Noerdlinger, 1968). This suggests that, unless reacceleration acts as a generator to produce perpendicular currents, equilibria will evolve to minimize particle faces; i.e., toward more nearly "force-free" equilibria. These facets can affect perturbations endangering jet stability.

Laboratory experiments injecting relativistic electron beams into dense plasma can illuminate the return current formation process. Spulak and Burns (1984) discussed the possibilities. I have carried out preliminary experiments with my graduate students, particularly James Means. Our system has $n_j/n_p \approx 10^{-2}$, $E_j \approx$ MeV, $R_j \approx 3$ cm, outer conducting wall radius $R = 10$ cm and length 1 m. We find that return currents appear at r >> R_j, and follow a profile of $I_{net} = I_j - I_r = I_j(1-\alpha r)$ outside the beam. About half the jet (beam) current returns in the plasma and half in the chamber walls. This means the screening length $c/\omega_p \sim$ cm is not the limit on dispersal of I_r. More experiments in this domain may help explore the development of extensive magnetic equilibria.

REFERENCES

Benford, G. (1978), Mon. Not. R. Astron. Soc. 183, 29.

Bicknell, G. V., and Henriksen, R. N. (1980), Astrophys. Lett. 21, 29.

Bridle, A. H. (1982), in Extragalactic Radio Sources, IAU Symposium No. 97, (ed. D. S. Heeschen and C. M. Wade, Reidel, Dordrecht), p. 121.

Bridle, A. H., and Perley, R. A. (1984) Annu. Rev. Astron. Astrophys. 22, 319.

Bridle, A., Perley, R., and Henriksen, R. 1986, Astron. J. 92, 534.

Camenzind, M. (1986), Astron. and Astrophys. 156, 137.

Chan, K. L., and Henriksen, R. N. (1980), Astrophys. J. 241, 534.

Eilek, J. A., Burns, J. O., O'Dea, C. P., and Owen, F. N. (1984), Astrophys. J. 278, 37.

Laing, R. A. (1981b), Astrophys. J. 248, 87.

Noerdlinger, P. (1968), Phys. Rev. Lett. 20, 146.

Siah, M., and Wiita, P. (1986), Astrophys. J. 300, 605.

Smith, M. D., Norman, M. L., Winkler, K.-H.A., and Smarr, L. L. (1985), Mon. Not. R. Astron. Soc. 214, 67.

Spulak, R. and Burns, J. (1985) in Physics of Energy Transport in Extragalactic Radio Sources, A. H. Bridle and J. A. Eilek, eds., Proc. 9th NRAO Workshop, NRAO, Greenbank, USA.

STABILITY OF MAGNETIC JET EQUILIBRIA

Gregory Benford
Department of Physics
University of California
Irvine, California 92717 USA

ABSTRACT. Magnetically confined plasma is notoriously unstable. Jets self-confined by magnetic fields can have advantages, however, if their streaming velocities exceed the Alfven speed. Traditional MHD modes (kink, pinch) have reduced growth rates in flowing jets. If the confining field threads a surrounding dense cocoon, this lowers the growth rates for wavelengths exceeding the jet radius, since motions must drag this cocoon mass along with the perturbation. The very straight jets of the well-aligned classical double radio sources imply remarkable sideways stability over distances of 1000 r_j. This implies either Mach numbers exceeding 50, quite dense jets, or stabilization by magnetic linking of jet and cocoon, which increases growth lengths by M_{cocoon}/M_j. Wriggling produces sideways shocks, which can also slow growth. B_z builds up from a seed field by velocity shear amplification, and thus may not influence stability until it becomes correlated over distances $> r_j$. Reflection modes ricochet around the interior, probably driving shocks which in turn may accelerate particles. Detection of a magnetically ordered cocoon would imply a major role for magnetic fields in the origin and propagation of jets.

Few jets are straight and serene, terminating in well-organized lobes. Instead, most turn and break into complex patterns.

Current-carrying beams are liable to many modes of instability. However, from ample study of solar coronal magnetic arches we know that when magnetic forces are significant, systems can evolve to minimize the net force on particles. This means $\vec{\nabla} \times \vec{B} = \mu\vec{B}$ and the equilibrium is a helix, with B_z strongest at small cylindrical radius (r) and B_Θ strongest further out. The magnetic helicity

$$K = \int \vec{A}\cdot\vec{B}\, dV$$

is conserved (Taylor, 1974); here, $\vec{A}$ is the vector potential. This minimum magnetic energy state allows two modes: an axisymmetric m=0, and m=1, where the perturbation goes as exp(imθ). Königl and Choudhuri (1985) showed that m=1 oscillations could be visible in jets as enhanced synchrotron emission which oscillates alternately to

W. Kundt (ed.), Astrophysical Jets and Their Engines, 205–210.

one side of a jet and then, downstream, to the other. This seems to fit observations in NGC 6251.

Generally, though, jets may not be able to evolve to a so-called "force-free" (actually, force-minimized) state. Lacking such knowledge of the equilibrium, one must consider general features which are insensitive to unknown internal features (for example, to the details of a magnetically confined equilibrium — $B_z(r)$, $B_\theta(r)$ -- or the exponent in the gas law of the confined plasma). Luckily, these affect long wavelength modes least, and large disturbances are most observable in radio maps.

LARGE SCALE INSTABILITIES

Fluid instabilities can be modified by a dynamically important $\vec{B}$. Generally B_θ destabilizes by augmenting sausage $m=0$, $\lambda \approx 2\pi R_j$) and kink (helical, $m=1$) modes. The fastest growing modes have $1 < \omega/kc_s < M$, with M the jet Mach number. If the Alfven speed greatly exceeds the internal sound speed, $v_A >> C_s$, all modes but the kink are stable. (Smarr, et al. 1984) Fluting ($m \geq 2$) can be suppressed by beam expansion but not pinching or kinking. Growth of pinches is most rapid at $\lambda \approx 2\pi R$, while kinks reach maximum growth at $\lambda \approx 6\pi M R_j$, and falls at higher λ. Pinches are easy to confuse with shock waves reflected from the jet walls and focusing into disk-shaped zones, where the added pressure presumably increases the synchrotron emission (Achterberg et al., 1983). Several attempts to use such models to deduce M or v_j (Ferrari, 1985) have met with plausible results ($1 << M \lesssim 10$).

It is difficult to see how jets of great length can maintain high M (or relativistic γ) in the face of entrainment, internal shocks, etc. Knots and flutes will persist but should not greatly thwart propagation. Kinks, though, should grow in a distance set by the phase velocity of the mode, v_p, so $L \approx v_p \omega_i^{-1}$ with ω_i the temporal growth rate. The phase velocity is a fraction of v_j (Hardee, 1983; Cohn, 1983). Beam expansion obeys (Chan and Henriksen, 1980)

$$\frac{dv_r}{dz} = 2\left[p_j + \frac{B_z^2}{8\pi} - \frac{B_\theta^2}{2n\pi}\right] (\rho_j R_j v_z)^{-1}$$

for slow motions. Since $B_z^2 \propto R_j^{-4}$, but both P_j and B_θ^2 scale as R_j^{-2}, an equilibrium with $p_j \approx B_\theta^2$ may form early in a jet's life and restrict pinch radial excursions. Here n = 4 for a surface jet current and n = 2 for current flowing uniformly throughout the jet, reflecting the greater stabilizing capability of distributed currents. Cohn (1983) showed that $M > 3$ pinching rates are too slow to affect all but the longest jets.

Conveniently, results for thermally confined jet calculations can often be used on magnetically confined jets, since for $M > 1$ or super Alfvenic flows the energy resides in fluid flow, not B^2. (See review by Ferrari, 1985). Even M87, which seems conspicuously unconfined by thermal pressure, can be studied this way (Hardee,

1985). Shia and Wiita (1986) found in numerical simulations that, in general, a self-field helps stabilize pinches. Interestingly, a B_θ dispersed into the volume around the jet was more stabilizing than if it is compressed into a small sheet around the jet. This favors thick cocoons carrying a jet return current.

The same is true of kink modes, because an ordered magnetic structure outside the jet will attach external ("cocoon") mass to the beam motion. The radius affected is about v_A/ω_i. For a disordered field in the cocoon, sound waves can force a blanket of mass around the slewing jet to move within a radius $\sim C_s/\omega_i$. The extra mass loaded on the sideways motion slows linear exponential growth but does not suppress it. However, when the sidewise excursion $\xi > \lambda/M$, with λ the observed wavelength, shocks can occur either in the external gas or in the jet. Such sidewise shock stabilization (Benford, 1981) slows kink growth to algebraic ($\xi \sim t^2$). The Centaurus A jet may display just such an effect (Clarke, Burns and Feigelman, 1986), complete with detailed edge brightening, filaments and streamers at 0".3 resolution. Hardee (1984) has argued that jet expansion can stop kink and other modes if $\dot{R}_j/R_j$ exceeds the (already shock-affected) growth rate. Such suppression must occur, since the "wiggles" in the Centaurus A jet seem to grow no more beyond the region of apparent sidewise shocks.

Sources such as NGC 6251 require a way to keep a jet stable against sidewise (kinking) instability for a distance $L \approx 500\ R_j$. Mass loading by a cocoon with mass M_c per unit length will allow a jet to survive

$$\frac{L}{R_j} \leqslant 18\pi \frac{M_c}{M_j} M^2\ N \sin^2\phi$$

where M is the Mach number, M_j the jet mass per unit length, ϕ the projection angle and N the maximum number of kink e-foldings tolerated ($N \leqslant 2$).

Consider a jet whose I_j is partially neutralized within a radius R_*, $I_r = -fI_j$. Beyond R_* a distant highly conductive outer boundary carries the remainder, $(1-f)\ I_j$. Neglecting the image forces in this distant region (Benford, 1978), we find kinking is stable if

$$\left(\frac{\lambda}{2\pi R_*}\right)^2 > \frac{\ln(\lambda/2\pi R_*)}{f} \quad .$$

This shows that long wavelengths can be suppressed if f is not too small; i.e., B_θ is not large. Beam expansion can steal free energy from the kink, stabilizing it if the expansion rate $\dot{R}_*/R_*$ is large, since the total kink growth rate (neglecting internal jet dynamics) is

$$\omega_i \approx \frac{\dot{R}_*}{R_*} - \frac{v_A}{R}\sqrt{\left(\frac{\lambda}{2\pi R_*}\right)^2 \ln\left(\frac{\lambda}{2\pi R_*}\right) - \frac{1}{f}} \quad .$$

FILAMENTATION

Consider a current-carrying plasma (a jet) penetrating another plasma which has high conductivity but no net current. Quickly the jet induces an opposite current in the plasma. A purely magnetic instability occurs, based on the fact that parallel currents attract. This filamentary mode is the magnetic analogy of the Jeans instability in gravitation, and begins with the growth of small pinches on a scale $\sim c/\omega_{pe}$. These nested pinches carry equal and opposite currents, reflecting the ability of the conductor to cancel the net current of the imposing flow jet. Linear growth at the rate $\approx \omega_{pj}(v_j/c)$, with ω_{pj} the jet plasma frequency, occurs far faster than any fluid MHD mode. Pinching rapidly approaches a nonlinear stage in which each pinch of size L* experiences a confining magnetic field B_* which perturbs the original v_z electron motion. When an electron oscillates laterally at a frequency

$$\omega_B \approx \left(\frac{2\pi e B^* v_j}{mc\ L^*}\right)^{1/2} ,$$

which eventually exceeds the linear growth rate, energy flow from jet kinetic energy into B_*^2 slows, but does not necessarily stop. Setting $\omega_B \cong \omega_{pj}(v_j/c)$ gives a small pinch size (Lee and Lampe, 1973, Benford, 1972):

$$L \approx 5\ 10^7 \mathrm{cm} \sqrt{(P/10^{-12}\ \mathrm{dyne/cm}^2)}(c/v_j)\ n_{-4}^{-1}$$

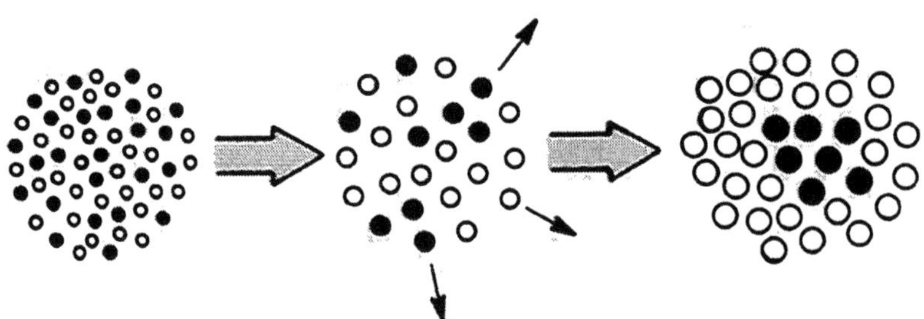

Fig. 1. Evolution of Filaments. Here solid dots are the filamented jet current and open dots the return current. A few of each type are ejected in the final nonlinear stages.

This is not the end of evolution, since pinches of the same current sense continue to attract each other, though mediated by a sea of oppositely directed current pinches. At this stage pinches grow by accretion and fluid MHD properties begin to matter. Since turbulence on a large scale at the jet-intergalactic medium is probably dominated by vortex structure, we can expect that the nonlinear pinch scale L >> L* will be the minimum size from which to estimate the appropriate fluid viscosity. No clear way appears to do this without numerical simulation, since here we face the knotty problem of connecting the far faster and smaller plasma modes to the larger fluid modes. We do know from purely noncollisional simulations of filamentation (Lee and Lampe, 1973) that the evolving pinches conglomerate into a central core of several pinches carrying a major fraction of the original jet current, I_j. Surrounding these now self-pressured pinches is a halo of oppositely directed current, a jostling crowd of somewhat smaller pinches. Some small pinches of both signs are ejected by the rapidly changing forces in the crowd. This ejection of "streamers" occurs when high-current beams propagate through plasma, and especially in neutral gases which the beam must first ionize to provide charge and then current neutralization necessary for its propagation. The figure suggests how return current finds its way to the outer region as required if the cocoon carries it. This occurs in laboratory experiments with return currents, in which the beams are filamentary unstable. The entire process is a way for I_j to propagate at a cost of thermalizing, since it reaches a Bennett equilibrium (Miller, 1982). This distribution has a density profile $n(r) = n_o[1+(r/R)^2]^{-2}$.

This suggests that jets penetrating the intergalactic plasma must filament if they carry a net current. Far from merely disrupting jets, filaments help transport current into an outer layer, where it returns to the source. "Streamers" should emerge and propagate into the surrounding gas. They will probably be most obvious near or in lobes, since they are aberrant paths and probably not replenished with jet material. This means they will do their own funneling through the gas and will age more swiftly, thus producing steep synchrotron spectra. This predicts pinched filaments which should appear as linear, curved features in lobes. Shocks seen sidewise will give the same appearance, but unlike magnetic filaments, should display considerable disruption when they meet. Since shock sheets should intersect frequently (unless they are onion-skin shells), the statistics of such shell crossings should distinguish these from the narrow filaments, which will seldom meet.

Our entire picture depends on the interpenetration of the incoming jet and the dense intergalactic plasma. Most fluid models assume that fluids do not interpenetrate in the sense that particle friction is important; instead, viscosity dominates, though there seems no useful way to estimate its size in jets. Further, magnetic fields could mediate the actual momentum exchange between jet and gas. None of this matters in electrodynamics, though. Inductive electric fields will generate nearby return currents, model-independently, since Maxwell equations demand it.

Filamentation instability points up a persistent lack in our astrophysical modeling. Small-scale ($\sim c/\omega_{pe}$), fast ($\sim \omega_{pe}$) plasma instabilies will grow to advanced nonlinear stages long before fluid models become plausible. Connecting the plasma scale to the fluid realm demands careful study of how "anomalous" plasma collisions enter into viscosity and thence the Reynolds number. If magnetic filaments provide the effective "particle" or eddy size for the onset of fluid analysis, this could tell us much about both pictures of jet dynamics.

Instability analysis, to be truly compared with jet observations, awaits such a synthesis.

REFERENCES

Achterberg, A., Blandford, R. D. and Goldreich, P. (1983), Nature 304, 607.

Benford, G. (1972), Phys. Rev. Lett. 28, 1242.

Benford, G. (1978), Mon. Not. R. Astron. Soc. 183, 29.

Benford, G. (1981), Ap. J. 247, 792.

Chan, K. L., and Henriksen, R. N. (1980), Astrophys. J. 228, 534.

Clarke, D., Burns, J. and Feigelman, E. (1986), Ap. J. Lett. 70, 289.

Cohn, J. (1983), Ap. J. 269, 500.

Ferrari, A. (1985), in Unstable Current Systems and Plasma Instabilities in Astrophysics, Proc. of 107th Symposium of IAU (eds.: M. R. Kundu and G. D. Holman), College Park, Maryland (1983), pp. 393-412.

Hardee, P. E. (1983), Astrophys. J. 26, 94; (1984), Astrophys. J. 287, 523; (1986) Astrophys. J. 303, 111.

Hardee, P. (1985), in Unstable Current Systems and Plasma Instabilities in Astrophysics, Proc. of 107th Symposium of IAU (eds.: M. R. Kundu and G. D. Holman), College Park, Maryland (1983), pp.439-443.

Kőnigl, A., and Choudhuri, A. R. (1985), Astrophys. J. 289, 173 (KC).

Lee, K. F. and Lampe, M. (1973), Phys. Rev. Lett. 25, 1390.

Miller, R. B. (1982), An Introduction to the Physics of Intense Charged Particle Beams, Plenum Press, New York.

Smarr, L. L., Norman, M. L., and Winkler, K.H.A. (1984), Physica 12D, 83.

DYNAMICAL EFFECTS OF LARGE-SCALE MAGNETIC FIELDS IN JETS

A. ACHTERBERG
STERREWACHT "SONNENBORGH"
ZONNENBURG 2
3512 NL UTRECHT
THE NETHERLANDS

ABSTRACT. The possible dynamical effect of large-scale magnetic fields in astrophysical jets is discussed. They range from collimation and refocussing of jets to interaction between adjacent current-carrying jets.

I. INTRODUCTION.

The interpretation of the non-thermal emission from extended extragalactic radio sources is usually in terms of of synchrotron emission of relativistic electrons with energy $\gamma m_e c^2$ in a magnetic field B at a frequency $\nu \propto \gamma^2 B$. Typical fieldstrengths derived from an equipartition (minimum energy) argument (e.g. [1]) range from 1 - 100 μG . The high degree of polarization of the emission (up to 40%) implies not too many field-loops along the line of sight. Indeed, in those jets which are resolved, the field is "ordered", but not necessarily unidirectional.

The gross direction of the magnetic field seems to be mostly perpendicular to the jet axis for weaker sources (including most two-sided sources) and mostly along the jet axis for the stronger (one-sided) sources, although there is a tendency for a transition to the perpendicular configuration far away from the nucleus, at least in the core of the jet [2,3]. This is to be expected on the grounds of conservation of magnetic flux, which predicts a scaling $B \propto 1/R^2$ and $B \propto 1/R$ where R is the radius of the jet. The fact that the longitudinal (poloidal) field drops much more rapidly than the azimuthal (toroidal) field explains the change of configuration observed in a number of jets. However, it also poses the question of the origin of the longitudinal field. If one scales the longitudinal fieldstrength observed in the outer regions of the source back to the galactic nucleus assuming simple flux conservation, the nuclear fieldstrength inferred turns out to be unreasonably high. Therefore, the longitudinal field must have been entrained along the way from the surrounding material, or must have been generated from the longitudinal field due to shear in the jet flow [e.g. 3], or

W. Kundt (ed.), Astrophysical Jets and Their Engines, 211–222.

through a dynamo mechanism operating due to small-scale turbulence. Consequently, the field probably does not have a simple "unidirectional" geometry, but consists of loose loops of field.

In this paper I will consider some of the dynamical consequences of the presence of a magnetic field, i.e. macroscopic effects. Some of the microscopic effects of magnetic fields, in particular for particle acceleration processes, have been dealt with in my other contribution in this Volume.

II. BASIC EQUATIONS OF JET MAGNETOHYDRODYNAMICS.

The dynamics of magnetic fields in jets can be described in the MHD limit, essentially since the gyroradius $r_G = 1.6 \times 10^9 \ \gamma/B_{\mu G}$ cm of electrons is absolutely microscopic compared with the transverse dimensions of the jet. Also, the high mobility of electrons along the field will short out any potential differences along a given field line. If there is no dissipation due to finite resistivity, the conservation of magnetic flux is most conveniently expressed as $B/\rho d\ell$ = constant [e.g.: 4,5]. Here ρ is the mass-density of the plasma and $d\ell$ a section of length along the field line. So if the distance between two fluid elements on the same fieldline changes, B/ρ changes in proportion. Using the equation for the components of a field-line in curvi-linear coordinates $h_i d\xi^i = (B^i/B)d\ell$, where $d\ell = \left(\sum_j (h_j \, d\xi^j)^2\right)^{\frac{1}{2}}$ is the line-element, this can be written as:

$$B^i(\boldsymbol{\xi},t)/\rho(\boldsymbol{\xi},t)h_i(\boldsymbol{\xi},t) = \left(B^j{}_o/\rho_o h_{jo}\right)\left(\partial\xi^i/\partial\xi^j{}_o\right) . \qquad (1)$$

This allows the calculation of the magnetic field from some initial value $\mathbf{B}_o$ at some position $\mathbf{x}_o = (\xi^1_o, \xi^2_o, \xi^3_o)$ at time t_o, once the flow field is known. The equation of motion, Maxwell's equations and mass conservation supplementing this read:

$$\rho \frac{d\mathbf{v}}{dt} = -\nabla p + \frac{1}{c}\,\mathbf{j} \times \mathbf{B} \ ; \ \mathbf{j} = \frac{c}{4\pi}\,\nabla \times \mathbf{B} \ , \ \frac{d\rho}{dt} = -\rho\,\nabla.\mathbf{v} \ . \qquad (2)$$

Here p is the gas-pressure and $\mathbf{j}$ is the electrical current density.

In a cylindrical jet with laminar flow at some constant velocity V (a reasonable approximation in strongly supersonic flow) which is slowly expanding laterally with distance with local radius R, mass conservation in the steady state gives $\rho V R^2 \sim$ constant , while the flux conservation law (1) gives $B_\phi/\rho R \sim$ constant, $B_z/\rho \sim$ constant. This simple model leads to a scaling $B_\phi \propto 1/R$, $B_z \propto 1/R^2$ for the azimuthal and longitudinal fields respectively, which was mentioned above. These equations are sufficient to understand the dynamics of magnetic fields as long as dissipative processes and dynamo action can be neglected. Most of the dissipation under astrophysical conditions is probably due to small-scale turbulence rather than ordinary (Coulomb) collisions. Of great interest is the possibility of dynamo action of the turbulence [5,6], which occurs if the turbulent fluid motion lacks reflectional symmetry (carries net helicity $\mathbf{v}.\nabla \times \mathbf{v}$).

Under such circumstances the **mean** magnetic field evolves according to the equation:

$$\frac{\partial \mathbf{B}}{\partial t} = \nabla \times \left(\boldsymbol{\beta} \times \mathbf{B} + \alpha \mathbf{B} - \lambda \nabla \times \mathbf{B}\right). \tag{3}$$

Here $\alpha \sim -\frac{1}{3}\langle\tilde{\mathbf{v}}.\nabla \times \tilde{\mathbf{v}}\rangle\,\tau$ and $\lambda \sim \frac{1}{3}\langle\tilde{v}^{\,2}\rangle\tau$ are the so-called "alpha coefficient" which is a measure of the dynamo action, and the turbulent diffusivity due to the turbulence of amplitude $\tilde{v}$ and correlation time τ . Also $\boldsymbol{\beta} \equiv \mathbf{v}/c$. . The brackets denote averaging over the turbulent velocity field ($\mathbf{v}_{tot} = \mathbf{v} + \tilde{\mathbf{v}}$, $\langle\tilde{\mathbf{v}}\rangle = 0$, and a similar relation for $\mathbf{B}$). These relations are approximate, and only valid for isotropic turbulence. In general, α and λ are tensors rather than scalars. The terms in the bracket in Eq. 3 are essentially the induction electric field due to the mean motion and the mean electric field $\langle\tilde{\mathbf{v}} \times \tilde{\mathbf{B}}/c\rangle = \alpha\mathbf{B} - \lambda\,\nabla \times \mathbf{B}$ induced by the turbulent fluctuations in the velocity and the magnetic field.

Such a dynamo probably operates in the accretion disk in the active nucleus where the jet is formed. There coriolis-forces in the turbulent differentially rotating disk impart helicity on the turbulence. Once the jet is formed and has propagated out of the nucleus it is hard to see how net helicity can be generated. The reason mainly is that the specific angular momentum $\ell = rv_\phi - \mu r B_\phi$ (μ = constant) is conserved along flowlines in the absence of torques. On these grounds one naively expects $\omega = v_\phi/r \propto 1/r^2$ to decay strongly as the jet expands, so that the flow itself carries little or no helicity. If the turbulence carries no **net** helicity, one still expects field amplification to occur [5], at least as long as the turbulence has locally some helicity. But the field will form loose loops on all scales larger than the dissipation scale, and a pre-existing large-scale magnetic field will not be significantly amplified as $\alpha \to 0$ in this case.

III. JET COLLIMATION, CONFINEMENT AND FOCUSSING.

The stress $c^{-1}\mathbf{j} \times \mathbf{B} = (1/4\pi)(\nabla \times \mathbf{B}) \times \mathbf{B} = (1/4\pi)(B^2\boldsymbol{\kappa} - \tfrac{1}{2}\nabla_\perp B^2)$ excerted by the magnetic field on the plasma (where $\boldsymbol{\kappa} = \mathbf{B}.\nabla\mathbf{B}/B^2$ is the curvature vector of the field and $\nabla_\perp$ the component of the gradient perpendicular to $\mathbf{B}$) can play a role in the initial collimation of jets near the accretion disk around a compact object [7,8]. This collimation is essentially caused by hoop stresses of order $B_\phi^{\,2}/4\pi r$ due to the toroidal field which furnish the inward radial force balancing pressure gradients (and centrifugal forces due to the plasma if the jet carries mechanical angular momentum). The inertia of the gas streaming along field lines anchored in the differentially rotating disk will wind up the field so that the toroidal (B_ϕ-) component will eventually dominate at large height from the disk. I will not look at the initial collimation in any detail here, but rather concentrate on the possibility of magnetic confinement once a jet has already formed.

III.1. Magnetic confinement of jets.

Some jets have a minimum energy density needed to explain the observed synchrotron emission which exceeds the typical pressure expected in the confining intergalactic medium. A number of these jets is probably free, expanding at a constant opening angle $\Theta \sim 1/M$, where M is the Machnumber of the flow in the jet. But some of the sources do seem to be collimated [1] continuously along their way.

If a jet carries a net electrical current, the associated magnetic hoop stresses can produce the needed collimation. The pressure in the core of the jet can be significantly increased above the thermal pressure in the surrounding medium [9]. In order to avoid charging up the nucleus, a return current must close the cirquit. As a simple illustration of how the confinement works, consider a jet with radius R with a constant axial current density j and total current $I = \pi R^2 j$. It is surrounded by a "cocoon" without any currents so that $B_\phi \propto 1/r$, while the return current is a surface current at the outer surface of the cocoon at $r = R_c$. (I will use cylindrical coordinates r, ϕ, z throughout this paper). Within the jet, the force balance (Eq.2) reads $\partial_r p + B_o^2 r/2\pi R^2 = 0$, where $B_o = 2I/cR$ is the magnetic field at the surface of the jet. This can be solved by $p(r) = p(0) - B_o^2 r^2/4\pi R^2$. Pressure balance at the jet-cocoon interface gives $p(R) = p_c$, where p_c is the thermal pressure in the cocoon which is constant since there is no electrical current in the cocoon. Likewise, balance of total pressure at the cocoon-surrounding medium interface gives $p_c + B_\phi^2(R_c)/8\pi = p_e$, where $B_\phi(R_c) = 2I/cR_c$ and p_e the thermal pressure in the surrounding medium. This shows the effect of the surface current, which screens out the toroidal field in the surrounding medium. Eliminating the pressure in the cocoon from these equations one finds: $p(0) = p_e + (B_o^2/4\pi)\left(1 - (R^2/2R_c^2)\right)$. The most effective confinement occurs when the cocoon is cold so that $p_c \ll B_\phi^2/8\pi$, p_e. Note that one expects a scaling $p \propto r^{-2\gamma}$ and $B_\phi^2/8\pi \propto 1/r^2$ as the cocoon expands to radius r, so that the magnetic pressure tends to dominate. Then $B_o^2/8\pi = p_e(R_c/R)^2$ and the average particle pressure in the jet (averaged over cross-section) is given by:

$$\bar{p} \sim 2p_e\left(R_c^2/R^2\right)\left(1 - \tfrac{1}{2}\,(R^2/R_c^2)\right). \qquad (4)$$

The maximum amplification of the thermal pressure in the jet over the pressure in the surrounding medium is achieved when $R_c \gg R$. This conclusion remains qualitatively the same if the return current is distributed over the cross-section of the cocoon, rather than being a surface current. If there is no cocoon so that the return current is a surface current on the jet ($R_c = R$) one has $\bar{p} = p_e$ and there is no net confinement. If there is also a longitudinal magnetic field $\bar{B}_z$ in the jet and a turbulent field component $\tilde{\mathbf{B}}$, one replaces $\bar{p}$ by $\bar{p} + \bar{B}_z^2/8\pi + \langle\tilde{\mathbf{B}}.\tilde{\mathbf{B}}/8\pi\rangle$.

The most important conclusion is that the magnetic collimation works well only for jets with extended cocoons. As is indicated by numerical calculations [e.g. 10] under-dense jets ($\rho/\rho_e << 1$) with high-Machnumber internal flow form extended cocoons. I will return to this point later. Also, The cocoon material consists mostly of jet material which has gone through the shock which terminates the jet (working surface) and which has been left behind, with some of the external material mixed in due to instabilities at the outer radius. This material is rather hot near the head of the jet (and the flow probably intrinsically time-dependent due to vortex shedding etc.). Magnetic confinement is probably most effective some distance back, where the cocoon material has cooled due to sidewise expansion. If this is true, the confining magnetic fields in the cocoon would not show up in the radio due to the expansion losses of the radiating electrons, except maybe in maps which have a very high dynamic range, allowing low-surface brightness extended emission to be seen.

III.2. Refocussing of jets: Hot Spots and Sco X-1.

Hoop stresses due to a toroidal field in a current carrying jet can also be used to refocus diverging jet flow, producing shocks near the focussing point where part of the kinetic energy in the flow is dissipated. This leads to particle acceleration which in turn gives rise to a "hot spot" in the radio emission of the jet. As such, it forms an alternative to a pure gas-dynamical model where internal shocks form when the jet pressure must suddenly adjust to the strongly varying pressure in the surrounding medium [11,12]. Such a model was proposed by Achterberg, Blandford and Goldreich [13] for the radio-lobes associated with the galactic X-ray binary Sco X-1. The situation is quite analogous to that of magnetic confinement, including the necessity of an extended cocoon where the return current runs. But here the hoop stresses balance the (inertial) centrifugal force due to the jet flow following curved flow-lines rather than a pressure gradient. Assuming for simplicity a cold jet with flow-velocity V and radius R, which slowly changes its cross-section ($dR/dz << 1$) with a locally self-similar toroidal field $B_\phi = 2Ir/cR^2$, the equation of motion for the jet flow $\rho V^2(d^2R/dz^2) = B_o{}^2/2\pi R$ $(B_o = 2I/cR)$ can be cast in the form of a particle in a potential well using mass conservation $\pi R^2 \rho V = \dot{M}$ = constant.

$$\tfrac{1}{2}\,\dot{R}^{\,2} + \Phi(R) = 0, \text{ with } \dot{R} \equiv V\frac{dR}{dz} \text{ and } \Phi(R) \equiv 2A^2\ln(R_{max}/R). \quad (5)$$

Here $A \equiv B_o(R)/\left(4\pi\rho(R)\right)^{\frac{1}{2}}$ is the Alfvén velocity, which is a constant by virtue of mass- and current conservation. This equation admits a solution in which the jet expands from some initial radius set by the boundary conditions near the nucleus to a radius R_{max} , than collapsing again, focussing to a point in a focussing distance $D \sim (V/A)R_{max}$. In reality, the contraction of the jet radius will be turned around by the increasing pressure in the compressed jet material, or through the formation of internal shocks.

If the pressure increase is adiabatic (no shocks), its effect can be modelled in a high-Machnumber flow by adding a term to the potential:

$$\Phi(R) = 2\,A^2 \ln(R/R_{max}) + s^2\left[(R/R_{max})^{2-2\gamma} - 1\right]/\gamma(\gamma - 1). \quad (6)$$

Here $\gamma = c_p/c_v$ and s is the internal sound velocity at the maximum expansion point $R = R_{max}$. With this potential the equation of motion (5) admits periodic solutions, in which the jet looks like a string of sausages, as was found by Chan and Henriksen [9] in a more elaborate calculation. If shocks are formed (which seems likely) so that the specific entropy of the jet material increases at each constriction point, similar solutions are still possible (provided the flow remains laminar on the large scales) but the focussing length after each constriction will be determined by the amount of heating at the shocks and the subsequent cooling due to expansion- and radiation losses.

This model is attractive for explaining the Sco X-1 radio lobes (at a distance of some 0.2 pc on either side of the X-ray source, assuming a distance to the source of 500 pc) since the observations put a stringent upper limit on the proper motion of the lobes of about 32 km/s [14]. The observed luminosity of the radio lobes is too large to be consistent with the standard interpretation of emission near the working surface at the head of the jet. The minimum-energy of the radiating electrons requires too large a density in the surrounding medium (in excess of 10^3 hydrogen atoms per cc) to be confined by ram-pressure, given the upper limit on the velocity of the lobes. In the magnetic focussing picture the distance between the nucleus and the hot spot is set at a <u>fixed</u> value by the boundary conditions, at least as long as the power input at the central source does not fluctuate. Behind the focussing point, the synchrotron surface luminosity fades rapidly ($S_\nu \propto \nu^{-\alpha} R^{-(7\alpha+6)/3}$ with α the spectral index) due to the combined action of electron expansion losses and the decay of the magnetic field $\propto 1/R$. It is also interesting to note that the <u>minimum</u> dissipation at the focussing point must be the kinetic energy of motion towards the axis. Since this motion is driven by the magnetic tension, the constriction velocity is always of the order of the Alfvén velocity. Dissipation of this component gives an energy density in particles near the focus of order $B^2/4\pi$, naturally leading to equipartition with the magnetic energy density as a lower limit for the particle energy density near the focus.

Recently, the discovery of a second set of lobes further out has been reported for Sco X-1 [15]. They are well aligned with the axis set by the original lobes, at distances of about 35 and 70 pc on either side of the central source. At this point it is not clear if these are secondary foci, with the different focussing length due to asymmetries in the confining medium or in the dissipation at the primary hot spots, or if we are seeing the end-points of the jets with their working surface. The fact that the weaker of the two secondary lobes is at the largest distance from the central source is consistent

with the latter interpretation. Assuming high Machnumber flow with the intrinsic kinetic luminosity $L_j = \pi R^2 \rho V^3/2$ set in the central source equal for both jets, the head of the jets advances into the surrounding medium with a velocity $V_h = \sqrt{\sigma} V/(1 + \sqrt{\sigma})$ with $\sigma \equiv \rho/\rho_e$ the density ratio between the jet and the surrounding medium, as follows from ram-pressure balance.
The power dissipated at the working surface will be of order $L \sim \pi R^2 \rho (V - V_h)^3 \sim L_j/(1 + \sqrt{\sigma})^3$. If some fixed fraction of this goes into relativistic electrons, this implies a scaling of the luminosity L with the distance $D = V_h T$ of the outer lobes $L \propto (VT - D)^3$ (T is the age of the source).

IV. INTERACTION WITH THE SURROUNDINGS: WIGGLES AND JET-JET INTERACTION IN 3C75.

I now turn to the question of how current-carrying jets interact with their surroundings or, as may be the case in the remarkable double-nucleus source 3C75 discovered by Owen et al. [16], with another jet. A central role in any discussion of electromagnetic interaction between a jet and its surroundings must be played by the return current and the question of extended cocoons. As long as one adheres to a pure MHD picture (but see [31] for a different point of view), the magnetic field goes where the plasma goes. Microscopic diffusion effects will allow the magnetic field to penetrate only very small distances into a neighbouring highly conducting plasma, at least under the conditions thought to be present in the intercluster- or intergalactic medium. This implies that the toroidal field due to a current-carrying jet will not penetrate into the surrounding medium (nor will the external field penetrate into the jet) if no extensive turbulent cocoon is formed where material from the surrounding medium and jet material (which has gone through the working surface) is mixed. Instead surface currents are formed at the jet surface which shield the field. This surface current automatically supplies the necessary return current. Under such circumstances, a jet is electromagnetically isolated from its surroundings in the sense that an external magnetic field $\mathbf{B}_e$ in the surrounding medium will not in general lead to a net Lorentz force $\mathbf{F}_L = \mathbf{I} \times \mathbf{B}_e/c$ on the jet as one might naively expect. The return (surface) current gets in the way, leading to a vanishing net current.

This situation can change when an extensive (turbulent) cocoon is formed, in which turbulent diffusion allows the return current to spread out, and enables the magnetic field in the surrounding medium to penetrate the cocoon. The evolution of the toroidal field B_ϕ in a turbulent cocoon is described by an equation:

$$\left(\frac{\partial}{\partial t} + \mathbf{U}_p . \nabla \right)\left(B_\phi/\rho r\right) = \frac{\lambda}{\rho r} \left(\nabla^2 - 1/r^2 \right) B_\phi \ . \tag{7}$$

This equation describes the spreading of the current $j_p = (c/4\pi r)\left(d(rB_\phi)/dr\right)$. Here I have neglected dynamo action and a possible poloidal magnetic field. $\mathbf{U}_p$ denotes the poloidal component of the fluid velocity. If the turbulent diffusion (left-hand-side of the equation) dominates so that $(\nabla^2 - 1/r^2)B_\phi \sim 0$, this allows a solution without surface currents on the jet (at radius R) or cocoon boundary, and with a constant (return) current density with magnetic field: $B_\phi(r) = (B_o/(R_c{}^2 - R^2))((R_c{}^2R/r) - (rR))$ in the cocoon.

If the jet remains intact but the exterior field penetrates the cocoon, the total force due to the exterior magnetic field is just the integral of the magnetic pressure over the exterior surface of the jet $\mathbf{F}_{em} = - \int d\mathbf{O}(B_e{}^2/8\pi)$. This gives rise to a net force which has the form of the Lorentz force [17], even though the exterior magnetic field does not actually penetrate into the jet where the current runs. In a heuristic way, the net "Lorentz Force" on a current-carrying jet due to an external field (with its current sources outside the jet) can in general be represented as:

$$\mathbf{F}_l = \frac{1}{c}(1-f)\ \mathbf{I} \times \mathbf{B}_e \quad . \tag{8}$$

Here $0 < f < 1$ is a factor which measures the fraction of the return current which shares the motion of the jet. If the cocoon is very thin (or collapses to a surface current) one will have $f \sim 1$, and there will be little or no electromagnetic force on the jet. If the cocoon is very thick, one expects $f \to 0$. On purely hydrodynamical grounds a jet which moves as a whole without breaking up drags along a roughly equal volume per unit length of surrounding fluid (cocoon material) (the so-called added mass, e.g. [18,19]).

The fact that extensive turbulent mixing seems to be a prerequisite for good interaction between jets and an external magnetic field probably means that only weaker sources (Fanaroff-Riley Class I sources) with only subsonic or moderately high-machnumber flow in the cocoon could experience net Lorentz-forces. The reason is that turbulent boundary layers spread only slowly in strongly supersonic flow (opening angles $\lesssim 1/M$, where M is the machnumber of the flow) and would not fill the cocoon.

It has been proposed [20] that the wiggles in some jets such as 3C465 are due to the interaction between a current-carrying jet and an external magnetic field. In view of the discussion above, this will only be effective if the jet has an extended cocoon, which in turn implies a high-Machnumber, under-dense jet. Also, the cocoon should not be dragged along with the jet. These restrictions make this particular mechanism for creating wiggles rather unlikely in general.

The discovery by Owen et al. [16] of the "interacting jets" in the double-active-nucleus radio source 3C75 has opened the question if the helical braiding of two of the jets (one from each active nucleus) suggested by the observations can be due to electromagnetic interaction between two current-carrying jets. For such a mechanism to work the jets must not only have an extended cocoon, but these cocoons must also have merged so that the jets "feel" each others magnetic field after turbulent mixing. If one assumes the jets move as a whole without their cross-section being deformed (the "firehose approximation", which is valid for phenomena with timescales slower than the sound crossing-time across the jet and lengthscales much longer than the jet radius) one can describe the motion of a single jet and its interaction with the surroundings with a variational principle [17]. The force balance in equilibrium is between the centrifugal force of the helical flow (partly balanced by magnetic tension, assuming of course super-Alfvénic flow) through the jet and the electromagnetic attraction between two currents. Assuming for simplicity two identical jets each carrying a current I in a helical pattern 180 degrees out of phase on a cylinder of radius R with a pitch-angle $dz/Rd\Theta = \cot\alpha$, the force balance reads:

$$\pi a^2\left(\rho V^2 - \frac{\langle B_{//}^{\,2}\rangle}{4\pi} + \frac{B_T^{\,2}}{16\pi}\right)\frac{\sin^2\alpha}{R} \sim (1-f)^2\,\frac{I^2}{cR}\,\cos 2\alpha \ . \quad (9)$$

Here the $\langle\rangle$ indicate an averaging over the jet cross-section and $B_T = 2I/ca$, where a is the jet radius. The term in brackets on the right-hand-side is the (negative) net tension due to the centrifugal force and magnetic forces. Note that the toroidal field B_T is destabilizing. As before, the factor f parametrizes the amount of screening of the mutual attraction due to return currents in the cocoon running within radius R. Terms of order $a^2/R^2 \ll 1$ where neglected. It turns out that such an equilibrium is unstable against perturbations which change the pitch-angle of the helix. The instability with a perturbation amplitude ξ is driven by perturbations in the centrifugal force $\propto (\pi a^2\rho V^2\cos^2\alpha\sin^2\alpha/R^2)\xi$ and resisted by the inertial force $\pi a^2(\rho + \rho_e)(\partial^2\xi/\partial t^2)$ The typical growth-time for the instability is of order $(\rho + \rho_e)^{\frac{1}{2}} R/\sqrt{\rho}V$, which shows the effect of the increased inertia due to the dragging around of an equal volume of surrounding material by the jet, so that the effective mass per unit length equals $\pi a^2(\rho + \rho_e)$ for small perturbations. If one makes the jet material in 3C75 sufficiently light $(\rho/\rho_e < 0.01)$, the observed helical structure could survive during the estimated age of the source (10^8 $(V_h/1000$ km/s$)^{-1}$ yr), simply because of the large inertia. This of course again assumes that there is good mixing between the cocoon material and the surrounding heavy medium. A cocoon consisting only of pure shocked jet material will in general have a lower density than the jet material due to the expansion behind the shock terminating the jet. Radiative cooling is too slow to affect this conclusion.

It is equally possible to explain the observed structure in 3C75 without invoking electromagnetic effects. If the jets are sufficiently under-dense, the observed wiggles could be due to the shedding of large-scale vortices in the wake of one of the jets if there is sufficient cross-flow over the jet. Since the gross-morphology of the jets is that of a head-tail source with the jets swept backwards due to motion of the nuclei through the cluster medium, this seems a likely scenario. A second jet which gets in the wake of the first jet experiences fluctuating hydrodynamical lift- and drag forces due to the fluctuating vorticity in the wake [17].

The typical amplitude ξ of the induced wiggles when compared with the overall radius of curvature R of the jets would be of order $0.2\,\tilde{C}\,(\rho_e U^2/\rho V^2)$. Here $\tilde{C}$ is the amplitude of the fluctuating drag- or lift coefficient for a cylinder (the force per unit length due to fluctuations in the wake equals $\rho_e U^2 a\tilde{C}$) in a wake with typical density ρ_e and mean streaming velocity U. V is the velocity of the jet material. Typically $\tilde{C} \sim 0.8$ for high- Reynoldsnumber flow. Wiggles in a single jet could be explained in a similar way if the jet sheds large-scale vortices. The observed amplitude in 3C75 requires $\rho/\rho_e \leqslant 0.05$. If on the other hand the jet material is much denser than that of the surrounding medium, the jet trajectories would be essentially ballistic, and could just reflect the orbiting motion of the bound active nuclei around each other [20].

V. DISCUSSION.

I briefly reviewed the possible dynamical effects of large-scale magnetic fields in astrophysical jets. Although the principles are well-understood, many of the details remain unclear. In particular this is true of the question of possible dynamo action in astrophysical jets. Simple numerical experiments [22] show that amplification of a seed field is indeed possible if the turbulence carries enough vorticity. Whether this will amplify the field all the way to equipartition values remains unclear. To properly appreciate the role of magnetic fields in the stability and confinement of jets will require extensive numerical work which is able to follow MHD instabilities into the strongly non-linear regime. Many papers have been written [e.g. 23-26] discussing the linear stability of magnetized jets. The conclusion is invariably that some perturbations are unstable. Unfortunately that does not answer the question whether these instabilities are dynamically important once they enter the non-linear regime. The observations at least imply that astrophysical jets are very stable and persist for great lengths. Given the rapid development in the area of large computers and computational physics, much progress can be expected to be made in the coming years. It may well be that the currently held believes in this area (or should one say prejudices) will ultimately prove to be incorrect or at best over-simplified.

I have also not looked in any detail at force-free magnetic fields, which have received some attention [27-29]. These fields satisfy the equation $\nabla \times \mathbf{B} = \mu\mathbf{B}$, with $\mathbf{B}.\nabla\mu = 0$. As the Lorentz-force vanishes in these fields, they do not have any direct dynamical effect. The attraction of the idea of force-free equilibria (which have been extensively studied in connection with magnetic structures in stellar coronae, see for instance [30]) lies in the fact that they are "minimum energy" solutions. The force-free field geometry proposed in [27] does bear a striking resemblance with what is observed in some jets. However, it is unclear how a force-free field can be maintained in a system with super-Alfvénic flow and/or particle pressure near equipartition, since it requires all body forces (right-hand-side of equation (2)) to vanish perpendicular to the magnetic field.

In particular, if the conditions in the jet perpendicular to B change on a typical scale L, the equilibrium will be nearly force-free provided $\mu L >> 4\pi\rho V^2/B^2$, $4\pi p/B^2$. Since for most realistic field configurations one has $\mu L = O(1)$, this implies that the force-free assumption is really only justified if the flow velocities are much smaller than the Alfvén velocity, and the thermal pressure is much smaller than the magnetic pressure. This is the case in the Solar Corona, but probably not in most jets. Also, the idea that in a slowly evolving jet dissipation of magnetic energy needed to maintain a force-free state with conserved magnetic helicity $\mathbf{B}.\mathbf{A}$ ($\mathbf{B} = \nabla \times \mathbf{A}$) can provide the energy of the observed synchrotron emission has been called into question [29,30].

ACKNOWLEDGEMENT. This work was supported by the Netherlands Organization for the Advancement of Pure Research (Z.W.O.) through a Huijgens fellowship.

REFERENCES.

1. Longair, M.S.: 1981, 'High Energy Astrophysics', Cambr. Univ. Press, p.296.
2. Bridle, A.H. and Perley, R.A.: 1984, Ann. Rev. Astron. Astrophys. **22**, 319.
3. Begelman,M.C.,Blandford,R.D. and Rees,M.J.: 1984, Rev. Mod. Phys. **56**, 255.
4. Roberts, P.H.: 1967, 'An Introduction to Magnetohydrodynamics' Longmans, London.
5. Zeldovich, Ya. B., Ruzmaikin, A. A. and Sokoloff, D.D.: 1983: 'Magnetic Fields in Astrophysics', Gordon and Breach, London, p75.
6. Moffat, H.K.: 1978, 'Magnetic Field Generation in Electrically Conducting Fluids', Cambridge Univ. Press, Cambridge, U.K.
7. Blandford, R.D. and Payne, D.G.: 1982, Mon. Not. Roy. astr. Soc. **199**, 833.
8. Norman, C.A. and Pudritz, R.E. : 1986, Astrophys. J. **301**, 571.
9. Chan, K.L. and Henriksen, R.N.: 1980, Astrophys. J. **241**, 534.
10. Norman, M.L., Winkler, K.A.H. and Smarr, L.:1985, in: 'Physics of Energy Transport in Extragalactic Radio Sources', A.H. Bridle and J.A. Eilek, eds., Proc. 9th NRAO Workshop, NRAO, Greenbank, USA.
11. Sanders, R.H.: 1983, Astrophys. J. **266**, 73.
12. Wilson, M.J. and Falle, S.A.E.G.: 1985, Mon. Not. Roy. astr. Soc. **216**, 971.
13. Achterberg, A., Blandford, R.D. and Goldreich, P.: 1983, Nature **304**, 607.
14. Fomalont, E.B., Geldzahler, B.J., Hjellming, R.M. and Wade, C.M.: 1983, Astrophys. J. **275**, 802.
15. Velusamy, T., Pramesh Rao, A. and Sukumar, S.: 1985, Mon. Not. Roy. astr. Soc. **213**, 735.
16. Owen, F.N., O'Dea, C., Inoue,M. and Eilek, J.A.: 1985, Astrophys. J. Lett. **294**, L85.
17. Achterberg, A.: 1986, Preprint, submitted to Astron. Astrophys.

18. Landau, L.D. and Lifshitz, E.M.: 1959, 'Fluid Mechanics', Pergamon Press, Oxford.
19. Newman, J.N.: 1977, 'Marine Hydrodynamics', M.I.T. Press, Cambridge, Ma.
20. Eilek, J.A., Burns, J.O., O'Dea, C.P. and Owen, F.N.: 1984, Astrophys. J. **278**, 37.
21. Yokosawa, M. and Inoue, I.: Publ. Astr. Soc. Japan **37**, 655.
22. De Young, D.S.: 1980, Astrophys. J. **241**, 81.
23. Ferrari,A., Trussoni, E. and Zaninetti,L.: 1981, Mon. Not. Roy. astr. Soc. **196**, 1051.
24. Ray, T.P.: 1981, Mon. Not. Roy. astr. Soc. **196**, 195.
25. Cohn, H.: 1983, Astrophys. J. **269**, 500.
26. Achterberg, A.: 1982, Astron. Astrophys. **114**, 233.
27. Königl, A. and Choudhuri, A.R.: 1985, Astrophys. J. **289**, 173.
28. Turner, L.: 1986, Astrophys. J. **305**, 668.
29. Königl, A. and Choudhuri, A.R.: 1986, Astrophys. J. **305**, 954.
30. Priest, E.R.: 1982, 'Solar Magnetohydrodynamics', D. Reidel Publ. Co., Dordrecht, Holland, Ch. 3.5.
31. Benford, G.: 1985. in: 'Physics of Energy Transport in Extragalactic Radio Sources', A.H. Bridle and J.A. Eilek, eds., p. 185. NRAO, Greenbank.

PARTICLE ACCELERATION IN ASTROPHYSICAL JETS

A. ACHTERBERG
STERREWACHT "SONNENBORGH"
ZONNENBURG 2
3512 NL UTRECHT
THE NETHERLANDS.

ABSTRACT. The different mechanisms for the acceleration of particles in astrophysical jets are discussed. The formation of power-law particle distributions is considered by the processes of Fermi acceleration, gyro-resonant acceleration and shock acceleration.

I. INTRODUCTION.

The non-thermal emission from extended extragalactic radiosources is commonly interpreted as the synchrotron radiation of relativistic electrons in a magnetic field. The observed frequency-dependence of the flux $F(\nu) \propto \nu^{-\alpha}$ suggests a power-law energy distribution of the relativistic electrons $N(E) \propto E^{-s}$, with $s = 2\alpha + 1$. The spectral index α lies in the range between 0.5 and 0.9 for 90% of the extended sources [1], corresponding to $2 < s < 2.8$. Jets and hot spots in radio lobes tend to have flatter spectra, whereas the diffuse radiolobes have steeper spectra ($\alpha > 0.8$). Recent reviews of both theory and observations can be found in Begelman, Blandford and Rees [2] and Bridle and Perley [1].

It is usually assumed that at least part of the energy of the relativistic electron population is generated locally by in-situ acceleration processes. Some of the arguments in favor of this picture are as follows:

1) The surface brightness of many jets decays much slower with jet radius than one would expect due to adiabatic cooling and radiation losses of a population of pre-accelerated electrons [3]. Also, many jets do not show the spectral-index gradients expected from expansion losses or synchrotron cooling of the radiating electrons.

2) Some of the larger diffuse radio lobes have linear dimensions that exceed the distance an electron of energy E can travel in a synchrotron loss time $t_{\frac{1}{2}}$. If one assumes that the material in the lobes is on average more or less stationary, this distance typically

W. Kundt (ed.), Astrophysical Jets and Their Engines, 223–236.

equals $\ell_{\frac{1}{2}} = \beta c t_{\frac{1}{2}} = 2500\ \beta E_{GeV}^{-1}\ B_{\mu G}^{-2}$ Mpc. Here $\beta = V_{str}/c$ is the bulk streaming speed in units of c. An electron of energy E radiates predominantly at a frequency $\nu_s = 16.1\ B_{\mu G} E_{GeV}^{2}$ MHz, so an electron radiating at 1.4 GHz in a 10 μG field will loose half its energy in a few Mpc travelling at the velocity of light. However, if there is a thermal plasma component in the lobes, plasma instabilities will reduce the bulk streaming speed of the relativistic electrons with respect to the thermal plasma to at most the ion-thermal speed $V_{ti} = 8 \times 10^7\ (T/10^6 K)^{\frac{1}{2}}$ cm/s if the thermal pressure exceeds the magnetic pressure, or the Alfvén speed $V_A = 2.2 \times 10^7\ B_{\mu G}(n/10^{-4}\ cm^{-3})^{-\frac{1}{2}}$ cm/s if the converse is true [4]. This reduces the actual value of $\ell_{\frac{1}{2}}$ by two-three orders of magnitude and puts it in the kpc range.

3) Observations of optical synchrotron emission in M 87 and more recently in the southern lobe of 3C33 [5] show electrons with energies of 100 GeV in a region several kpc in extent. The corresponding synchrotron lifetime is about 100 yr, so that in-situ acceleration is absolutely necessary to maintain the emission for any length of time.

Particle acceleration in jets or radio lobes is usually thought to be due to two possible mechanisms: **stochastic acceleration** due to turbulent motion in a magnetized plasma or low-frequency plasma waves and MHD waves (stochastic Fermi acceleration), or **shock acceleration** (regular Fermi acceleration) near shocks in the flow. In principle both mechanisms are capable of accelerating charged particles to a power-law distribution under the right circumstances. One expects stochastic acceleration to be important when the jet flow is dissipated gradually, for instance due to "friction" at the interface between the jet material and the surrounding medium. Fluid instabilities such as the Kelvin-Helmholtz instability could convert part of the kinetic energy of the directed jet flow into turbulent motions (e.g.[6-10]) in a boundary-layer, which gives rise to a turbulent viscosity.

The mechanism of shock acceleration will be important when shock transitions occur in the jet flow. Such shocks occur at the working surface, where the endpoint of a supersonic jet impinges on the intergalactic (or cluster-) medium [11]. Internal shocks in a jet occur when the jet overtakes dense clouds [12], is forced to adjust to a steep pressure gradient in the surrounding medium [13,14] or is forced to bend through a large angle [15]. Also, Kelvin-Helmholtz (surface) instabilities could be stabilized by the formation of internal shocks, as seems to be indicated by numerical calculations [16-18]. One expects localized "hot spots" in the radio emission near these shocks, as are often seen in jets. I will consider these mechanisms in some detail.

II. STOCHASTIC ACCELERATION: A UNIFIED APPROACH.

Fermi [19] proposed the first serious model of stochastic acceleration of charged particles in a turbulent medium. In this mechanism, particles gain energy during collisions with randomly moving, magnetized clouds. Assuming elastic reflection in the rest-frame of a cloud, a net energy gain results because the frequency of head-on collisions $\nu_+ = (v + V)/L$ exceeds that of the overtaking collisions $\nu_- = (v - V)/L$. Here v, V and L are the particle velocity, cloud velocity and inter-cloud distance respectively. The typical momentum gain per collision equals $\Delta p_{\pm} = \pm\ \gamma m V$ (γ is the particle Lorentz factor, m its mass). So the net energy gain per unit time equals:

$$\frac{dE}{dt} = (\nu_+ - \nu_-)v\ \Delta p \simeq \frac{V^2}{Lc}\frac{v}{c}\ E \ \rightarrow \alpha E \quad (\text{when } v \rightarrow c). \tag{1}$$

"Fermi acceleration" has become the generic name for stochastic acceleration processes in which at any given time a particle undergoes (fractionally) small energy changes due to interaction with clouds or an ensemble of plasma waves with random phases. In what follows, I will use the term "turbulence" loosely, meaning either fully developed MHD turbulence characterized by large fluctuations, or "wave turbulence" where an ensemble of waves propagates in a well-definded background plasma that supports them. Such processes can be visualized as a random-walk (diffusion process) of an individual particle in momentum. As such it is described by a Fokker-Planck collision term in the equation describing the evolution of the momentum distribution $f(\mathbf{x},\mathbf{p},t)$: $\left.\partial_t f(\mathbf{x},\mathbf{p},t)\right)_{\text{Fermi}} = (1/p^2)\partial_p\left(p^2 D_p \partial_p f(\mathbf{x},\mathbf{p},t)\right)$. [20,21]. The mean energy change is related to the momentum diffusion coefficient D_p through the relation $\langle dE/dt\rangle = (1/p^2)\partial_p\left(p^2 D_p v\right)$. At the same time, two individual particles starting out with the same energy E will diffuse with respect to each other in energy by an amount $\Delta E = (v^2 D_p t)^{\frac{1}{2}}$ in a time t.

The general equation describing the evolution of the particle distribution $f(\mathbf{x},\mathbf{p},t)$ in a plasma flowing with velocity **u** is given by:

$$\partial_t f + \boldsymbol{\nabla}.\left(\mathbf{u}f - \mathbf{D}.\boldsymbol{\nabla}f\right) = \frac{1}{p^2}\ \partial_p\left(\left[\frac{p}{3}\ \boldsymbol{\nabla}.\mathbf{u} + \beta p^2\right]p^2 f + p^2 D_p \partial_p f\right) + S(\mathbf{x},\mathbf{p},t). \tag{2}$$

This equation assumes that the relativistic particles are coupled to the thermal plasma through scattering by plasma waves. This allows only a slow spatial diffusion of the particles with respect to the plasma, characterized by the diffusion tensor **D**, and keeps the distribution nearly isotropic in momentum space. The first term on the right-hand-side gives the momentum changes due to divergence in the flow (expansion losses), synchrotron radiation ($\beta = 4e^4B^2/9m^4c^6$, corresponding to the synchrotron loss averaged over the particle pitch-angle) and stochastic acceleration respectively.

The source term $S(\mathbf{x},\mathbf{p},t)$ describes injection of particles from the thermal pool into the acceleration process. The precise form of the momentum diffusion coefficient D_p depends on the acceleration mechanism under consideration. Specific examples will be considered below.

For a full description of the acceleration process, an equation decribing the generation of the turbulent motions or plasma waves must be added. In general, it will have the form:

$$\partial_t W_k + \nabla . \mathbf{S}_k = Q_k + L_k - (\sigma^b{}_k + \sigma^r{}_k) W_k \ . \tag{3}$$

Here $W_k dk$ is the energy density contained in the turbulent motions or waves in a wavenumber range k,k+dk, $\mathbf{S}_k$ is the energy flux in that range (for waves just the group velocity x energy density), Q_k a source term describing the generation of the turbulence or waves and L_k a term describing the non-linear interactions between turbulent eddies and waves. The function of L_k is to describe the redistribution of the energy over different length scales, for instance in a turbulent cascade where large eddies degrade into smaller and smaller eddies, until viscous dissipation destroys them [22,23]. As such it does not destroy any energy, and the integral over all wavelengths should vanish: $\int dk\ L_k = 0$. The last term on the right-hand-side describes the linear damping of the turbulence by the thermal plasma (e.g. through viscous dissipation) and by the relativistic particles, characterized by damping rates $\sigma^b{}_k$ and $\sigma^r{}_k$ respectively. The energy gained by the relativistic particles at the expense of the turbulence should equal the energy lost by the turbulence to the particles, so one can appeal to a principle of "detailed balancing". Writing the momentum diffusion coefficient as an integral counting the contributions of turbulence at each wavelength k, $D_p = \int dk\ \tilde{D}_p(k)$, this implies the following relation:

$$\sigma^r{}_k W_k = \int dp\ 4\pi p^2\ E \frac{1}{p^2} \partial_p \left(p^2 \tilde{D}_p(k) \partial_p f \right) = \int dp\ 4\pi p^2\ f < \frac{dE}{dt} >_k , \tag{4}$$

where $<dE/dt>_k \equiv (1/p^2)\partial_p \left(p^2 \tilde{D}_p(k) v \right)$, after partial integration.

This set of equations in principle describes the dynamics of relativistic particles undergoing stochastic acceleration, and of the (wave-) turbulence.

III. FORMATION OF POWER-LAW DISTRIBUTIONS.

I now consider the mechanisms proposed for the formation of power-law electron distributions which seem to be implied by the observations. Such a mechanism should also explain the relatively narrow range of values for the slope of the distribution observed. (For relativistic particles the conversion between the energy distribution N(E) and the momentum distribution f(p) is simply $N(E) = 4\pi p^2 f(p) dp/dE \cong 4\pi p^2 f(p)/c$.) It is interesting to note that galactic cosmic rays have a similar distribution (for a review see [24] and references therein).

The radio observations usually cover frequencies due to electrons in the energy range 100 MeV-10 GeV. When non-thermal optical emission from hot spots is observed, such as in M87 or 3C33, it shows some steepening, as is expected from synchrotron losses. So the acceleration mechanism should explain the scale-free (self-similar) power-law behaviour over at least two decades in electron energy. One way to achieve this is to assume that the losses are "parasitic" and the particle distribution is totally determined by the escape of particles from the source and the energy dependence of the acceleration mechanism. Another possibility is to exactly balance energy losses and gains of the electrons in such a way that a power-law results. I will consider both cases below.

III.1 FERMI TYPE ACCELERATION.

As a simple model of fully-developed MHD turbulence, where a turbulent cascade maintains a wide range of scales, consider an ensemble of fluid eddies with some velocity-size distribution $V(\ell)$. Defining the energy spectrum of the turbulent velocity field by $\frac{1}{2}\rho V^2(\ell) = \int_{2\pi/\ell}^{\infty} dk\ W_k$ in the continuum limit, the momentum diffusion coefficient leading to the acceleration rate (1) is given by:

$$D_p = \frac{1}{3}\sum_{\ell} \frac{V^2(\ell)}{c\ell}\ p^2 = \frac{p^2}{6\pi c\rho}\int dk\ k\ W_k\ . \qquad (5)$$

Here I assumed isotropic turbulence and a plasma mass-density ρ . Existing models for fully developed turbulence give $W_k \propto k^{-m}$, with m = 5/3 (Kolmogorov law) [22] when the kinetic energy in the turbulence is much greater than the magnetic energy in the fluctuations, and m= 3/2 (Kraichnan law) [23,24] if there is equipartition between the two. In either case, the integral in (5) is dominated by the largest k (smallest eddies), while most of the energy in the turbulence resides in the largest eddies. The largest wavenumber available for the acceleration process is either determined by the scale at which the turbulence is dissipated, or by the requirement that the eddy size should exceed the electron gyro-radius $r_G = 1.6 \times 10^9\ \gamma/B_{\mu G}$ cm in the magnetic field B carried by the eddy, whichever criterion gives the largest scale. Eddies on scales smaller than the gyration radius scatter a particle only over a small angle (differential scattering) rather than reflecting it, and the energy gain per collision is correspondingly smaller.

The simplest model neglects expansion and synchrotron losses in Eq. (3), and approximates the propagation of the electrons through the source by a simple, energy independent escape term, i.e.: $(\partial_t f)_{esc} = -f/T$. Assuming particles are injected into the source at some rate R at energy E_o, balancing the particle acceleration against escape for energies $E > E_o$ and using Eq. (5) for D_p gives an energy distribution:

$$N(E) \propto E^{-s} \;;\; s = \tfrac{1}{2}\left(9 + \frac{16}{\alpha T}\right)^{\frac{1}{2}} - \tfrac{1}{2} \;,\; \left(\alpha = \frac{2}{3\pi\rho c}\int dk\; k\; W_k\right). \quad (6)$$

The parameter αT which determines the slope of the spectrum follows from the total energy balance in the acceleration region between the accelerated particles and the turbulence [25,26]. The relevant equations read:

$$\frac{dU}{dt} = RE_0 + \alpha U - \frac{U}{T} \quad ; \quad \frac{dW}{dt} = \varepsilon - \alpha(1+X)U \;. \quad (7)$$

Here $X \equiv \int dk\; \sigma^b(k) W_k/\alpha U$ is the fraction of the turbulent energy dissipated by the thermal plasma. The first equation for the energy density $U = \int dp\; 4\pi p^2\; Ef(p)$ of the electrons can be obtained by multiplying Eq.(2) with energy and integrating over momentum. The second equation for the total energy $W = \int dk\; W_k$ of the turbulence follows from Eq. (3), using $\int dk\; L_k = 0$, detailed balancing (Eq.4) and assuming no turbulent energy escapes from the region. The quantity $\varepsilon = \int dk\; Q_k$ is the energy injected into the turbulent cascade at large scales due to instabilities. In a steady state ($d/dt = 0$) and when dissipation by the thermal plasma can be neglected ($X \ll 1$) one finds $\alpha T = \varepsilon/(\varepsilon + RE_0)$. According to Eq.(6) the slope of the distribution equals $s = \tfrac{1}{2}\left(16 + 25(RE_0/\varepsilon)\right)^{\frac{1}{2}} - \tfrac{1}{2}$, depending only on the ratio of energy input into turbulence and particles. The observations suggest $RE_0/\varepsilon \lesssim 1$. So in this model the feedback between the particles and the accelerating turbulence determines the slope of the resultant particle distribution.

This same model applies to acceleration by small-amplitude compressive magnetosonic waves [27,28]. The physics in this case is rather different. It supposes a well-ordered background magnetic field B with small fluctuations in the field strength $\delta B \ll B$ superposed. For an ensemble of plane waves with $\delta B = \int d\kappa\; b(\mathbf{k},\omega)\; \exp\,(i\mathbf{k}.\mathbf{x} - i\omega t)$ where $d\kappa \equiv d^3\mathbf{k}d\omega/(2\pi)^3$, particles tend to interact with those waves which have a stationary phase as seen from the unperturbed orbit of the particle. This leads to a resonance condition $\omega - k_{\parallel}v_{\parallel} + n\,\Omega = 0$ ($n = 0, \pm 1, \pm 2, \ldots$) . Here $k_{\parallel}$ and $v_{\parallel}$ are the components of the wave-vector and particle velocity along the magnetic field, and $\Omega = eB/\gamma mc$ the gyration frequency. The interaction with magnetosonic waves takes place at the **Cherenkov resonance** $\omega - k_{\parallel}v_{\parallel} = 0$. In that case, the particle momentum perpendicular to the field $p_{\perp}$ is a constant of motion, and the parallel momentum changes according to $dp_{\parallel}/dt = -\,(p_{\perp}v_{\perp}/2B)(\partial B/\partial s)$ [28], where s is the pathlength along the ambient field. Scattering is necessary to redistribute the gain in parallel momentum over the other two degrees of freedom. Using $dp/dt = (p_{\parallel}/p)(dp_{\parallel}/dt)$ and introducing the cosine of the pitch angle $\mu \equiv p_{\parallel}/p$, the momentum changes according to $dp/dt = \tfrac{1}{2}\mu(1-\mu^2)(c\mu\partial\ln B/\partial s)p$. The momentum diffusion coefficient then scales according to $D_p \equiv \tfrac{1}{2}\langle\Delta p^2/\Delta t\rangle \propto p^2$, as in classical Fermi acceleration. As such it can be considered to be wave-turbulence version of the stochastic Fermi mechanism. The cascade which redistributes wave energy over a range of wavelengths is now due to wave-wave interactions between the MHD wave-modes.

III.2 GYRORESONANT ACCELERATION.

A somewhat different approach to the formation of power-law spectra has been taken by Eilek and Henriksen [29]. They employ gyro-resonant interactions ($\omega - k_{\parallel}v_{\parallel} = \pm \Omega$) with Alfvén waves with magnetic field perturbation **b(k)**. The energy gain exactly balances synchrotron losses at each energy, leading to a power-law (self-similar) behaviour. Expansion losses and particle escape from the source are neglected. It is assumed that the Alfvén waves are generated by the Lighthill-mechanism in the turbulent cascade. A turbulent eddy at scale ℓ with velocity $V(\ell)$ generates waves with frequency $\omega = k V_A \sim V(\ell)/\ell$, which determines the wavenumber k as a function of ℓ, given $V(\ell)$. The power transferred from the turbulence to the Alfvén waves in an oktave dk = k equals:

$$kQ_k = \eta\ (k\ell)\ \rho V^2(\ell)\times(V(\ell)/\ell) \sim \eta\ (V(\ell)/V_A)(V(\ell)/\ell)\times\rho V^2(\ell). \quad (8)$$

where η is a numerical factor of order unity. The relevant equations for the particles and waves, c.f. Eqs. (2,3) then are:

$$(1/p^2)\partial_p\left(p^2(\beta p^2 f + D_p\partial_p f)\right) = 0\ ;\quad Q_k k - \sigma^r{}_k k W^A_k = 0. \quad (9)$$

Here $W^A{}_k dk = \left(\langle|\mathbf{b}(\mathbf{k})|^2\rangle/4\pi\right)dk$ is the energy density between k, k+dk in Alfvén waves. The momentum diffusion coefficient in this case is given by (assuming isotropic Alfvén turbulence with magnetic field amplitude **b(k)** [e.g. 20]:

$$D_p = \gamma^2 m^2 V_A{}^2\ \Omega\ \frac{\pi}{8}\ \frac{k\ \langle|\mathbf{b}(k)|^2\rangle}{B^2}\Big|_{k\ \sim\ eB/cp} \quad (10).$$

With no preferred energy scale in the system, the energy balance is obtained through an argument of detailed balancing, at some energy E ~ pc: synchrotron radiation emitted = energy lost by emitting electrons = energy gained by same electrons from Alfvén waves = energy lost by waves to electrons = energy gained by waves from fluid turbulence. With a synchrotron emissivity per unit frequency $j(\nu)d\nu$ at a frequency $\nu_\gamma = eB\gamma^2/(2\pi m_e c)$ for an electron with Lorentz-factor γ, and using the resonance condition $k \sim eB/\gamma m_e c^2 \equiv k_\gamma$, this energy balance can be written as:

$$j(\nu_\gamma)d\nu_\gamma = \frac{4}{3}\ \sigma_T c\ \frac{B^2}{8\pi}\ \gamma^2 N(E)dE = \left\langle\frac{dE}{dt}\right\rangle_k N(E)dE = \sigma^r{}_k\frac{\langle\mathbf{b}(\mathbf{k})^2\rangle}{4\pi}dk_\gamma = Q(k_\gamma)dk_\gamma.$$

Here σ_T is the Thompson cross-section. Using Eq.(10) and the definition of the <u>turbulent</u> energy density W_k, it follows that the energy distribution of the accelerated particles has a slope $s = (15 - 7m)/(3 - m)$ when $W_k \propto k^{-m}$. For a Kolmogorov spectrum ($m = 5/3$) this gives $s = 2.5$, close to the median value inferred from the observations.

This particular mechanism however does have a few disadvantages. First of all, the fraction of the turbulent energy channeled into relativistic electrons is by necessity small: if too much of the turbulent energy is radiated into Alfvén waves the turbulent cascade would break off and not maintain the broad-band fluctuation spectrum needed to generate the resonant Alfvén waves with several decades in wavelength. The typical transfer time at a scale ℓ in which an eddy is degraded into smaller eddies is of order of the eddy turn-over time $\ell/V(\ell)$. So the energy transport rate in the cascade is of order $\frac{1}{2}\rho V^2(\ell) \times V(\ell)/\ell$. Compared with the Lighthill-radiation rate (Eq. (9)), this means that a fraction $V(\ell)/V_A$ of the turbulent energy at each scale is lost to Alfvén waves. So one requires $V(\ell)/V_A \ll 1$, i.e. the turbulence takes place in a strong magnetic field with an amplitude much less than the average Alfvén velocity. Only this fraction of the turbulent energy is available for electron acceleration, in contrast to the "classical" Fermi model considered above, where in principle a large fraction of the turbulent energy can be used for particle acceleration.

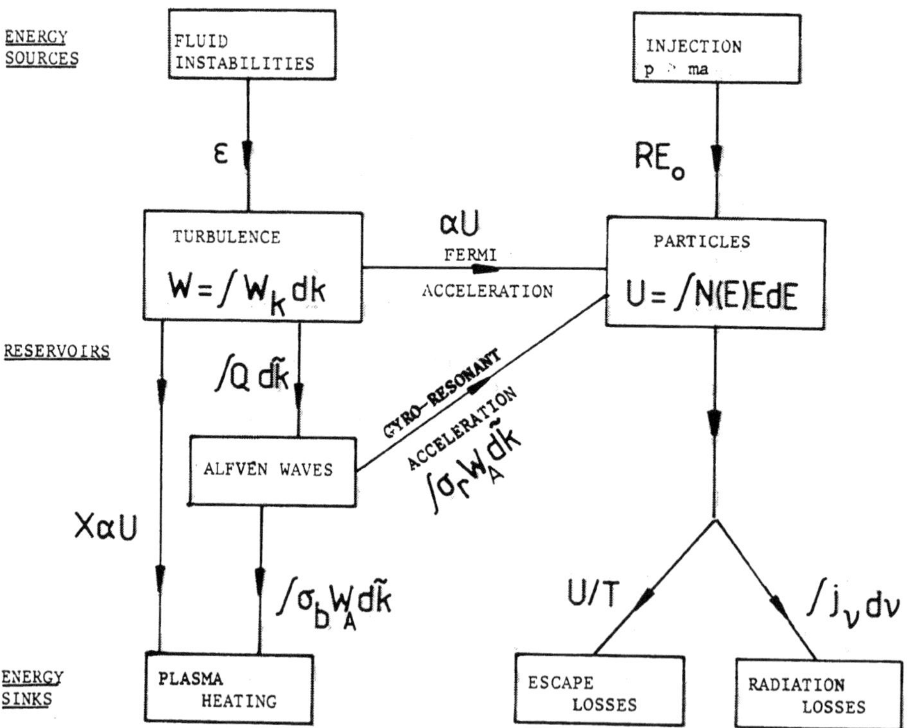

Figure 1. A flow-diagram of the different acceleration mechanisms which can lead to a power-law distribution of relativistic electrons. Energy flow-rates are shown. The notation is as defined in the paper.

A second problem is that the model does not allow losses of the waves to the thermal plasma or to a population of (resonant) relativistic protons. These losses in principle scale differently with energy and would destroy the scale-free behaviour needed to explain a power-law distribution of the accelerated electrons. Some of these problems could be avoided if the majority of the positive (relativistic) charge carriers in the jet material are positrons, which are subject to the same synchrotron losses as electrons.

A third problem concerns the fact that the synchrotron-loss timescale in some sources is rather long, and it is not quite clear that the energy balance described above can be maintained if the source evolves strongly on a similar timescale.

The possible ways of using turbulence to create a power-law of relativistic electrons discussed here are schematically shown in Fig.1 in the form of a diagram showing the energy flow rates in the system.

III.3. ACCELERATION BY SHOCKS.

Much theoretical effort in recent years has been devoted to the process of diffusive particle acceleration near shocks propagating through a magnetized plasma [30-33]. In this mechanism, scattering on irregularities in the magnetic field (Alfvén waves) traps energetic particles near the shock front. If the scattering mean-free-path is much larger than the shock thickness, the particles see the shock as a discontinuity in the velocity of the scattering centers. Its function is to create a converging flow between the scattering centers on either side of the shock. Particles trapped between these converging scattering centers are accelerated (regular Fermi acceleration) during repeated shock crossings, until they are convected away into the downstream region and escape.

This mechanism is very attractive since, in its simplest form where one neglects the back-reaction of the accelerated particles on the pre-shock flow (the test particle limit), it naturally leads to a power law distribution for the accelerated particles. At each shock crossing, a particle on average experiences a fractional momentum change $\Delta p/p = 4\,(U_1 - U_2)/3v$. Here U_1 (U_2) is the velocity of the scattering centers in front of (behind) the shock. The chance of escape per shock crossing into the downstream region is roughly $4U_2/v$. With n denoting the number of shock crossings, one has in a steady state (for $p > p_o$ where p_o is the injection momentum):

$$\frac{d\ln N}{d\ln p} = \frac{(d\ln N/dn)}{(d\ln p/dn)} = -\frac{3}{r-1} \quad ; \qquad (r \equiv U_1/U_2). \quad (11)$$

This corresponds to a distribution $f(p) = dN/4\pi p^2 dp \propto p^{-\sigma}$, , with $\sigma = 3r/(r-1)$. In a strong, purely hydrodynamical shock the compression equals $r = 4$, leading to a slope $\sigma = 4$, or equivalently $N(E) \propto E^{-2}$ for relativistic particles. The time between subsequent shock crossings can be estimated by balancing the diffusion away from the shock with the advection of particles into the shock. With a diffusion coefficient $D_1 = \lambda_1 v/3$ $(\lambda_1 = v/\nu$ is the scattering mean-free-path in the upstream medium), a particle diffuses a distance

$L = (Dt)^{\frac{1}{2}}$ in a time t, seen from a frame tied to the upstream flow. In a time U_1/L a fluid element at distance L in front of the shock is overtaken by the shock. So the time a particle diffuses before being overtaken is of order $t = L/U_1$,and the typical distance it diffuses ahead of the shock in that time equals $L = D_1/U_1$. The typical cycle time between subsequent shock crossings equals L/c, which gives a time t_{acc} in which the energy of a given particle is doubled of order $(L/c)(p/\Delta p) \sim 3D/U^2$. A more precise calculation yields [34]:

$$t_{acc} = \frac{3}{U_1 - U_2} \left((D/U)_1 + (D/U)_2\right) . \quad (12)$$

The test-particle limit predicts very high efficiencies for the acceleration. This means that the back-reaction of the accelerated particles on the pre-shock flow must be taken into account. The pressure gradient of the precursor of accelerated particles with scale-length L decelerates the incoming flow, regulating the shock strength, and creating a region of slow compression in front of the viscous shock proper.

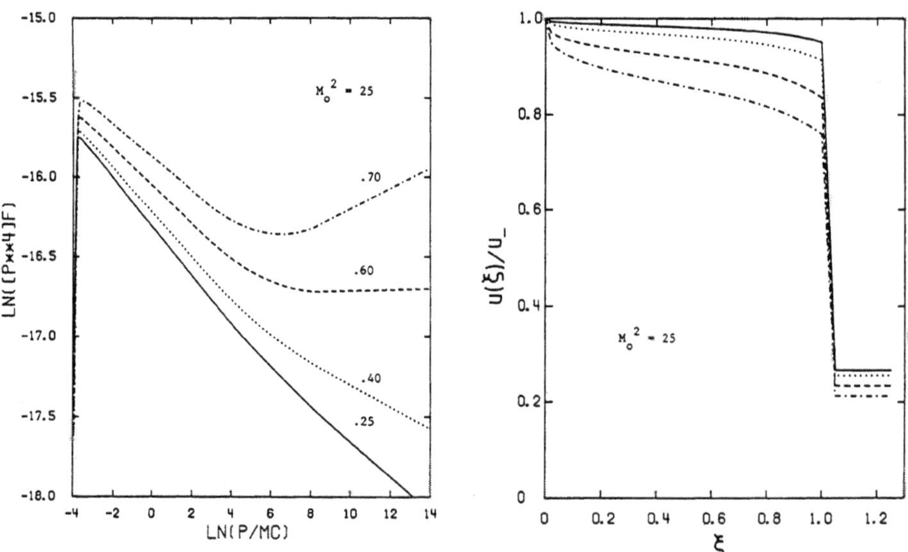

Figure 2. The particle distribution $\ln(p^4 f)$ as a function of ln(p/mc) (left panel), and the normalized velocity in the shock precursor as a function of $\xi = \exp(x/L)$ where x is the distance in front of the shock and L a normalizing distance (right panel). The Machnumber of the shock squared $M_o^2 = 25$, and the curves are labeled by $(P_{seed}/P_{gas}) \times 100$. P_{seed} is the pressure of seed particles far ahead of the shock and P_{gas} the gas pressure. For low seed-particle pressure, the deceleration of the flow in the precursor is small, and the particle distribution is close to the test-particle power-law.

A further complication is that one expects that the diffusion coefficient increases with particle energy. An often used (but essentially unproven) hypothesis is that the mean free path in strong MHD wave-turbulence ($b(k)k/B = O(1)$) will roughly scale as the particle gyroradius, so that $D \sim c^2/\Omega \propto \gamma$. This means that particles of higher energy on average get further upstream, and consequently sample a higher velocity contrast between the upstream and downstream medium since the pressure residing in particles with lower energy has not yet had a chance to decelerate the flow. This makes the problem very non-linear, and except in a perturbation analysis [35,36] one has to resort to Monte Carlo calculations [37] or to numerical solution of the equations [38]. These show that high efficiencies can be reached even for relatively low Machnumbers (more than 20% of the total incoming momentum flux being converted into particle pressure). The particle distribution tends to be harder at high energies, as long as losses are neglected (e.g. fig.2).

If one adds synchrotron losses , the net effect is to create a bump in the distribution near the momentum p^* where the energy gain due to shock acceleration balance the loss. Here electrons tend to pile up. This is followed by a rapid fall-off of the distribution for $p > p^*$ [39]. The value of p^* follows from equating the acceleration time t_{acc} (Eq. 13) with the synchrotron loss-time $t_{\frac{1}{2}} = 1/\beta p$, which yields at the shock:

$$p^* = 1/\beta t_{acc} \sim (m_e c/3) \frac{U_1 - U_2}{(D/U\tau)_1 + (D/U\tau)_2} , \qquad (13)$$

where $\tau = 6\pi m_e c/(\sigma_T B^2)$. Assuming a spatial diffusion coefficient $D = fr_G c = 5 \times 10^{19} f\gamma B_{\mu G}^{-1}$ cm^2/s with f a number of order unity, one typically finds: $\gamma^* \sim 4\times10^4\ B_{\mu G}^{-3/2}(U/1000\ \mathrm{km/s})\left((r-1)/fr\right)^{\frac{1}{2}}$.

IV. OPEN QUESTIONS AND DISCUSSION.

I have shown above that there are a number of mechanisms which, under the right conditions, are capable of accelerating electrons to a power-law energy distribution. Yet there are still a number of essentially unsolved questions which stand in the way of a detailed comparison between theory and observations. The most important ones can be summarized as follows.

The role of surface instabilities for particle acceleration in jets is not completely clear. We do not know whether Kelvin-Helmholtz instabilities in astrophysical jets saturate by converting the free energy in the relative motion between the jet material and the surrounding medium into small-scale turbulence which provides anomalous viscous damping in a boundary layer, or by the formation of shocks in the non-linear regime which then furnish the necessary dissipation. Since shocks due to corrugations in the jet boundary can penetrate the whole jet (and a surrounding cocoon), in situ acceleration by the shocks would not be confined to a turbulent boundary layer. This would avoid the problem of edge-brightening of jets, which is not observed in most sources.

The composition of the jet material is not really known. In particular, it is not clear whether the positive charge carriers in the jet material are protons, positrons or a mixture of both. Models in which the "central engine" in the galactic nucleus consists of a black hole with a magnetized accretion disk driving a relativistic electron-positron wind [40,41,42] due to pair creation by γ-photons from the disk predict that the majority of the positive charge carriers are positrons, with the fraction of protons probably determined by the (poorly understood) process of entrainment of galactic and intergalactic material by the jet. Also, it is not clear if such a relativistic pair-plasma can escape from the active nucleus without catastrophic radiation losses. Wave properties in an electron-positron plasma are quite different [43] from that in a hydrogen plasma. To my knowledge, little or no work has been done on the theory of stochastic particle acceleration in such a (relativistic) electron-positron plasma.

It is also not clear whether there is an energetically important "thermal" plasma in the jets capable of supporting the (low-frequency) plasma waves needed for stochastic particle acceleration and shock acceleration. If the properties of the thermal plasma are not known, it is difficult to estimate how much of the available energy in plasma waves goes into relativistic electrons, and how much is dissipated by the thermal plasma (the "partitioning problem").

Injection of electrons into the process of shock acceleration or gyroresonant acceleration is not well understood. This problem arises since Alfvên waves in a hydrogen plasma propagate only at frequencies $\omega < \Omega_i = eB/m_i c$, where m_i is the proton mass. The resonance condition gives $p = eB/ck = m_i V_A(\Omega_i/\omega) > m_i V_A$. So electrons must be accelerated up to a momentum $m_i V_A$ before they can be scattered or accelerated by Alfvên waves. The constraint is also valid for Fermi-acceleration by small- amplitude plasma waves, since there the gyroresonant scattering is needed to redistribute the gain in p over all degrees of freedom [28]. Fermi acceleration by strong turbulence provides its own scattering and does not have such an injection problem. In shock acceleration gyroresonant scattering is needed to trap the energetic particles near the shock. One therefore needs some pre-acceleration of the electrons. For shock acceleration this question is closely connected to the poorly understood dissipation in quasi-parallel (magnetic field direction within 45° of the shock normal) shocks. It is quite possible that electron acceleration near shocks is at least partly due to magnetosonic waves or other plasma waves generated by (nearly thermal) ions reflected at the shock in the manner proposed by Galeev [44,45]. At higher energies, the electrons would then be picked up by the shock acceleration process decribed above.

A related problem is that all the mechanisms treated here work equally well for protons as for electrons at the same momentum. Unless the injection process for the electrons is well-understood, it remains difficult to decide what fraction of the energy in the turbulence or shocked flow is expended in accelerating (invisible) protons to supra-thermal energies.

It remains difficult to give strong **observational predictions** which would enable one to distinguish between the different acceleration mechanisms. Generally, one expects that the stochastic Fermi acceleration process would lead to rather diffuse regions of synchrotron emission, whereas shock acceleration would start rather suddenly, being confined to the immediate surroundings of the shock.

There has been some work (e.g. by Bicknell [46-48]) to describe fully turbulent, low-Machnumber jets in which the spreading of the jet is caused by turbulent dissipation (Reynolds stresses) in the bulk plasma. This heats the relativistic electron population through Fermi acceleration. If one assumes that most of the pressure of the jet material resides in the radiating electrons, one can derive a surface brightness-jet radius relation which can be compared with the observations. Work of this nature, (also numerical), investigating different scenarios of particle acceleration, is clearly needed to further our understanding of the conditions in astrophysical jets and make the connection with the observations.

ACKNOWLEDGEMENT.
This work was supported by the Netherlands Organization for the Advancement of Pure Research (ZWO) through a Huijgens Fellowship.

REFERENCES.

1. Bridle, A.H. and Perley, R.A.: 1984, Ann. Rev. Astron. Astrophys. **22**, 319.
2. Begelman,M.C.,Blandford,R.D. and Rees,M.J.: 1984, Rev. Mod. Phys. **56**, 255.
3. Bridle, A.H.: 1982, in: 'Extragalactic Radio Sources', IAU Symp. 97, D.S. Heeschen & C.M. Wade,eds.,p.121, Reidel Publ. Co., Dordrecht, Holland
4. Achterberg, A.: 1981, Astron. Astrophys. **98**, 161.
5. Meisenheimer, K. and Röser, H.J.: 1986, Nature **319**, 459.
6. Achterberg, A: 1979, Astron Astrophys. **76**, 276.
7. Eilek, J.A.: 1979, Astrophys. J. **230**, 373.
8. Benford, G., Ferrari,A. and Trussoni, E.: 1980, Astrophys. J. **241**, 98.
9. Bicknell, G.V. and Melrose, D.B.: 1982, Astrophys. J. **262**, 511.
10. Pelletier, G. and Zaninetti, L.: 1984, Astron. Astrophys. **136**, 313.
11. Blandford, R.D. and Rees, M.J.: 1974, Mon. Not. R. astr. Soc. **169**, 395.
12. Blandford, R.D. and Königl, A.: 1979, Astropys. J. Lett. **20**, L15.
13. Königl, A.: 1980, Phys. Fluids **23**, 1083.
14. Sanders, R.H.: 1983, Astrophys. J. **266**, 73.
15. Icke, V.: 1986, Preprint.
16. Norman, M.L., Smarr, L., Winkler, K.-H.A. and Smith, M.D.: 1982, Astron. Astrophys. **113**,285.
17. Norman, M.L., Winkler, K.-H.A. and Smarr, L.: 1983, in: "Astrophysical Jets", A. Ferrari and A.G. Pacholczyk, eds., D. Reidel Publ. Co., Dordrecht, Holland.

18. Norman, M.L., Winkler, K.-H.A. and Smarr, L.: 1985, in: "Physics of Energy Transport in Extragalactic Radio Sources", A.H. Bridle and J.A. Eilek, eds., Proc. 9th NRAO Workshop, NRAO,Greenbank,USA.
19. Fermi, E.: 1949, Phys. Rev. **75**, 1169.
20. Skilling, J.: 1975, Mon. Not. R. astr. Soc. **172**, 557.
21. Melrose, D.B.: 1980, 'Plasma Astrophysics', Vol. 2, Ch. 8, Gordon and Breach Science Publishers, London.
22. Landau, L.D. and Lifshitz, E.M.: 1959, Fluid Mechanics, Ch. III, Pergamon Press, London.
23. Kraichnan, R.H.: 1965, Phys. Fluids **8**, 1385.
24. Achterberg, A.: 1984, Adv. Space Res. **4**, 193.
25. Burn, B.J.: 1975, Astron. Astrophys. **45**, 435.
26. Achterberg, A.: 1979, Astron. Astrophys. **76**, 276.
27. Kulsrud, R,M and Ferrari, A.: 1971, Astrophys. Space Sc. **12**, 302.
28. Achterberg, A.: 1981, Astron. Astrophys. **97**, 259.
29. Eilek, J.A. and Henriksen, R.N.: 1984, Astrophys. J. **277**, 820.
30. Krymsky, G.F.: 1977, Dokl. Acad. Nauka SSSR **234**, 1306.
31. Axford, W.I., Leer, E. and Skadron, G.: 1976, EOS **57**, 780.
32. Bell, A.R.: 1978, Mon. Not. R. astr. Soc. **182**, 147.
33. Blandford, R.D. and Ostriker, J.P.: 1978, Astrophys. J. Lett. **221**, L29.
34. Lagage, P.O and Cesarsky, C.J.: 1983, Astron. Astrophys. **125**, 249.
35. Heavens, A.F.: 1983, Unpublished Thesis, Univ. of Cambridge, U.K.
36. Blandford, R.D.: 1980, Astrophys. J. **238**, 410.
37. Ellison, D.C and Eichler, D.: 1984, Astrophys. J. **286**, 691.
38. Achterberg, A.: 1986, Preprint, submitted to Astron. Astrophys.
39. Webb,G.M.,Drury, L.O'C. and Biermann, P.: 1984, Astron Astrophys. **137**, 185.
40. Blandford, R.D. and Znajek, R.L.: 1977, Mon. Not. R. astr. Soc. **179**, 433.
41. Phinney, E.S.: 1983, in "Astrophysical Jets", A. Ferrari and A.G. Pacholczyk, eds., p. 201, D. Reidel Publ. Co.,Dordrecht, Holland.
42. Blandford, R.D.: 1986, in "Quasars", G. Swarup and V.K. Kapaki, eds.,IAU sump. 119, p.359, D. Reidel Publ. Co.
43. Sakai, J. and Kawata, T.J.: 1980, J. Phys. Soc. Japan **49**, 747.
44. Galeev, A.A.: 1985: in: 'Advance in Space Plasma Physics', B. Buti, ed., p. 273, World Scientific, Singapore.
45. Galeev, A.A.: 1985, ibid. p. 401.
46. Bicknell, G.V.: 1984: Astrophys. J. **286**, 68.
47. Bicknell, G.V.: 1986: Astrophys. J. **300**, 591.
48. Bicknell, G.V.: 1986: Astrophys. J. **305**, 109.

OPTICAL SYNCHROTRON EMISSION FROM RADIO HOT SPOTS

P. R. Hiltner
Max-Planck-Institut für Astronomie
Königstuhl 17
D-6900 Heidelberg 1
FRG

ABSTRACT. *Optical polarization measurements of the following 3 objects are reported: The southern hot spot of 3C33; the western hot spot of Pic A; and the jet of 3C273. The work has been done by Klaus Meisenheimer and Hermann-Josef Röser[1,2,3] in 1983-1985 using a CCD and a Savart plate (two dimensional, imaging polarimeter).*

1. GENERALITIES

The intriguing properties of extended extragalactic radio sources[4,13] (ERS) - such as jets and lobes of very different size and shape, hot spots which may or may not be present in different numbers, the spectral characteristics of these structures and of the central object - after 20 years of investigation pertinaciously defy all efforts for a coherent understanding. This state of affairs only reflects the fact that the ERS constitute a new and exceedingly complex class of physical objects, and underlines the necessity of fully exploiting the information contained in the radiation (i.e. photometry, spectroscopy and polarimetry combined) over as great a range of frequencies as possible in order to reduce the number of the various, often conflicting, models.

We have embarked on an optical polarization survey of the so-called hot spots occurring in the lobes of ERS. They constitute a very interesting subgroup of the structures connected with ERS in so far as they obviously represent a prominent seam between the energyflow from the core and the ambient intergalactic medium. A full understanding of the physical processes in these hot spots would therefore help clarify a lot of the riddles of the ERS. Yet in frequency ranges other than the radio (and in marked contrast to it) it is only in the last few years that data on some of these hot spots have been collected[1-3,5,6]. This may partly be due to the grave observational problems which require detectors of very high quantum efficiency and in general long integration times. 5 hours at a 2.2-m telescope are needed to reach an accuracy of 3% in polarization at R=22 mag.

W. Kundt (ed.), Astrophysical Jets and Their Engines, 237–243.

What does one hope to see? Radio observations tell us that the radiation from the hot spots is highly linearly polarized, 30% or more, with a nonthermal (power law) spectrum[4, 13]. The most natural explanation is synchrotron radiation of relativistic electrons in a partially ordered magnetic field. The data indicate magnetic field strengths around 10^{-4} G in the hot spots (assuming minimum, i.e. near-equipartition, of energy). This in turn implies a lower bound for the Lorentz factor γ of the emitting electrons of 10^4 (take λ=1 cm), which have a synchrotron lifetime of about 10^4 yr. Now, since the synchrotron cutoff frequency $\nu_c \sim \gamma^2 B_\perp$, the detection of optical synchrotron radiation would imply an increase in the Lorentz factor by $(\nu_{opt}/\nu_{radio})^{1/2}$, (the minimum total energy and the magnetic field would only for a very flat spectrum be affected sensitively). This would be an important discovery already per se, all the more so since a decrease of the corresponding synchrotron lifetime proportional to $(\nu_{opt}/\nu_{radio})^{-1/2}$, down to ≈ 100 yr, would follow. This would add to the question of in-situ acceleration provided that one is not willing to assume that γ-factors of $\approx 10^6$ can be maintained over a 1 Mpc-distance from the core to the hot spots[7, 8]. Or, inversely, once the possibility of effective in-situ acceleration is accepted, the optical synchrotron radiation could be used as an effective tracer of regions where physical processes like shock acceleration take place.

In the following the results of the observations of the hot spots in 3C33 and Pic A and of the jet of 3C273 are described to some detail.

2. OBSERVATIONAL RESULTS

2.1 The southern hot spot of 3C33[1]

3C33 is a typical example of a powerful, edge-brightened, double radio source. At 5 GHz it extends over 4' on the sky at a redshift of the central D galaxy of z=.0592. The southern hot spot shows a beautiful bow shock structure[10] with a flux density of 700 mJy[9, 16]. The measurements have been done at the 2.2-m telescope of the Max-Planck-Society on La Silla, in the red at λ=660 nm.

The resolution of the images is such that not only the hot spot but also a backward tail and a NW-wing emerges, tracing the bow shock structure of the radio image. Therefore a polarization map could be constructed by determining the polarization at 11 different positions across the proper hot spot, the NW-wing and the backward tail. The degree of polarization varies from (29.2±2.4)% in the hot spot to about 45% in the NW-wing and 26% in the backward tail (off the hot spot the brightness level decreases to > 24 mag/arcsec^2).

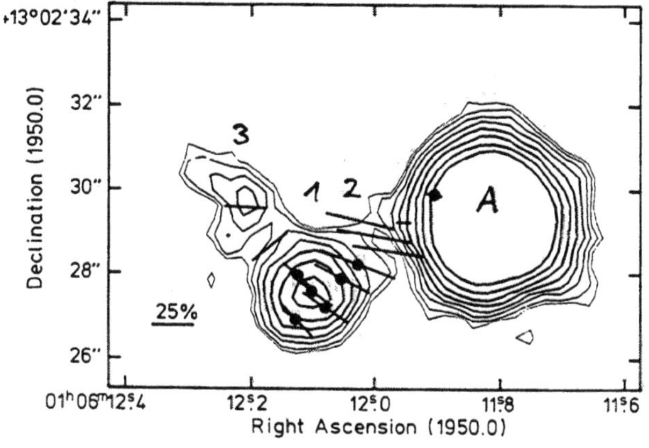

Fig. 1: Map of optical brightness and polarization within 3C33 south (1.5" resolution). The vectors show the fractional amount of linear polarization integrated over an effective circular beam of 2.0" FWHM. Positions where the amount of polarization exceeds 3σ are indicated by a central dot.
1, hot spot;
2, north-west wing;
3, backward tail;
A, field star.

Compared to the radio 6 cm data[11] a rotation in PA by 7.2° and a depolarization of 25%, both of which are not present at 2 cm, are indicated for the hot spot. It should also be mentioned that the positional offset between the radio hot spot and its optical counterpart is less than 0.8", and that there is a remarkable agreement between both frequency ranges as well for the polarization structure as for the intensity contours.

Of outstanding interest is of course the determination of the spectral index of the hot spot and of an eventual cutoff frequency. The data lead to $F_\nu = 4.2 \pm .8$ μJy at 660 nm. A more recent blue exposure indicates preliminarily $F_\nu \approx 1.5$ μJy at 423 nm. These data, together with an $\alpha = -.89 \pm 0.5$ powerlaw established at 5 GHZ[11], yield a cutoff frequency of $\nu_c = (1.3 \pm .1) \cdot 10^{14}$ Hz. A near-infrared measurement at 864 nm, which is now available but not yet reduced, is expected to support this conclusion. In contrast to this, the spectrum in the backward tail is considerably more extended, there being no evidence of a high frequency cutoff below $4.5 \cdot 10^{14}$ Hz.

From these data the following physical parameters for the hot spot can be inferred (assuming its volume to be 1 kpc^3):

minimum energy magnetic field	: $B_{me} = .64$ mG
Lorentz factor at B_{me} and ν_c	: $\gamma = 2.2 \cdot 10^5$
synchrotron lifetime	: $\tau_{syn} = 90$ yr
synchrotron luminosity (10^8–10^{14} Hz)	: $L_{syn} = 5.1 \cdot 10^{42}$ erg/s

2.2 The western hot spot of Pic A[2]

Pic A, a double lobed southern radio source, is remarkable because it is a very powerful and a highly polarized one. At 5 GHz it extends over 7' on the sky, and the central object, which is no distinguished radio source, is a D galaxy at z=.0342. The flux density of the western hot spot is 284 Jy[12]. The data have been taken at the 3.6-m telescope on La Silla, in the blue at $\lambda = 425$ nm with the Savart plate, and three direct exposures and two spectra in the red (660 nm) and the blue.

A higly polarized object of B=19 mag stands out on the Savart frame, but since no exposure with a different orientation of the Savart plate could be taken, it is only possible to deduce a lower limit of 30% for the polarization. If the polarization angle is assumed to be more or less the same as in the radio, polarization up to 60% would result.

The direct exposures reveal a secondary brightness maximum about 4" towards the central galaxy, and a filamentary structure inclined at about 70° to the source axis. The positional offset with respect to the VLA-position[12] is 3" to the NW.

The spectrum lacks any significant emission or absorption feature. Together with the direct images a radio-to-optical spectral index of α=-1±.1 can be deduced. There is no cutoff in or below the optical waveband. In fact it is the most fascinating aspect of this particular source that there might be no cutoff up to the X-ray domain! Radiation from the hot spot region can be isolated (though with a mediocre signal-to-noise ratio and very low spatial resolution) on an EINSTEIN IPC-frame. There is good positional agreement with the hot spot and the estimated X-ray flux falls very close to what is expected from an extrapolation of the radio-to-optical spectral slope. Should this X-ray feature really relate to the hot spot and be of synchrotron origin, a minimum energy magnetic field of $\leq$5.3 mG and a Lorentz factor exceeding $5 \cdot 10^6$ would be the consequence, giving Pic A a sound chance of becoming a record holder!

2.3 The jet of 3C273[3]

3C273 is notorious for being the closest and the first detected quasar. It is also one of the rare objects exhibiting an optical jet. The quasar has a redshift z=.158. The flux density of the hot spot at 5 GHz is 5.2 Jy[3]. Data have been taken at the 90" telescope on Kitt Peak and the 2.2-m telescope on Calar Alto in the blue, red and IR-waveband (423 nm, 613 nm, 864 nm). The polarization was determined only for the red exposures owing to the poor S/N-ratio in the other wavebands.

The polarization varies considerably along the jet, and its morphology is made up of at least 4 different regions: (a) the gap, from the quasar core out to a distance of ≈10"; (b) the plateau region, extending out to 17.5" with a formal value of the polarization of (3.7±2.5)%; (c) the optical knot from 17.5" to 19.6" which shows a polarization of (12.7±2.8)% perpendicular to the jet axis and which essentially terminates the optical jet; and (d) the radio hot spot at 21.2", which is optically detectable only in the infrared and has a radio polarization along the jet axis. The corresponding flux distribution for the three colours is visualized in figure 2.

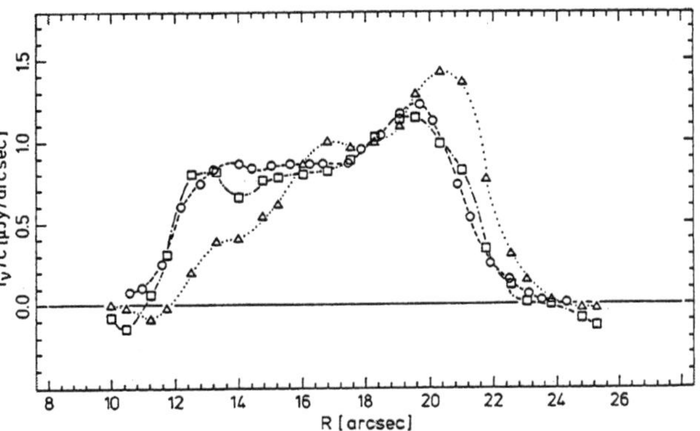

Fig. 2: Flux distribution along the jet for the 3 colours. Normalization constants are for the Blue: (□) C=2.17, the Red: (o) C=4.57 and the Near infrared: (Δ) C=4.73.

As for the spectral shape two points can be made from Fig. 2:

(i) in the plateau region there is an excess of short wavelength radiation which, together with the low polarization in the optical and the high polarization in the radio, indicates a diluting radiation component of non-synchrotron origin. Such an interpretation is also suggested by the observed EINSTEIN X-ray flux and by reports on optical line emission near the jets of other sources[14].

(ii) beyond the optical knot the spectrum steepens abruptly.

Since the diluting component does not seem to influence the optical knot very much (polarization at 5 GHz is essentially the same as in the red), it can be attempted to be explained by a synchrotron spectrum. From Fig. 2, however, it is obvious that no such spectrum can explain the knot and the hot spot simultaneously. Therefore the hot spot is isolated by subtracting the knot, which has been modelled with the point spread function of 2.4" seeing. One thus can deduce a cutoff spectrum with $\alpha=-1.04$, $\nu_c=7\cdot10^{13}$ Hz and a corresponding minimal energy magnetic field of $B_{me}=1.5$ mG for the hot spot. For the optical knot only a lower limit for the cutoff frequency of $\nu_c \geq 3\cdot10^{4}$ Hz can be derived, with $\alpha=-.96$ (radio-to-optical in both cases) and $B_{me}=.4$ mG. The inferred Lorentz factors are $\gamma=10^5$ for the hot spot and at least four times as much for the optical knot. The corresponding synchrotron lifetimes, on the order of 100 yr, do not even suffice for the particles to travel the distance from the knot to the hot spot (15,000 Lyr for $H_o=75$ km s^{-1} Mpc^{-1}), so again the question of in-situ acceleration is raised.

The gaps in these measurements can be hoped to be filled by the excellent new data taken last year at the 2.2-m telescope on La Silla, which are currently being reduced.

3. OUTLOOK

The reported observations provide the first direct evidence for optical synchrotron radiation from radio hot spots of ERS. However, revealing so different properties, they can be regarded only as a beginning: a far more extended data base is required if one wishes to come to conclusions not only concerning individual sources.

We are therefore systematically investigating the most luminous sources of the well defined sample of 166 3CR-objects as given by Jenkins et al[15]. Of course, our selection criterion according to the 5 GHz flux of the hot spot in a 1" aperture can only give a hint on a promising source since it is the cutoff frequency which ultimately decides whether the hot spot is optically detected or not.

So far, we are confident of having been successful with a number of hot spots additional to those I just have reported on, while some others were not detectable in the optical waveband or simply proved to be most likely chance coincidences with unrelated field objects. Since all of these data are not yet properly analyzed and/or still incomplete, I shall only make a few preliminary remarks on them here.

Positive candidates are 3C6.1, 3C20, 3C111 and 3C330, where weak optical features are found in the vicinity of the hot spot. In particular for the 3C20 and 3C111 hot spots a very good positional coincidence has been established and the polarimetry which is available for 3C20 shows the supposed optical counterpart to be possibly polarized.

No optical feature near the hot spots is discernible on very deep direct exposures of 3C123 and 3C454.3, a suspected candidate for 3C86A is 2" off and shows no sign of high polarization, and for 3C327 a chance coincidence with a galaxy is being evidenced by the appearance on the Savart exposure.

As for Cyg A, which we have also observed extensively getting very good polarization data, our reduction procedure is severly hampered by the lack of an appropriate deep direct exposure needed in view of the very crowded field.

As a conclusion I may say that a very exciting and promising, yet longterm and time consuming investigation is being carried out, the scientific impact of which can not yet be fully estimated. However, it is already clear from the existing data that a wide range of cutoff frequencies is encountered, which contain crucial information on the individual sources and their statistics. This we hope to substantiate by extending the observed frequency range also to the IR.

References

1 Meisenheimer K., Röser H.-J.: Nature **319**, 459 (1986)
2 Röser H.-J., Meisenheimer K.: Astrophys. J. **314**, (1987)
3 Röser H.-J., Meisenheimer K.: Astron. Astrophys. **154**, 15 (1986)
4 Bridle A.H., Perley R.A.: Ann. Rev. Astron. Astrophys. **22**, 319 (1984)
5 Simkin S.M.: Astrophys. J. Lett. **222**, L55 (1978)
6 Crane P., Tyson J.A., Saslaw W.C.: Astrophys. J. **265**, 681 (1983)
7 Webb G.M., Drury L.O'C., Biermann P.: Astron. Astrophys. **137**, 185 (1984)
8 Kundt W.: J. Astrophys. Astr. **5**, 277 (1984)
9 Spinrad H., Djorgovski S., Marr J., Aguilar L.: PASP **97**, 932 (1985)
10 Rudnick L.: Preprint 1986
11 Rudnick L., Saslaw W.C., Crane P., Tyson J.A.: Astrophys. J. **246**, 647 (1981)
12 Prestage R.M.: PhD thesis, Edinburgh 1985
13 Miley G.: Ann. Rev. Astron. Astrophys. **18**, 165 (1980)
14 Breugel W.v.: Can. J. Phys. **64**, 392 (1986)
15 Jenkins C.J., Pooley G.G., Riley J.M.: Mem. R.A.S. **84**, 61 (1977)
16 Hargrave P.J., McEllin M.: MNRAS **173**, 37 (1975)

GENERAL RELATIVISTIC EFFECTS ON COLLIMATION OF A JET

R.C.Kapoor
Indian Institute of Astrophysics
Bangalore 560034, India

Collimated beams (jets) with opening angles of order of a few degrees observed in extragalactic energetic sources are generally explained as having been produced by accreting supermassive black holes or supermassive magnetized rotators in centers of galaxies (For a review see Wiita 1985). Considering the first alternative, several workers have attempted to understand jet structure and collimation in a general relativistic framework in recent times. Our purpose here is to point out certain intrinsic gravitational effects that have been hitherto overlooked but which may significantly influence collimation processes if operative in the immediate vicinity of the black hole (say $r_o \lesssim 20$ m; $m = GM/c^2$).

First it is deflection in particle trajectories in the spacetime around the black hole which effectively widens the beam. A study of equations of geodesic motion in the Schwarzschild spacetime reveals that due to spacetime curvature, a relativistic beam opens up by a factor ~ 2; particle density and effective luminosity in the beam (L/θ_o^2) are reduced by a factor $\sim 10^{-1}$. Thus in the local rest frame, the beam is narrower, and starts out with a density and effective luminosity by an order of magnitude larger than one would usually assume. Further, if there is a distribution in particle energies in the jet, the decollimation effect will try to seggregate its multigamma constituents near the black hole. The low gamma component of the jet will tend to be confined closer to the wall of the funnel as it suffers decollimation more substantially. Beams with subrelativistic particle velocities may be decollimated appreciably even for $r_o > 20$ m. In this regime Newtonian considerations would necessarily be an under evaluation of the decollimation effect. A detailed calculation is given in Kapoor (1986).

Another interesting effect which might be important is twisting of the beam caused by dragging of inertial frames in case the black hole is Kerr. The dragging modifies particle trajectories in the jet as it will exert a torque on particles at given r_o, which therefore tend to acquire an angular velocity of cumulative dragging, ω, in the direction of rotation of the black hole. If we consider rotation of the black hole to be 'slow', $\omega = 2\,(am/r^3)$, 'a' being its angular momentum parameter, the net amount of twist would be

W. Kundt (ed.), Astrophysical Jets and Their Engines, 245–246.

$$\Phi_d > \int \omega \, dt \approx \frac{am}{r_0^2} , \qquad (1)$$

assuming that the jet axis and rotation axis of the black hole coincide. For $a \simeq 0.1$ m, $r_0 \simeq 6$ m, $\Phi_d > 0^\circ.2$. This is small but obviously the twist would be increased for larger opening angles and for the low gamma component of the jet. Moreover, in an active galactic nucleus, the black hole is expected to have a large angular momentum so that a much larger amount of twist in the jet near the black hole can result. This work is in progress.

While curvature effect is stronger than the rotational effect, decollimation and/or twist do not jeopardize an initial symmetry in the jet about the jet axis. Given this, spacetime curvature would overwhelm collimation effects near the black hole brought about by pressure, magnetic field etc and serve as an input to the collimation processes which could prevail over gravity in regions >> 20 m. To conclude, effects of decollimation and twist due to spacetime curvature and dragging of inertial frames will play a nontrivial role in questions relating to stability of a jet in a general relativistic framework.

Kapoor,R.C. 1986 Astrophys. Letters (in the press).
Wiita,P.J., 1985, Physics Reports, 123, 117.

SOME STUDIES ON GIANT RADIO GALAXIES

L. Saripalli and Gopal Krishna
T.I.F.R.
P.O. Box 1234
I.I. Sc. Campus
Bangalore 560 012 / India

Giant Radio Galaxies, of diameters D of a few megaparsec (defined here as $D \gtrsim 1.5$ Mpc; $H_o=50$ km/sec Mpc; $q_o=0$), form only a small fraction (2-3%) of the sources in metre wavelength samples (3C,6C). We study such sources because they offer a much higher linear resolution for the same distance and also because they could represent radio galaxies under an extreme set of conditions. One may also hope to study thereby a sample of radio sources predominantly rid of projection effects. Moreover, the radio lobes of these galaxies extend so far away from the parent galaxy that they are virtually under the influence of the intergalactic medium, and hence can be used as diagnostic tools.

Systematic searches for such objects have concentrated on the northern sky. The total number reported so far is only ~15. Clearly, for any collective study of such a class of objects, one needs to augment the sample. We have started a systematic search concentrating mostly on lower declinations. The method adopted by us is different from that usually followed. Briefly, we formed pairs of radio sources from the Molonglo Catalogue with separations ranging from 15' to 45', with at least one source listed as extended. The optical fields of each of these pairs were examined on the Palomar sky prints. Those pairs we called 'good candidates' had a bright ($m \lesssim 16$) elliptical galaxy roughly midway between the pair and with neither of the sources constituting the pair coincident with any optical object. A subset of 20 such good candidates was observed at Ooty (327 MHz) and Effelsberg (1.4 GHz). We have been able to establish one such candidate as a giant radio galaxy, of size 2.5 Mpc (Saripalli et al., 1986), and there is one other very probable case currently under investigation.

There are many properties common to the giant radio sources. Most of these galaxies are of the FR II class morphology, with their powers falling in a narrow range of 2 orders of magnitude ($10^{25} \lesssim P_{408} \lesssim 10^{27}$ W/Hz). They have uniform magnetic fields over hundreds of kiloparsecs. The minimum pressures in their lobes are very similar; they are mostly within a factor of 2 of $2 \cdot 10^{-14}$ dynes/cm^2 (Waggett et al., 1979, comment upon the possibility of a selection bias). Two other properties of concern are the prominence of radio cores and the frequent occurrence of asymmetry in their radio structures.

W. Kundt (ed.), Astrophysical Jets and Their Engines, 247–249.

Some of the most prominent radio cores found in 3CR radio galaxies belong to the giant radio sources. In 3C236, the largest of all radio galaxies, and NGC 6251, the cores contribute 60% of the total flux at 5 GHz. Because the giant radio galaxies constitute the extreme end of the linear size distribution of radio source samples, they are not expected to be much removed from the plane of the sky. Hence a distribution of their core fractions (defined here as CF = core flux at 5 GHz / total flux at 408 MHz) is expected to represent the intrinsic distribution.

Comparing the core fractions among the 14 giant radio galaxies with those found for a metre-wavelength selected sample of 93 (presumably randomly oriented) radio galaxies having the same range in z and luminosity (Ferretti et al., 1985), we find that the two samples have a similar median (0.5% for the GRCs, 0.2-1% for the comparison sample - the range being due to the upper limits being assigned the value 0, and when they are taken as they are, respectively), see figure 1. This is unexpected because (1) the comparison sample is likely to have sources whose cores are Doppler boosted, (b) the giant radio galaxies are predominantly of the FR II type while a sizeable fraction of the comparison sample is of FR I strucure which are known to have prominent radio cores, and (c) the giant radio galaxies are old systems and are therefore expected to have weak cores - at a late stage of activity. The distributions of absolute V-magnitudes of the GRGs and the companion sample have been found to be the same.

From such a comparison it can be surmised that sustained nuclear activity is one of the causes for the formation of large radio structures.

In order to investigate the cause of the frequent occurrence of asymmetric radio structure and to explore why they grow to such large sizes, we are studying the optical environments of each of these sources. In this study we have looked at all the neighbouring galaxies within 2 magnitudes of that of the parent galaxy, in a region of radius equal to the largest angular size of the source. We find a tendency for a higher galaxy concentration on the side which extends less from the parent galaxy. In other words, there seems to be a correlation between the galaxy distribution around the giant galaxy and its radio asymmetry.

This finding supports the idea that the advance of a jet as well as its radio power can be significantly influenced by the ambient gaseous medium. It has been earlier inferred from the radio asymmetries observed in normal-sized radio galaxies (e.g. Ekers, 1982; Mackay, 1971). In the case of giant radio galaxies whose sizes are comparable to the separations between groups and poor clusters of galaxies, the asymmetry of the environment could perhaps arise due to the increased probability for one of the jets to encounter a group or a poor cluster. Evidence for a substantial gaseous medium in such systems has come from X-ray observations (Biermann and Kronberg, 1983).

This study is being carried out for all the known giant radio galaxies, and a quantitative analysis is in progress.

REFERENCES

Biermann, P. and Kronberg, P.P., in 'Clusters & Groups of Galaxies', ed. Mardirossian, F. et al., 495, 1983

Ekers, R.D., IAU Symp. 97, 465, 1982

Ferretti, L., Giovanni, G., Gregorini, L., Parma, P. and Zamorani, G., Astron. Astrophys. 139, 55, 1984

Mackay, C.D., Mon. Not. Roy. Astr. Soc. 154, 209, 1971

Saripalli, L., Gopal Krishna, Reich, W., and Kühr, H., Astron. Astrophys. 170, 20, 1986

Waggett, P.C., Warner, P.J., Baldwin, J.E., Mon. Not. Roy. Astr. Soc. 181, 465, 1977

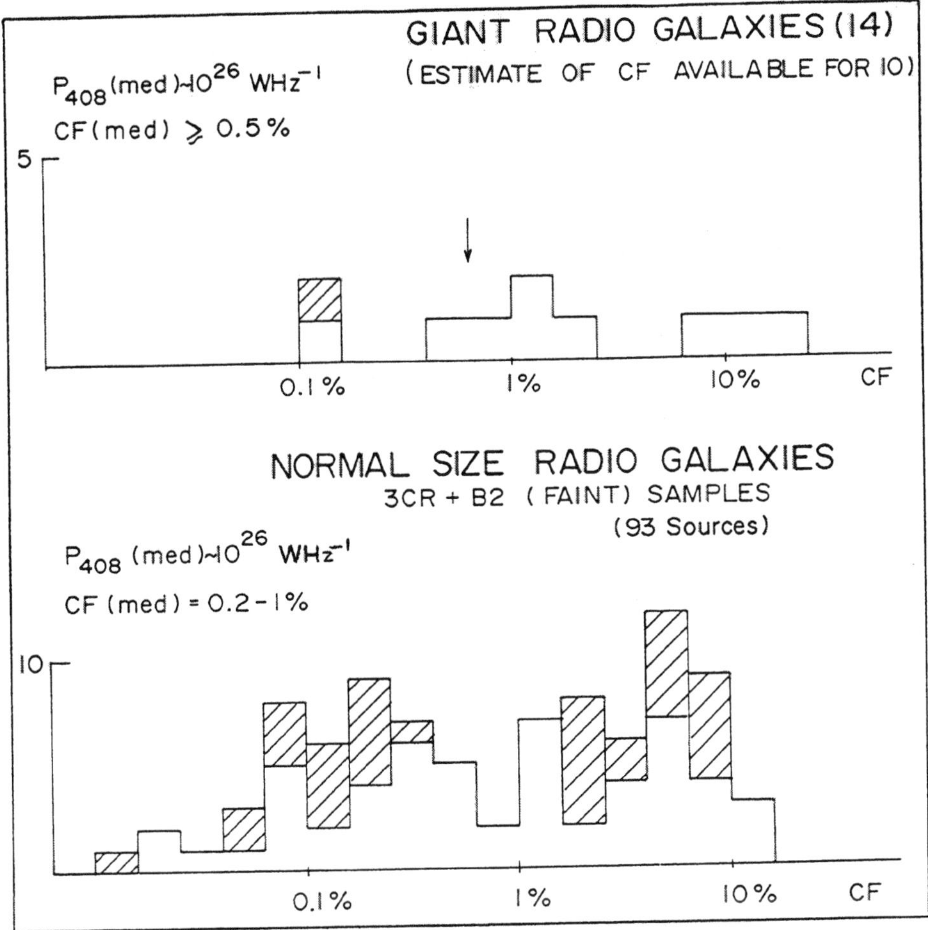

Figure 1: Histograms of core fractions CF:= core flux at 5 GHz / total flux at 408 MHz.

members of the jet set

SUBJECT INDEX

GPSR Compliance
The European Union's (EU) General Product Safety Regulation (GPSR) is a set of rules that requires consumer products to be safe and our obligations to ensure this.

If you have any concerns about our products, you can contact us on

ProductSafety@springernature.com

In case Publisher is established outside the EU, the EU authorized representative is:

Springer Nature Customer Service Center GmbH
Europaplatz 3
69115 Heidelberg, Germany

www.ingramcontent.com/pod-product-compliance
Ingram Content Group UK Ltd.
Pitfield, Milton Keynes, MK11 3LW, UK
UKHW021830190726
13853UKWH00003B/1270

* 9 7 8 9 4 0 0 9 3 9 2 8 8 *